Flood Risk Manageme

NATO Science Series

A Series presenting the results of scientific meetings supported under the NATO Science Programme.

The Series is published by IOS Press, Amsterdam, and Springer in conjunction with the NATO Public Diplomacy Division

Sub-Series

I.   **Life and Behavioural Sciences**   IOS Press
II. **Mathematics, Physics and Chemistry**   Springer
III. **Computer and Systems Science**   IOS Press
IV. **Earth and Environmental Sciences**   Springer

The NATO Science Series continues the series of books published formerly as the NATO ASI Series.

The NATO Science Programme offers support for collaboration in civil science between scientists of countries of the Euro-Atlantic Partnership Council. The types of scientific meeting generally supported are "Advanced Study Institutes" and "Advanced Research Workshops", and the NATO Science Series collects together the results of these meetings. The meetings are co-organized bij scientists from NATO countries and scientists from NATO's Partner countries – countries of the CIS and Central and Eastern Europe.

Advanced Study Institutes are high-level tutorial courses offering in-depth study of latest advances in a field.
Advanced Research Workshops are expert meetings aimed at critical assessment of a field, and identification of directions for future action.

As a consequence of the restructuring of the NATO Science Programme in 1999, the NATO Science Series was re-organised to the four sub-series noted above. Please consult the following web sites for information on previous volumes published in the Series.

http://www.nato.int/science
http://www.springer.com
http://www.iospress.nl

Series IV: Earth and Environmental Sciences – Vol. 67

Flood Risk Management: Hazards, Vulnerability and Mitigation Measures

edited by

Jochen Schanze

Leibniz Institute of Ecological and Regional Development (IOER)
Member of the Dresden Flood Research Center (D-FRC),
Dresden, Germany

Evzen Zeman

DHI Hydroinform a.s.
Prague, Czech Republic

and

Jiri Marsalek

National Water Research Institute,
Burlington, ON, Canada

 Springer

Published in cooperation with NATO Public Diplomacy Division

Proceedings of the NATO Advanced Research Workshop on
Flood Risk Management - Hazards, Vulnerability and Mitigation Measures
Ostrov, Czech Republic
6 –10 October 2004

A C.I.P. Catalogue record for this book is available from the Library
of Congress.

ISBN-10 1-4020-4597-2 (PB)
ISBN-13 978-1-4020-4597-4 (PB)
ISBN-10 1-4020-4596-4 (HB)
ISBN-13 978-1-4020-4596-7 (HB)
ISBN-10 1-4020-4598-0 (e-book)
ISBN-13 978-1-4020-4598-1 (e-book)

Published by Springer,
P.O. Box 17, 3300 AA Dordrecht, The Netherlands.

www.springer.com

Printed on acid-free paper

TABLE OF CONTENTS

v

Part 3. Flood forecasting

Part 4. Vulnerability and flood damages

Part 5. Mitigation measures

PREFACE

After a series of disastrous flood events during the recent years flood risks are in the forefront of public concerns. World-wide statistics indicate continuously increasing flood damages, and losses of human lives remain at unacceptably high levels. Many of these concerns have manifested themselves during recent extreme floods in Central Europe, like in the Vltava-Elbe river basin in August 2002 with catastrophic damages in the Czech Republic (2 billion EURO) and in Germany (9 billion EURO). In science and among professionals, there is a growing recognition that inundations by extreme floods cannot be totally avoided and maybe their occurrence will increase due to climate change. Accordingly, the previous paradigm of flood protection has to change to a societal flood risk management. This paradigm shift especially requires more comprehensive and continuous approaches considering all natural and societal factors of flood risks. Research and practice on flood risk management therefore depends on an enhanced collaboration of professionals in different fields, administrative sectors and regions or countries.

Against this background, the NATO Advanced Research Workshop (ARW) on 'Flood Risk Management – Hazard, Vulnerability and Mitigation Measures' aimed at discussing and advancing the understanding of an integrated and sustainable flood risk management. Therefore, it referred to the main risk factors and their theoretical and methodological investigations, like weather forecasting, climate change, flood propagation modelling, vulnerability assessment, design of risk reduction measures as well as the development of management strategies and instruments. In addition, it reflected practical experience from recent floods in Central Europe and elsewhere, considering both long-term as well as flood event measures, like flood warning, evacuation, etc. In this process, interdisciplinary and transboundary co-operation issues played an important role.

The workshop was held in Ostrov (near Decin), Czech Republic, close to the Czech-German border, from September 29th to October 3rd, 2004. Forty-three participants with backgrounds in natural sciences, social sciences, engineering and practical flood risk management represented 14 countries. The workshop covered sessions on 'flood hazard modelling', 'flood forecasting', 'modelling of vulnerability', 'flood risk mitigation' and 'historical floods and transboundary issues'. The proceedings provide the full texts of most of the formal oral presentations, and furthermore present the final conclusions of the ARW, which were announced during a reception by The Mayor of the City of Dresden.

| Jochen Schanze | Evzen Zeman | Jiri Marsalek |
| Dresden, Germany | Prague, Czech Republic | Burlington, Canada |

ACKNOWLEDGEMENT

The Advanced Research Workshop (ARW) on 'Flood Risk Management – Hazards, Vulnerability and Mitigation Measures' was directed by Jochen Schanze, Leibniz Institute of Ecological and Regional Development (IOER) as Member of the Dresden Flood Research Center (D-FRC), Dresden, Germany, and Evzen Zeman, DHI Hydroinform a.s., Prague, Czech Republic. Both were assisted by Jiri Marsalek, National Water Research Institute (NWRI), Environment Canada, Burlington, Canada, as member of the organising committee.

The ARW was sponsored by an award of NATO under the Programme Environmental and Earth Science & Technology. The co-directors and members of the organizing committee especially thank Dr. Alain H. Jubier and Dr. Deniz Beten for their personal assistance in preparing the workshop and the proceedings.

The institutes of the co-directors provided additional financial and staff resources. The local organisation of the workshop was predominantly provided by DHI Hydroinform a.s. headed by Evzen Zeman with extensive contributions from Pavlina Nesvadbova and other staff. The typescript of the proceedings was prepared by Jochen Schanze supported by Alfred Olfert, Katrin Vogel and Margitta Wahl at the Leibniz Institute of Ecological and Regional Development (IOER).

The co-directors thank all participants for their presentations at the workshop and their papers contributed to this book. The exchange of various scientific perspectives and experiences from different regions of the world made the workshop particularly fruitful in terms of the future improvement of flood prevention by the civil societies.

PART 1

FLOOD RISK MANAGEMENT

Chapter 1

FLOOD RISK MANAGEMENT – A BASIC FRAMEWORK

JOCHEN SCHANZE

Leibniz Institute of Ecological and Regional Development (IOER), Member of the Dresden Flood Research Center (D-FRC), Dresden, Germany

Keywords: Flood risk management, risk analysis, risk assessment, risk reduction, flood risk system, vulnerability, management strategies, flood research, water policy.

1. INTRODUCTION

Floods are one of the most threatening natural hazards for human societies (e.g. WBGU 1999). This is evident from the increase in damages in the last 50 years due to a series of extreme floods (Munich Re Group 2003). Recently, the tsunami in South East Asia caused 220,000 deaths which makes it probably one of the most disastrous floods. During the International Decade of Natural Disaster Reduction (IDNDR) from 1990 to 1999 it was appreciated that the previous paradigm of "flood protection" was inappropriate (UNDRO 1991, Plate 1999). Absolute protection is both unachievable and unsustainable, because of high costs and inherent uncertainties. Instead, risk management has been recommended as being more suitable and this paradigm is now receiving growing attention within flood research (e.g. Plate 1999, Schanze 2002, Hall et al. 2003, Hooijer et al 2004). Also currently environmental and regional policies in many countries are starting to shift from flood protection to flood risk management (e.g. Budapest Initiative 2002, EU 2004).

Flood risk management deals with a wide array of issues and tasks ranging from the prediction of flood hazards, through their societal consequences to measures and instruments for risk reduction. Due to this variety of aspects, management of flood risks needs systematisation and integration. This chapter provides definitions of central terms, the systematisation of tasks and components and a basic framework for flood risk management. Based on this, there are challenges for research and practice which arise especially from an integrated risk based approach.

J. Schanze et al. (eds.), Flood Risk Management:
Hazards, Vulnerability and Mitigation Measures, 1–20.
© 2006 *Springer.*

2. TERMS AND CONCEPTS

2.1. Flood risk

Floods can be defined as a temporary covering of land by water outside its normal confines (FLOODsite-Consortium 2005; cf. Munich Re 1997). They happen in small and large river basins, in estuaries, at coasts and locally. Beside these general conditions, floods can be systematised according to the cause of events, such as winter rainfall floods, summer convectional storm induced floods, snow-melt floods, sea surge and tidal floods, tsunamis, rising ground water floods, urban sewer floods, dam break or reservoir control floods. (cf. Penning-Rowsell and Peerbolte 1994, enhanced). A special type of flood is the highly dynamic flash floods. Each flood event can be characterised by features such as water depth, flow velocity, matter fluxes, and temporal and spatial dynamics. Flooding in most cases is a natural phenomenon which, for example, in natural floodplains cannot be classified as a threat. Nevertheless, floods in intensively used catchments are often influenced by man through land use, river training etc.

The probability of the occurrence of potentially damaging flood events is called *flood hazard* (cf. ITC 2004). Potentially damaging means that there are elements exposed to floods which could, but need not necessarily, be harmed (FLOODsite-Consortium 2005). The flood hazard encompasses events with various features. For instance, a building in a floodplain can be threatened by a 50-year flood, with a water level of 1 metre and by a 100-year flood, with a water level of 1.5 metres. Moreover, these events may be associated with different transport capabilities regarding debris, sediment and other (e.g. toxic) substances with varying impacts on man and the environment.

Damage by flood hazards depends on the *vulnerability* of exposed elements. The term vulnerability refers to inherent characteristics of these elements which determine their potential to be harmed (Sarewitz et al. 2003). It can be understood as a combination of susceptibility and societal value (FLOODsite-Consortium 2005) and expressed by direct and indirect effects which are tangible or intangible (Messner and Meyer, this issue). In contrast to the societal value, which is independent from the hazard, susceptibility indicates the process of damage generation (cf. Penning-Rowsell et al. 2003). It depends on both the type of flood event with its features and the constitution of the elements at risk. Three basic areas of flood vulnerability can be distinguished according to the principle of sustainability: social and cultural, economic and ecological vulnerability. *Social and cultural vulnerability* refers to loss of life, health impacts (injuries), loss of vitality, stress, social impacts, loss of personal articles, and loss of cultural heritage. *Economic vulnerability* alludes to direct and indirect financial losses by damage to property assets, basic material and goods, reduced productivity, and relief efforts. *Ecological vulnerability* comprises anthropogenic pollution of waters, soils and ecological systems with their biota (cf. Messner and Meyer, this issue).

Flood risk emerges from the convolution of flood hazard and flood vulnerability (WBGU 1999, ISDR 2004). It can be defined as the probability of negative consequences due to floods and depends on the exposure of elements at risk to a flood hazard (cf. ibid.). The general understanding of the term risk dates from the initial risk research (e.g. Knight 1921). In terms of floods it is interpreted as harm to flood-prone elements with a specific vulnerability ("elements at risk") due to probable flood events with their features. It should not be confused, therefore, with risk in terms of reliability, which plays a major role for quantifying the safety of structural works for flood protection (Plate 1999; see below).

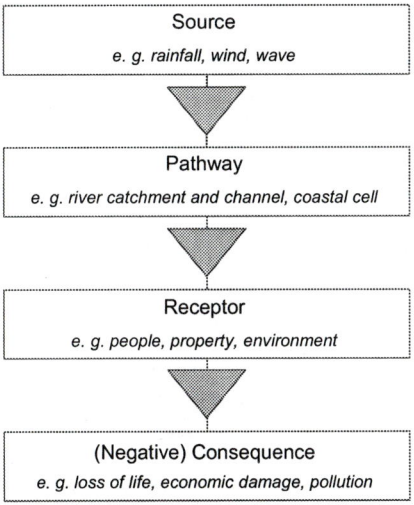

Figure 1. Source-Pathway-Receptor-Consequence-Model (ICE 2001, modified)

In order to describe flood risk the conceptual Source-Pathway-Receptor-Consequence-Model (SPRC-Model) has been proposed (ICE 2001; see Figure 1). It shows a simple causal chain ranging from the meteorological and hydrological events either in inland or at coasts (sources) through the discharge and inundation (pathways) and the physical impacts on elements at risk (receptors) to the assessment of effects (consequences). The chain links 'source', 'pathway' and 'receptor' refer to the physical process, whereas the assessment of the '(negative) consequence' is a matter of societal values.

In terms of flood risk, 'source' and 'pathway' represent the flood hazard. 'Source' is determined by the probability (p) of flood events with a certain magnitude and other features (m). Early warning (w) and the retention capacity of the source areas of inland floods (t) can be considered as two risk reduction factors. The 'pathway' can be described by the inland discharge or coastal overflow and inundation (i) with various attributes (a) and interventions for flood control (c). 'Receptor' and '(negative) consequence' state the vulnerability, whereas 'receptor' specifies the susceptibility (s) with interventions to strengthen resistance and resilience (r). 'Consequence' stands for the harm to values (v; damage) with interventions to decrease or to compensate them (d). Accordingly, flood risk can be expressed by the following function:

Flood risk = f $((p,m,w, t)_{source}, (i, a, c)_{pathway}, (s, r)_{receptor}, (v, d)_{consequence})$

In reality the causal chain of the SPRC-Model occurs for each element at risk and each flood hazard. Moreover, complex interrelations exist between pathways, interventions for flood control and the exposure of vulnerable elements. In some cases the interrelations consist of multiple feedbacks. A system which is assumed to include all related elements and processes is called here a "flood risk system". For inland floods it refers to river catchments, for coastal floods to coastal cells as areas which are hydraulically connected. The overall risk associated with a flood risk system can be described as the sum of risks of all individual elements.

2.2. Flood risk management

The term management is used in at least two different ways in the literature on floods, either excluding or including risk analysis. The first understanding is based on the hydrological reliability of existing flood defence structures. Management is interpreted, therefore, as decisions and actions undertaken to mitigate the remaining risk above flood protection design standards. In the past, the remaining risk has been assessed by scientific investigations. Dealing with flood risks in this case means carrying out flood risk analysis and then flood risk management (e.g. Marsalek 1999, Hooijer et al. 2004, Oumeraci 2004). The second understanding defines management as decisions and actions undertaken to analyse, assess and (to try to) reduce flood risks. In this case flood risk management covers the risk analysis, risk assessment and risk reduction (Plate 1999, Sayers et al. 2002, Hall et al. 2003). Both concepts are real alternatives and can hardly be combined.

Originally the term "management" comes from business economics. It is defined as all those activities which control the decisions and actions of an actor, an organisation, or a set of organisations (network) effectively and efficiently. Such activities include planning (data gathering, analysing, goal setting, evaluation of options, and so forth), organising, directing, staffing, monitoring, controlling and learning (Weihrich and Koontz 1992). In this sense "management" is already used in European Water Policy. One example is provided by the requirements of River Basin Management Plans (Art. 13) specified in Annex VII of the European Water Framework Directive (2000/60/EC). Beside others, they encompass (i) the *analysis* of pressures and impacts, (ii) the *assessment* of the water status and (iii) a programme of *measures*. Similar demands can be found in the "Guidelines for an integrated Management of Coastal Zones" (2002/413/EC).

Against this background, it is recommended that *flood risk management* should be defined as 'holistic and continuous societal analysis, assessment and reduction of flood risk'. 'Holistic' refers to the flood risk system which should be considered as comprehensive as possible. The term 'continuous' expresses the need for an ongoing assessment of flood risks, their dynamic change and effects of reduction activities. Analysis, assessment and interventions for risk reduction as common elements of a development model are dedicated to the 'managing entity'. As far as the management of the flood risk system is concerned, society, represented by politicians, experts and individuals, could be seen as the 'management entity' (cf. Parker 2000; see below). These represent-atives perceive flood risks and assess whether a certain flood risk is tolerable or not (Adams 1995) and decide on risk reduction interventions.

Flood risk management takes place as a decision-making and development process of actors. According to potential interventions on the flood risk system these actors represent various fields (e.g. water authority, spatial planning authority), adjacent areas (e.g. multiple municipalities) and different levels (e.g. local, regional, national, international). The decision-making and development process varies depending on the political, administrative, planning and cultural systems.

During the continuous management process different modes of management can be distinguished: the pre-flood, the flood event respec-tively flood management (Rosenthal and Hart 1998) and the post-flood modes. These terms have been introduced for the systematisation of flood risk reduction activities (Kundzewicz and Samuels 1997; see below). According to the comprehensive understanding of flood risk management, the modes also extend to risk analysis and risk assessment. Especially for risk analyses, a differentiation between these modes is already common as 'long term' and 'operational' approaches with differing methodological re-uirements. Plate (2002) quite similarly distinguishes a (project) planning level with an embedded project design level and an operational level.

The *pre-flood mode* is determined by the aim of reducing flood risks in the long term. It can be characterised by the availability of time and resources. Flood risk may be investigated in detail and with some pre-cision. Alternative developments of preventive or preparatory interven-tions may be intensively discussed, also in informal institutional regimes. In contrast, the *event management mode* is influenced by the nature of the flood event. Response times are short and the resources are limited to prepare actions in comparison with the pre-flood phase. The dynamic course of flood risk has to be estimated in the short-term. Decisions need to be taken immediately and strongly formalised. The *post-flood mode* is dedicated to recovery and the avoidance of further negative consequences. Its organisational structure and temporal dynamic may be classified as something leading from event management to the pre-flood mode.

Some authors describe a temporal circle of flood risk management, beginning with a flood event which is followed by the post-flood and then the pre-flood phase (e.g. DKKV 2003). The image of a circle can serve to support public awareness referring as it does to the recurrence of floods and related risk reduction. For research and flood risk management practice it should be remembered that previous actions influence future ones. The modes, therefore, more correctly follow a kind of helix. Moreover, the modes can partially overlap in real-world conditions (e.g. inclusion of preventions measures in recovery activities).

3.   TASKS AND COMPONENTS

Based on the definition of flood risk management, three tasks with specific components can be used for structuring the management activities (see Figure 2). The main tasks are:
- risk analysis,
- risk assessment and
- risk reduction.

Risk analysis provides information on previous, current and future flood risks, *risk assessment* deals with their perception and evaluation and *risk reduction* is dedicated to interventions with a potential to decrease the risks. To achieve the aims of each task certain components are required. They range from hazard determination to the specification of post-flood interventions. Alternative systematisations have been introduced in the literature (e.g. Yadigaroglu and Chakraborty 1985, UNDRO 1991), but none of them fit the demands of the comprehensive and actor-oriented understanding of flood risk management introduced here.

Figure 2. Tasks and components of flood risk management

3.1.   Risk analysis

Flood risk analysis ascertains past, current or – based on proposed activities or uncontrollable trends (global change) – future risks. It is based on the determination of the flood hazard, the flood vulnerability and the flood risk itself. To describe the flood risk system, multiple methods from natural and social science are required. Based on the SPRC-Model in principle meteorological, hydrological, hydraulic, economic, social science and ecological methods should be included. This leads to the challenge of how to gather and integrate the knowledge from all these fields. In general, there is already a rich experience of deterministic modelling combined with statistical analysis of the hydro-meteorological processes. Real probabilistic approaches are less frequent. As far as the socio-economic field is concerned some approaches deal with direct economic losses (e.g. Roos 2003). Knowledge and methods on indirect consequences (e.g. Vrouwenfelder et al. 2003) as well as on social (e.g. Tapsell et al. 2002) and ecological impacts (e.g. Geller et al. 2004) are still in their infancy. Really rare are especially more comprehensive approaches which try to integrate many of these fields. Recently, there have been increasing efforts to develop complex model systems (e.g. FORESIGHT, FLOODsite, VERIS Elbe). Moreover, tiered methodologies for various levels of risk analysis can be found (Sayers et al. 2002).

Despite the required efforts to develop more integrated approaches, however, knowledge on flood risk systems and the impacts of flood risk reduction interventions will inevitably remain incomplete (Hall et al. 2003). Consideration of uncertainty, therefore, plays an important role for flood risk analysis (Hall 2003, Apel et al. 2004, FLOODsite-Consortium 2005). Two kinds of uncertainty may be differentiated (ibid.): eleatory and epistemic uncertainty. *Eleatory uncertainty* refers to the fact that quantities are inherently variable over time, space or subjects and objects. *Epistemic uncertainty* results from the limited knowledge of the elements and processes of the flood risk system. This becomes clearer against the fact that mostly only (easily or sufficiently) quantifiable aspects are considered for risk analysis. The limited heuristic construction thus unavoidably leads to a restricted view of the real flood risk system.

In principle, eleatory and epistemic uncertainty can be seen as the main reason for the understanding of flood risk management as a continuous process. The inevitable restriction of scientific knowledge on the flood risk system requires periodic repetitions of the risk analysis to reflect inherent changes of the system and to examine assumed effects of risk reduction interventions. Nevertheless, uncertainty should not only be seen as an unavoidable issue. There are methods available for its quantification (e.g. Monte Carlo Simulation etc.; cf. Hall 2003). These methods allow a consideration of the degree of uncertainty which then can be seen as valuable additional information for flood risk management.

Main results from flood risk analysis are currently – printed or digital – risk maps in the long term and real time flood forecasting and warning systems as operational procedures. *Risk maps*, which reduce the flood risk system to a static and 2-dimensional format, provide information on the flood probability, the water level, flow velocity, sediment transport etc.. Approaches are widespread (Walz et al. 2005). Their reliability depends on the accuracy and the specification of the remaining uncertainty of the risk analysis. More flexible, interactive and web-based risk map systems do partly exist and are in further development.

Real time flood forecasting and warning systems provide information on ongoing flood events (e.g. Parker and Fordham 1996). Their aim is to enhance the lead time for preparatory activities, like reservoir control, evacuation etc. They mainly consist of meteorological and hydrological modelling (Cluckie and Hajjam 2001). Full risk based approaches, which would include precise information on potential damage are also not available up to now. Vulnerability normally is considered in terms of critical peak discharges, but not as expected damage. Nevertheless, the importance of the existing systems for flood (event) management is broadly acknowledged (Penning-Rowsell et al. 2000, Sorensen 2000).

3.2. Risk assessment

The results of scientific analyses of flood risks can be assessed quite differently by the society depending on individual and collective perception as well as on the weighting of the tolerability of certain risks. Applying this distinction does not mean to value a (more precise) 'objective' risk analysis from a (less valid) 'subjective' assessment (cf. Wildavsky 1993). Instead, it acknowledges the complementary need of information describing the flood risk system with available theories and methods on the one hand and the existence of individual and collective perception and weighing of risk as a societal behaviour on the other hand. In the following, risk perception and risk weighing are considered in a two step approach of risk assessment.

Risk perception refers to the fact that the overall view on the kind and magnitude of risk of individuals and groups involved in flood risk management depends on their individual and collective backgrounds. For example, individuals who have already experienced an extreme flood event will most probably have a different perception of flood risks than other individuals. There are psychological-cognitive and sociological-cultural approaches of explaining risk perception (cf. Weichselgartner 2002). The psychological-cognitive approaches focus on factors which influence (non-)rational awareness and decisions (cf. Slovic 2000). Among others they are interested in the relevance of individual and collective values, feelings, experiences and perspectives while perceiving 'real' risks. Because of the comparison of a rational and presumable correct flood risk with the subjective awareness and decisions, these approaches match with the understanding of an 'objective' information from scientific risk analysis as a basis for societal risk perception. Nevertheless, the potential for appropriate interpretation of factors of risk perception (Gethmann 1993) are still critically discussed.

The sociological-cultural approaches summarise a multifarious field. The area of risk communication research may be distinguished from cultural anthropological investigations (Weichselgartner 2002). Risk communication research stresses the meaning of communication between experts and the public and deals with conditions, issues and improvement options for a proper knowledge transfer (Wiedemann et al. 1990). More recent research enhances this passive role of communicating an objective risk to the negotiation of the acceptability or tolerability of risk (Adams 1995). Cultural anthropology assumes that all perception of risk is determined by the culture of the society (cf. Tompson 1980, Wildavsky 1993). It ascribes the selection process during risk perception to the cultural organisation types within the society.

Within the constraints of this article a detailed discussion of the various theories is impossible. As an overall position, risk perception here is understood as the 'construction of risk as the individual or collective imagination of a probable negative consequence'. It is based on values, feelings, experiences and perspectives which are influenced by the culture of a society. It is not interpreted as a gradual awareness of a somehow specified 'real' risk.

The risk of whole flood risk systems, however, cannot be directly perceived by the society. Risk analyses provide, therefore, information and serve as a step in a comprehensive risk perception. From this point of view risk perception could also be interpreted as an overall framework for risk analysis. Due to different paradigms - risk analyses rely on scientific principles, risk perception expresses real-world attitudes - the proposed distinction between risk analysis and risk perception seems to be appropriate.

Risk weighing takes into account the fact that the perception of risk does not already include decisions on how to deal with this risk. It emphasises that risk always depends on options of behaviour (WBGU 1999). Risk refers to options with the probability of negative consequences, whereas opportunities define options with the probability of positive consequences and both are closely related. In principal, it is impossible to totally avoid risk. Otherwise lots of opportunities are missed (WBGU 1999). Wildavsky (1984) stated: "No risk is the highest risk at all." Weighing risk therefore means to decide on a degree of risk as probable 'costs' which can be tolerated in relation to certain opportunities as 'benefits'.

Luhmann (1997) accepts the term risk only if disadvantages can be related to at least one aspect of a decision. To include risk as causal process of probable negative consequences - like specified under risk analysis - a distinction between a 'causal risk' and a 'normative risk' as a decision option is needed. The 'normative risk' refers to the tolerability of a formally quantified 'causal risk'. A normative risk alone would not make sense if there are no real specifications of the probable negative consequences. The same is true the other way round. Causal risk would not be sufficient as a basis for risk management since it does not refer to prerequisites for decision making.

Risk assessment altogether covers the perception of the causal interrelations and the weighing of the tolerability of risks. The overall question is: How much risk is tolerable considering expected 'costs' and 'benefits'? The weighing should encompass the manifold benefits of floods or flood-prone areas. Floods at least provide habitats for various species and serve as resources for irrigation - not only in sub-tropical regions (cf. Green et al. 2000). Especially urban flood-prone areas moreover are of high economic value due to access to waters and navigation as well as through location, often close to the city centres.

In recent literature as one main concept for risk assessment the so-called ALARP-Principle ('as low as reasonable practicable') is being used (e.g. Oumeraci 2004). The engineering principle assumes that certain vulnerability already exists and the capabilities for risk reduction are limited. Against the background of the previous explanations it becomes clear that risk assessment needs to include both 'costs' (risks) and 'benefits' of use (opportunities). 'Costs' cover both, the negative con-sequences and the efforts for risk reduction (see below). Accordingly damages should not be weighed with the reduction efforts only, like it is proposed by the ALARP-Principle. Instead, overall costs should be compared with benefits as positive consequences (opportunities).

Undoubtedly, there are intangible costs, like loss of life or injuries, which cannot be weighed against benefits. In this respect a 'prohibition area' of costs will be broadly accepted. The same is the case for a 'normal area' with unavoidable minor risks. The 'border area' between both, the prohibition and the normal area, is crucial, therefore, and has to be negotiated within flood risk management (WBGU 1999).

Although the overall approach of risk assessment appears to be an exercise in cost-benefits-analysis, (cf. Brent 1996), in terms of sustainability, the manifold direct and indirect tangible and intangible 'costs' additionally require multi-criteria approaches. They allow the inclusion of a wide range of criteria and fuzzy-linguistic scales of thresholds. Appropriate GIS-based MADM-methods are available from other fields of environmental research and water management (e.g. Thinh et al. 2004), but have not yet been sufficiently exploited in assessing multiple flood risks. Evaluation of these risks becomes more complex if costs and benefits from other areas of (water) policy should be included as is required by integrated river basin and flood risk management plans.

3.3. Risk reduction

If risks have been assessed as not tolerable, measures and instruments are applied for risk reduction. *Measures* are defined as 'interventions based on direct physical actions', *instruments* are 'interventions based on mechanisms which lead to measures indirectly or influence human behaviour' (Olfert and Schanze 2005). Up to now predominantly the terms structural measures for interventions of flood defence and non-structural measures for all other interventions have been used (e.g. White 1975, Kundzewicz 2002, Hooijer et al. 2004). The reason for that may result from the traditional engineering perspective of flood risk management. The societal point of view, which seems to be crucial for the consideration of the drivers of interventions, calls for further specification of the rather collective term non-structural measures. Instead, the term 'instrument' is already quite common to designate activities in various fields of financial and legal policies.

Measures can be differentiated as permanent or temporary. *Permanent measures* are 'direct physical interventions which lead to durable change of the physical conditions of the flood risk system', whereas *temporary measures* are 'direct physical interventions to reduce the risk during ongoing flood events' (Olfert and Schanze 2005). Permanent measures encompass engineering work for flood control (cf. Penning-Rowsell and Peerbolte 1994) as well as constructive efforts to improve flood resistance and resilience of buildings and infrastructure. They also include activities to increase the retention capacity aiming to reduce the peak discharge and to reduce matter losses from landscapes (e.g. suspended and bed load, debris) in upstream catchments and on the floodplains. Temporary measures comprise demountable flood protection (e.g. sand bags and movable flood barriers), the relocation of people, animals or goods on-site or off-site (evacuation) and the securing of objects and goods (e.g. locking gas valves). Moreover, recovery of flood damages (e.g. removal of toxic sediment accumulation) is limited in time and, therefore, temporary.

Instruments for flood risk management may be structured according to the kind of mechanism as regulatory, financial and communication instruments (Olfert & Schanze 2005). *Regulatory instruments* are interventions deploying legal or other normative mechanisms and come from the policy sectors of water management (e.g., flood protection acts), spatial planning (e.g. legally binding land use plans) and environmental protection and nature conservation (e.g. Strategic Environmental Assessment) (cf. NHV 2004). *Financial instruments* are supporting (e.g., incentives for flood proofing of built structures) or restrict flood risk related activities (e.g. zoning of insurances). *Communication instruments* are based on the transfer of knowledge (e.g. media, brochures, literature), informal cooperation (e.g. collaboration of relevant actors) as well as education, preparation, warning and instructing (e.g. flood warning systems). Instruments are partly used for programming, regulating, implementing and sometimes even funding the realisation of measures.

Measures and instruments fit certain coefficients of the SPRC-model (see above): Retention for instance is oriented to the factor 't' of sources, flood control to the factor 'c' of pathways, improvement of constructive resistance and resilience to the factor 'r' of receptors and compensation the factor 'd' of consequences. In accordance to the modes of flood risk management they can be systematised as pre-flood, flood event and post-flood interventions (Kundzewicz and Samuels 1997):

- *Pre-flood* interventions cover *prevention*: a decrease of the magnitude of floods and vulnerable elements in flood-prone areas, *protection*: structural protection of existing vulnerable elements and *preparedness*: behavioural preparation for probable flood events. Examples are water and matter retention in the source areas of river floods, risk conscious spatial development in flood-prone areas including zoning and building construction (prevention), design of flood control measures by the construction of reservoirs, polders and dykes (protection) and the preparation of people at risk (preparedness) (cf. DKKV 2003).

- *Flood event management* consists of forecasting and *warning* of an ongoing event as a basis for flood defence and providing information to people at risk, *flood control* by operative management of the discharge and water level, *flood defence* based on flood protection structures and *emergency response* as mitigation of damages and harm by evacuation and rescue. Examples are flood forecasting and warning by official and unofficial flood warning systems (warning; Parker and Handmer 1998), flood control by operative management of reservoirs and polders (defence) as well as risk mitigation by evacuation and rescue (emergency response).

- *Post-flood* interventions encompass *recovery* as relief and reconstruction. An example is the coverage of flood damages by insurance. The effects, effectiveness and efficiency of measures and instruments vary widely. *Effects* are manifold, intended and unintended (side-effects) and depend on the flood risk system. *Effectiveness* expresses the attainment of intended effects. *Efficiency* of measures and instruments refers to the ratio of effectiveness and required efforts or resources. For instance, cost-efficiency of flood protection measures means the ratio of the reduced damage potential to the costs of the measures. In this case it needs information on capital investments or opportunity costs. In general, effectiveness depends on the functionality of measures respectively mechanisms of instruments, the impacts on the flood risk system and the monetary and non-monetary efforts. They not only differ depending upon the type of intervention but also with respect to the site-specific conditions and implementation.

In order to consider the interdependencies between parallel or temporally consecutive measures and instruments in a flood risk system, it is valuable to consider the formulation of *strategic alternatives*. They describe alternative concepts for interventions based on combinations of single options from various policy fields. A broader view may be provided by the additional inclusion of natural and societal trends. Natural trends for instance include the ongoing climate change (c.f. IPCC 2001), societal trends include population change, increase in GDP, etc. Together, both are designated here as qualitative-quantitative *scenarios* (cf. UNEP 2003). These scenarios can be understood as conceptualised futures for the flood risk system.

4. THE FLOOD RISK MANAGMENT PROCESS

Management of the flood risk system requires further consideration of the linkages between the tasks and components introduced in chapter 3 as well as their application by representatives of the society. Links between risk analysis, risk assessment and risk reduction are the prerequisite for a consistent approach to steering a flood risk system. Due to widely varying methods and tools incorporating these into an operation system is challenging.

It should be realised that the application of tasks and components by the representatives of the society within flood risk management will normally not follow only a formal textual logic. Instead, various aspects of societal behaviour influence decisions and their implementation. The usability and real usage of scientific contributions to risk analysis, risk assessment and risk reduction in societal discussions and decisions are, therefore, of special importance for understanding flood risk management (cf. Plate 2002).

Figure 3 shows a basic framework for flood risk management considering both the linkages and orientation to the societal decision-making and development process. As the main element the latter is located at the top. Here, decisions are made and actions taken by the actors of flood risk management. All tasks and components introduced before are oriented to this societal decision-making and development process. This means that decision makers somehow have to deal with risk analysis, risk assessment and risk reduction. To do this, they receive information from or are supported by scientists and experts. Based on this information they take decisions and actions are preformed.

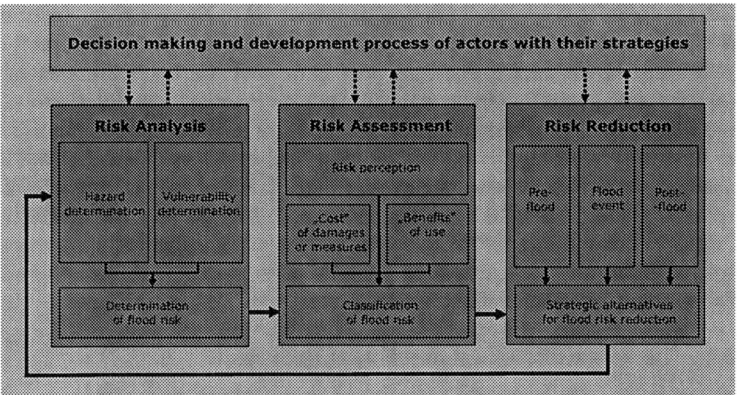

Figure 3. Basic framework of flood risk management

The framework is drawn as a logical contiguity. It does not include a temporal dimension. To do this would lead to further specifications which will not be the subject of this chapter (Sayers et al. 2002, Schanze 2005). In the following paragraphs the textual aspects and the societal decision-making and development process are dealt with in more detail.

4.1 Optimisation and monitoring of flood risks

Tasks and components of flood risk management have already been introduced above. Now there is the question of how they jointly contribute to the overall issue of flood risk management. In this respect all tasks and components can be seen as linked with each other within an optimisation loop. Optimisation refers to steering the flood risk system to a state of risk which is consistent with the tolerable risks and requires minimum effort. With it, risk analysis provides the simulation of the flood risk system. Risk assessment then indicates requirements for the improvement of the flood risk system. Risk reduction develops options and alternatives in terms of effects and efforts to solve those risks that are not tolerable. These options and alternatives – ideally covered with relevant trends in scenarios – afterwards are ex ante-simulated by methods of risk analysis.

For this optimisation a consistent approach is required to make risk analysis, risk assessment and risk reduction operational. Model systems for the simulation of the flood risk system, assessment tools and measures and instruments need, therefore, a joint parameterisation. This is not only needed for the targeted effects but also for side-effects to ensure that all the consequences of certain decisions are displayed. Due to the manifold effects of flood management actions the necessity of using multi-criteria tools for risk assessment is obvious.

Scenarios and their ex ante analysis, especially in the long-term, often have a high degree of uncertainty. Attainment of intended effects and the avoidance of unintended side-effects, therefore, can only be partly estimated. Monitoring of the real development of the flood risk system, therefore, plays an important role. It needs continuous data collection of the SPRC-factors. Moreover, information on the maintenance of physical measures over a long period is required (Penning-Rowsell and Peerbolte 1994). This should also include ex post evaluations of the performance of measures and instruments during previous events (Olfert and Schanze 2005). The monitoring of the real system dynamic and the performance of interventions provide sources of knowledge to improve the validity of estimated effects and to adapt interventions.

4.2 Decision-making and development process

As a basis for the framework it has been assumed that the actors of the decision-making and development process of flood risk management may behave as a 'managing entity'. In fact, many decisions, at least within catchments but presumably also in coastal cells, are made by fragmented actors (Bressers and Kuks 2004). This means that there may be parallel decision-making and development processes which are not or only partially linked. At least in certain sectors, administrative units or at administrative levels actors are behaving as 'managing entities' and analysing, assessing and reducing flood risks.

Because of the central importance of the decision-making and development processes a further understanding of their societal conditions is required. Especially, the prerequisites of an effective risk management including the requirements for integration of fragmented actors and decisions into a catchment-wide or coastal cell-wide approach are meaningful. Knowledge of these issues is still limited. Basic insights are already available from strategy and spatial planning research. Hutter (this issue) defines three dimensions of strategy development in flood risk management, the content, the context and the process dimensions. He emphasises that the *contents* regarding the tasks and components of flood risk management are only one dimension. To understand societal decision-making, the context of decision-makers as well as the process of formulating and implementing strategies has to be considered.

The *context* dimension reflects conditions that actors of flood risk management treat as given in a specific situation. It comprises external (political, economic, legal) and internal (e.g., cultural) factors. The *process* dimension refers to the fact that the procedure of flood risk management can vary widely. In research, often a linear process model is assumed with its emphasis on a top-down, well-structured, analytical, and sequential procedure. The formulation and implementation of strategies under real-world conditions is, however, mostly not as linear. Understanding these processes and testing alternatives may contribute to improving the effectiveness of flood risk management (cf. Hutter; this issue). Especially continuity including implementation, ex post evaluation and reformulation of strategies with learning from previous decisions could be of special importance in terms of the real dynamic of the flood risk system and uncertainties in predicting the impacts of decisions.

Without further specification in this chapter it becomes evident that flood risk management cannot be seen as restricted to the content dimension. It is based on the societal conditions of decision-making. In particular, the process dimension seems to be essential for the implementation of strategies. It provides the context for monitoring of the development of the real flood risk system and allows for the adaptation of previous aims and targets. It could be one prerequisite in terms of a co-evolutionary development between the societal decisions and the real dynamic and change of the flood risk system (cf. Schellnhuber 2001). Furthermore, context and process conditions can be expected to be crucial for an integration of fragmented decisions.

5. CHALLENGES FOR RESEARCH AND DEVELOPMENT

The comprehensive and continuous perspective of flood risk management shows a wide range of relevant aspects from physical processes to societal decision-making. Beside further knowledge on single items, integration appears to be a central challenge. Accordingly, in the literature on flood risk management the adjective "integrated" is sometimes used. In some cases integration of structural measures and non-structural instruments is meant (Hooijer et al. 2004), in other cases temporal integration, including pre-flood, flood event and post-flood activities for risk reduction (Kundzewizc and Samuels 1997, DVVK 2003). Originally the term 'integration' meant (i) restore something as a whole (lat.) or (ii) involvement in a larger entity (Brockhaus-Enzyklopädie 1989).

Transferred to flood risk management, integration can be interpreted as an inclusion of all relevant aspects in an overall understanding of flood risk management. Due to the variety of aspects, several dimensions should be distinguished. One dimension refers to the physical processes of the flood risk system with its spatial extent, another to the continuous timeframe including the pre-flood, flood event and post-flood modes. Further dimensions can be seen in the dimensions of strategies as well as in sectors, levels and adjacent areas of decision-making.

Against the background of the definition of flood risk management introduced here, all these dimensions are in principle relevant. Accordingly, an explicit use of the term 'integrated' seems not to be required anymore. Nevertheless, the theoretical, methodological and technological knowledge of integration becomes of high importance for the research and practice of flood risk management. This ranges from the development of comprehensive model systems for the simulation of the flood risk system (partly encompassing their technological integration in decision support systems), the quantification of uncertainties of such approaches to the usability of the results under the contexts and in the processes of societal decision-making. The awareness of the necessity of such approaches seems to grow consistently in research and practice. Beside efforts to improve knowledge on partially natural and societal processes, progress regarding integration can already be based on basic insights and further developed in parallel.

A second main challenge for research on and the practice of flood risk management is the temporal continuity of dealing with the flood risk system. On the one hand it requires a consideration of the change of such systems based on inherent trends of natural or societal factors as well as intended and unintended interventions. Predictions or scenarios with ex ante analyses of such changes in the long term are unavoidably associated with uncertainties. Accordingly, the uncertainties should be anticipated and quantified. Ex post analyses hereby can play an important role.

The dynamic of the flood risk system and its limited predictability on the other hand needs to be reflected while formulating and implementing strategies. The decision making and development process should, therefore, be based on procedures which ensure flexibility to adapt aims and interventions. This also requires the ability to learn from previous decisions and to be prepared for unforeseeable developments.

6. CONCLUSIONS AND OUTLOOK

The chapter has shown requirements and possibilities to understand and deal with flood risk management as a continuous and holistic issue. Main tasks and components have been indicated and briefly explained. A basic framework provides a mental map for the systematisation of the main components and their relations. It provides a means of navigation navigating the scientific and practical contributions that are required for flood risk management (cf. Tonn et al. 2000). Moreover, it makes it possible to indicate whether all relevant aspects of site-specific flood risk management are being considered.

The introduced framework stresses the meaning of the societal decision-making and development process as the managing procedure. Dedication of all tasks and components to this process appeared to be crucial. Furthermore, it is evident that flood risk management is not just restricted to information and options on the flood risk system. Instead, it happens within a societal context as a process of formulation and implementing strategies. Hereby, global change and unavoidable uncertainties of knowledge require a co-evolutionary understanding of steering and learning.

Multiple dimensions of flood risk management ask for further research efforts to describe and make partly operational the framework of integration. Accordingly knowledge can be expected to be crucial in contributing to more comprehensive and effective flood risk management. In this respect research is currently being carried out within the Integrated Project FLOODsite under the 6th European Framework Programme. Results are dedicated to the EU Action Programme (EC 2004). Further research on an integrated approach is ongoing in the European Member

States, as in the RIMAX Programme in Germany and the EPSRC Flood Risk Management Project in the United Kingdom. Together these efforts will support the change from the paradigm of flood protection to the more fuzzy but realistic paradigm of flood risk management.

7. ACKNOWLEDGEMENT

This contribution has been funded by IOER. The author especially thanks Prof. Jim Hall, University of Newcastle, and Prof. Roger Bettess, HR Wallingford, for their review and colleagues from the FLOODsite project team (contract-no. GOCE-CT-2004-505420) for fruitful discussions.

References

Adams J. (1995) *Risk*. UCL Press, London.

Apel H., Thieken A.H., Merz, B. and Blöschl, G. (2004) Flood risk assessment and associated uncertainty. *Natural Hazards and Earth System Sciences* 4, 295-308.

Brent R.J. (1996) *Applied Cost-Benefit Analysis*. Cheltenham.

Bressers H. and Kuks S. (2004) *Integrated Governance and Water Basin Management*. Conditions for Regime Change and Sustainability, Kluwer, Dordrecht.

Cluckie I.D. and Hajjam S. (2001) An Overview of Real Time Flood Forecasting. Griffith R.J., Cluckie I.D., Austion G.L. and Han D. (eds) *Radar hydrology for real time flood forecasting*. European Commission, Directorate General for Research, Brussels

DKKV (Deutsches Komitee für Katastrophenvorsorge e.V.) (ed.) (2003) *Hochwasservorsorge in Deutschland. Lernen aus der Katastrophe 2002 im Elbegebiet* (Flood prevention in Germany. Learning from the disaster 2002 in the River Elbe basin). Bonn.

Gethmann C.F. (1993) Zur Ethik des Handelns unter Risiko im Umweltstaat (About the ethics of bahaviour under risk in a state of the environment). Gethmann C.F. and Kloepfer M. (eds.) *Handeln unter Risiko im Umweltstaat*. Springer, Berlin, 1-54.

Geller W., Ockenfeld K., Böhme M. and Knöchel A. (eds.) (2004) *Schadstoffbelastung nach dem Elbe-Hochwasser 2002* (Pollution after the River Elbe flood 2002). http://www.ufz.de/data/HWEnd1333.pdf.

FLOODsite-Consortium (2005) Language of Flood Risk. www.floodsite.net.

Hall J.W. (2003) Handling uncertainty in the hydroinformatic process. *J. of Hydroinformatics* 5 (4), 215-232.

Hall J.W., Meadowcroft I.C., Sayers P.B. and Bramley M.E. (2003) Integrated Flood Risk Management in England and Wales. *Natural Hazard Review*, ASCE 4 (3), 126-135.

Hooijer A., Klijn F., Pedroli B. and Van Os A. (2004) Towards Sustainable Flood Risk Management in the Rhine and Meuse River Basins: Synopsis of the Findings of IRMA-SPONGE. *River Research and Applications* 20, 343-357.

Hutter G. (2005) Strategies for flood risk management – A process perspective. (this issue).

18

Hutter G. and Schanze J. (2004) Potenziale kooperativen Lernens für das Hochwasser-risikomanagement – am Beispiel der Vorsorge gegenüber Sturzfluten im Flussgebiet der Weißeritz (Potentials of co-operative learning on flood risk management). Felgentreff C. and Glade Th. (eds.) *Naturgefahren / Naturrisiken – Von der Analyse natürlicher Prozesse zur gesellschaftlichen Praxis* (Natural hazards / natural risks – from analysis of natural processes to societal practice). Potsdamer Geographische Schriften, Potsdam, 63-87.

ICE - Institution of Civil Engineers (ICE) (2001) Learning to Live with Rivers. Final Report of the ICE's Presidential Commission the Review the Technical Aspects of Flood Risk Management in England and Wales, London, http://www.ice.org.uk/rtfpdf/iceflooding.pdf.

ITC - International Institute for Geo-Information Science and Earth Observation (2004) Application 1. http://www.itc.nl/ilwis/applications/application01.asp.

ISDR - International Strategy for Disaster Reduction (2004) Terminology: Basic terms of disaster risk reduction. http://www.unisdr.org/eng/library/lib-terminology-eng%20home.htm (17.06.04).

Knight F. (1921) *Risk, uncertainty and profit*. New York, Harper & Row.

Kundzewicz Z. and Samuels P.G. (1997) Real-time Flood Forecasting and Warning. Conclusions from Workshop and Expert Meeting. Proceedings of Second RIBAMOD Expert Meeting, Published by DG XII, European Commission, ISBN 92-828-6074-4.

Luhmann N. (1997) Die Moral des Risikos und das Risiko der Moral (The risk of morale and the morale of risk). Bechmann, G. (ed.) *Risiko und Gesellschaft*. Westdeutscher Verlag, Opladen, 327-338.

Marsalek J. (2000) Overview of flood issues in contempory water management. Marsalek J, Watt W.E., Zeman E. and Sieker F. (eds.) *Flood Issues in Contemporary Water Management*. NATO Science Series, Kluwer Academic Publishers, Dordrecht, 1-16.

Meadowcroft I.C., Sayers P.B. and Hall J.W. (2002) Risk, Performance and Uncertainty in Flood and Coastal Management - A defining Review. Proceeding of the DEFRA Conference of River and Coastal Engineering and Management.

Messner F. and Meyer V. (2005) Flood damage, vulnerability and risk perception – challenges for flood damage research. (this issue).

Munich Re (1997) *Flooding and insurance*. Munich.

Munich Re Group (2005) Annual Review: Natural Catastrophes 2004. Knowledge Series Topics geo. Munich Re Group, http://www.munichre.com/publications/302-04321_en.pdf, 56 p.

NHV – Netherlands Hydrological Society (2004) Water in the Netherlands – managing, checks and balances. *nhv-special* 6, Utrecht, The Netherlands.

Olfert A. and Schanze J. (2005) Identification and ex-post evaluation of existing pre-flood measures and instruments - A theoretical framework. FLOODsite Report No. T12-05-01, Leibniz Institute of Ecological and Regional Development (IOER), Dresden.

Oumeraci H. (2004) Sustainable coastal flood defences: Scientific and modelling challenges towards an integrated risk-based design concept. 19 p. (unpublished).

Parker D.J. (2000) Managing Flood Hazards and Disasters: International Lessons, Directions and Future Challenges. Parker D. J. (ed.) *Floods*. Volume II, Routledge, London, 287-306.

Parker D.J. and Fordham M. (1996) An evaluation of flood forecasting, warning and response systems in the European Union. *Wat. Res. Manag.* 10, 279-302.

Penning-Rowsell E. and Peerbolte B. (1994) Concepts, Policies and Research. Penning-Rowsell E. and Fordham M. (eds.) *Floods across Europe. Flood Hazard Assessment, Modelling and Management*. Middlesex University Press, London, 1-17.

Penning-Roswell E., Tunstall S.M., Tapsell S.M. and Parker D. (2000) The Benefits of Flood Warnings: Real but Elusive and Politically Significant Water and Environmental Management. *Journal of the Institution of Water and Environmental Management* 14 (1), 7-14.

Penning-Rowsell E., Johnson C., Tunstall S., Tapsell S.M., Morris J., Chatterton J., Coker A. and Green C. (2003) The Benefits of flood and coastal defence: techniques and data for 2003. Flood Hazard Research Centre, Middlesex University.

Plate E. J. (1999) Flood risk management: a strategy to cope with floods. Bronstert A., Ghazi A., Hladny J., Kundzewicz Z.W. and Menzel L. (eds) Proceedings of the European Meeting on the Oder Flood 1997. Ribamod concerted action, European Communities, Office for Official Publications of the European Communities, Luxemburg, 115-128.

Plate E. J. (2002) Flood Risk and Flood Management. *Journal of Hydrology* 267, 2-11

Roos W. (2003) Damage to buildings. DC1-233-9, Delft Cluster, Delft.

Rosenthal U. and Hart P. (1998) *Flood Response and Crisis Management in Western Europe.* Springer, Berlin and others.

Sayers P.B., Hall J.W. and Meadowcroft I.C. (2002) Towards risk-based flood hazard management in the UK. Proceedings of the Institution of Civil Engineers. *Civil Engineering* 150, 36-42.

Schanze J. (2002) Nach der Elbeflut – die gesellschaftliche Risikovorsorge bedarf einer transdisziplinären Hochwasserforschung (After the River Elbe flooding - societal risk prevention needs a transdisciplinary flood research). *GAIA* 11 (4), 247-254.

Schanze J. (2005): Perspektiven für ein flussgebietsbezogenes Hochwasserrisikomanagement (Perspectives for a catchment-based flood risk managment). Köck W. (ed.) *Leipziger Schriften zum Umwelt- und Planungsrecht.* Nomos-Verlag, Baden-Baden, 167-177

Schellnhuber H.-J. (2001) Die Koevolution von Natur, Gesellschaft und Wissenschaft – eine Dreiecksbeziehung wird kritisch (Coevolution of nature, society and science – a triangle relation). *GAIA* 10 (4), 258-262.

Slovic P. (2000) *The Perception of Risk.* Earthscan Publications, London.

Sorensen J.H. (2000) Hazard Warning Systems: Review of 20 Years of Progress. *Natural Hazards Review* 1 (2), 119-125.

Tapsell S.M., Penning-Rowsell E.C., Tunstall S.M. and Wilson T. (2002) Vulnerability to flooding: health and social dimensions, *Phil. Trans. R. Soc. Lond.* 360, 1511-1525.

Thinh N. X., Walz U., Schanze J., Ferencsik I. and Göncz, A. (2004) GIS-based multiple criteria decision analysis and optimization for land suitability evaluation. Wittmann J. and Wieland R. (eds.) Simulation in Umwelt- und Geowissenschaften. Workshop Müncheberg 2004. Aachen, Shaker, *Berichte aus der Umweltinformatik 88*, 208-223.

Tompson M. (1980) Aesthetics of risk: culture on context, Schwing R.C. and Albers W.W. (eds.) *Societal risk assessment: How safe is safe enough?* Plenum Press, New York, 273-285.

Tonn B., English M. and Travis C. (2000) A Framework for Understanding and Improving Environmental Decision Making, *J. of Environmental Planning and Management,* 43(2), 163-183.

UNDRO (1991) Office of the United Nations Disaster Relief Coordinator: Mitigation natural diasters: phenomena, effects and options. A Manual for policy makers and planners, United Nation, New York.

Vrouwenvelder T., van der Veen A. and Stuyt L. (2003) Methodology for flood damage evaluation, DC1-233-3, Delft Cluster, Delft.

Walz U., Etter J. and Dransch D. (2005) Actor-oriented flood risk maps as support for societal decision making, Proceedings of the International Conference "Cartograpic Cutting-Edge Technology for Natural Hazard Management", Dresden (in print).

WBGU – Wissenschaftlicher Beirat der Bundesregierung für Globale Umweltveränderungen (1999) *Welt im Wandel – Strategien zur Bewältigung globaler Umweltrisiken* (World in change – Strategies for dealing with global environmental risks). Jahresgutachten 1998, Springer, Berlin and others.

Weihrich R. and Koontz H. (1992) Management, McGraw-Hill, New York.

Wiedemann P.M., Rohrmann B. and Jungermann H. (1990) Das Forschungsgebiet „Risiko-Kommunikation" (Research area „risk communication), Jungermann H., Rohrmann B. and Wiedemann P.M. (eds.) Risiko-Konzepte, Risiko-Konflikte, Risiko-Kommunikation. *Monografien des Forschungszentrums Jülich* 3, Jülich, 1-9.

Wildavsky A. (1984) Die Suche nach einer fehlerlosen Risikominderungsstrategie (Search for an errorless risk mitigation strategy). Lange S. (ed.) *Ermittlung und Bewertung industrieller Risiken*. Berlin, Heidelberg, New York, Springer, 224-233.

Wildavsky A. (1993) Vergleichende Untersuchung zur Risikowahrnehmung: Ein Anfang (Comparative study on risk perception, a beginning). Bayerische Rückversicherung (ed) *Risiko ist ein Konstrukt: Wahrnehmung zur Risikowahrnehmung*. Knesebeck, München, 191-211.

White G.F. (1975) Flood Hazard in the United States: A Research Reassessment. University of Colorado, Institute of Behavioral Science, Boulder, Colorado.

Yadigaroglu G. and Chakraborty S. (1985) Risikountersuchungen als Entscheidungsinstrument (Investigations on risk as decision-making instrument). TÜV Rheinland publication, Essen.

PART 2

FLOOD HAZARD MODELLING

Chapter 2

A EUROPEAN PERSPECTIVE ON CURRENT CHALLENGES IN THE ANALYSIS OF INLAND FLOOD RISKS

Paul G. SAMUELS

HR Wallingford, United Kingdom

Keywords: Floods, risk analysis, uncertainties, flood risk management, EU research policy.

1. INTRODUCTION

1.1. The threat of floods

Flooding poses a threat to many millions of the citizens of Europe. In the Netherlands more than half of the population live below mean sea level. In the UK about 10% of the population lives in areas of flood risk; and in Hungary about a quarter of the population live on the floodplain of the Danube and its tributaries. Although the greatest loss of life in Europe in recent decades occurred from the 1953 North Sea floods, which caused about 2500 deaths across the UK, Netherlands, Belgium and Germany, flooding remains the most widely distributed natural hazard in Europe leading to significant economic and social impacts. Concentrations of fatalities in river floods are associated with flash flooding events, such as Vaison-la-Romaine (1992), and the mudflows at Sarno (1997). Recent internationally and nationally significant floods have occurred, for example, in Slovak Republic (1993 and 1997), the Czech Republic (1997, 1998 and 2002), Poland and Germany (1997, 2002), Britain (1998, 2000), Austria (1991, 1997, 1999, 2002), Eastern Slovakia and Hungary (1998 and 2001) and France (2002 and 2003).

Flooding, however, is not just a European problem but a key facet of global water management. For example, Wang and Plate (2002) discuss the recent situation in China and White (2001) describes the international issues on flooding in rivers as presented as part of the World Water Vision at the Hague conference on water in March 2000.

J. Schanze et al. (eds.), Flood Risk Management:
Hazards, Vulnerability and Mitigation Measures, 21–34.
© 2006 *Springer.*

1.2   Flood damages

European scale information on flood risk and damage is not readily accessible. However, the national scale of importance of flood and coastal defence activities has been documented for England and Wales as preventing annual average damages of approximately €4 Billion, with the value of assets at risk of river and coastal flooding being about €300 Billion (Purnell, 2002). In the Netherlands estimates of the possible damage due to flooding vary from €300 to €800 Billion. In Austria there are over 10,000 catchments which experience "torrential" flash flood hazard and they cover an area of 46,921 km². Between 1973-2002 over 4,500 events occurred, affecting over 40% of the communities at risk, (just under 20% of all communities in Austria). In 2002 alone, these flood damages amounted to €163 Million. The 1997 flooding on the Odra river in Poland and the Czech Republic caused damages of about €4.5 Billion, the evacuation of over 200,000 people and over 100 deaths (Bronstert *et al*, 1999).

The potential for flood damage is also increasing on many rivers and coastal plains arising from social and economic development bringing pressures on land use. Failures of flood defence at any particular locality are rare because of the low probability of loading at the design standard, but internationally failures appear to be frequent. This is in part a statistical scale affect but could also be an indicator that climatic extremes are increasing in frequency and severity through the impacts of global change. In dry decades there is a natural political tendency to reduce expenditure on the provision of flood defences in favour of other pressing social objectives, leading eventually to increased risk of failure as infrastructure deteriorates (Bronstert *et al*, 1999). The past decade of significant floods in many parts of Europe has now reversed any such tendency for complacency as evidenced by the preparation in 2003 of a "Best practices document on flood prevention in Europe" (available from the ACTIF project library: http://www.actif-ec.net/library_page1.asp).

1.3   Flood risk

In the assessment and analysis of flood risk it is important to remember that "risk" is entirely a human issue. Floods are part of the natural hydrological cycle and are random; the risk arises because the human use and value of the river flood plains conflicts with their natural function of conveyance of water and sediments. The impact of floods on people can be dramatic even if there are no fatalities; pictures of rescue make good television material with stories told of bravery and fortunate escapes. However, the aftermath of a flood is distressing with personal possessions ruined, houses deep in sewage-contaminated mud and businesses destroyed. Cleaning up, community recovery and economic restoration takes months and for many there remains the fear that the next time it rains, the experience will be repeated.

In analysing the risk posed by floods, it is helpful to classify floods as broadly flash floods (or torrential floods) and lowland floods (or plains floods). Flash floods occur typically in the headwaters of a river system, with intense rainfall over steep topography. The spatial scale of the meteorological events which cause a flash flood can be small (e.g. under 100 km^2 for the UK "Boscastle" flood in August 2004). This leads to a rapid response of the river with large discharges and rapid increases in depth occurring without warning. Indeed the inability to provide effective warning was recommended by the ACTIF cluster project as the defining characteristic of a flash flood (see the document on "Future EC research needs under the EC sixth framework for flood and drought forecasting" available in the ACTIF library: http://www.actif-ec.net/library_page1.asp). Flash floods are often accompanied by substantial erosion, transport and deposition of sediments, and the generation of debris which may block river structures. The key characteristics of "plains" floods compared with flash floods are that they rise more slowly, cover greater spatial extent, last longer – days or weeks – and warning can be given to allow communities and individuals at risk to evacuate and take action to reduce the potential damage. The Rhine and Meuse floods of 1994 in Germany and the Netherlands are typical of plains flooding. Obviously, during a large-scale event there can be flash flooding in the catchment headwaters and plains flooding further downstream such as occurred in the Oder flood of 1997 and the Elbe flood of 2002.

2. FLOOD DEFENCE AND FLOOD RISK MANAGEMENT

2.1 Floods in history

Rivers and their floods have shaped human civilisation for millennia. Early narratives include the Mesopotamian Epic of Gilgamesh telling of a great flood and the Biblical description of the Deluge or Flood in the time of Noah. The prosperity of the great Egyptian civilisations depended upon the periodic flooding of the Nile to provide nutrients and the first measurement of water level being invented, now known as the "Nilometer", in the times of the Pharaohs to forecast the year ahead. About 2700 years ago, in the ancient civilisation of Sparta, there was the king-god Eurotas of Laconia, who controlled water and drained the marshes; now a Greek river bears his name.

The European documentary evidence for flood and drainage works stretch back at least for 800 years with legislation in the UK in about 1215 (Purnell, 1993) and 13th century records of flood defence works being constructed along the River Danube to protect communities like Szigetköz and Csallóköz (Zorkóczy, 1993). By the 14th century ring dikes had been constructed in the Netherlands creating polders where land and settlements were protected from flooding (Klaassen, 1998). In 17th century England, the Dutch Engineer Vermuyden undertook the construction of embankments, perhaps 100 km of new channel and wind pumps, thus changing the drainage patterns and flood levels in many thousand square kilometres of land. By the end of the 19th century the courses of many of the main rivers in Europe had been altered and embankments constructed for flood protection, see for example Belz *et al* (2001) for works on the Rhine. Thus the 21st century management of rivers is constrained by and must respond to the legacy of many generations of interventions in the natural system.

2.2   Policy approach and philosophy

The language used by the professional community involved in flood management reflects differing national and social attitudes to flooding and these have changed over the past 50 years. In Hungary the concept of "flood fighting" portrays the struggle between the community and the natural forces of the river and this is similar to the notion in many countries of providing flood defence, protection or prevention. All these terms indicate a philosophy of human control over nature and the protection of property against the "common enemy" of the elements. The Dutch approach to flood management has been strongly influenced by the 1953 disaster, with a statutory safety standard being maintained. Ensuring this involves both policy decisions on the acceptable standard and design and management activities to deliver that standard (Jorrissen, 1998).

However, with the adoption of Sustainable Development as an overarching policy commitment in the mid 1990's, the emphasis has shifted in the UK from defence *per se* to the broader action of flood risk management. In the summer of 2004 a consultation was launched in the UK on the future strategy for flood risk management called "Making space for water" (DEFRA, 2004). The title of this initiative is close to the concept of "room for the rivers" from the IRMA-SPONGE project (NCR, 2002), which considered flood management in the trans-national Rhine and Meuse basins.

The concept of flood risk management now joins together activities of several professional communities: hydrologists, hydraulic engineers, economists, social scientists, ecologists, and planners. The concepts used in these professions and between them internationally do not share a common terminology; thus there is a need to develop an accepted language of risk to facilitate communication of ideas and implementation of sustainable risk management measures. There has been progress in this regard,

- the summary report of the IRMA-SPONGE project (NCR, 2002) includes a multi-lingual glossary
- the UK flood risk management research programme has produced 9 page glossary at the in front of its initial risk and uncertainty review (DEFRA, 2002)
- the **FLOODsite** Integrated Project in the EC Sixth Framework Programme has an action to develop a common language of risk (see www.floodsite.net).

For the purposes of this paper, the exposure of a community or enterprise in a particular area to flood risk is taken as a combination of two factors, the *probability* of flood hazard in the area and the vulnerability of the area to undesirable *consequences* of a flood (Gendreau and Gilard, 1997). A similar definition was adopted in the IRMA-SPONGE project (NCR, 2002). Sometimes, this may be measured in simple terms as (DEFRA, 2002):

$$Risk = probability \times consequence$$

but this hides considerable complexity in the common valuation of consequences as well as describing the probability of flooding which may depend on many factors (event rainfall, landslides, bank failure, antecedent conditions, blockage of structures by debris etc.).

3. PRINCIPLES OF FLOOD MANAGEMENT PRACTICE

3.1 The RIBAMOD principles

It is generally recognised that the management and mitigation of flood risks requires a holistic, structured set of activities, approached in practice on several fronts, with appropriate institutional arrangements made to deliver the agreed standard of service to the community at risk. These are summarised a set of principles (Kundzewicz and Samuels, 1998), which were developed during the EC Concerted Action RIBAMOD.

Pre-flood activities include:
- *Flood risk management* for all causes of flooding
- *Disaster contingency planning* to establish evacuation routes, critical decision thresholds, public service and infrastructure requirements for emergency operations etc.,
- *Construction of flood defence infrastructure,* both physical defences and implementation of forecasting and warning systems,
- *Maintenance of flood defence infrastructure*
- *Land-use planning and management* within the whole catchment,
- *Discouragement of inappropriate development* within the flood plains, and
- *Public communication and education* of flood risk and actions to take in a flood emergency.
- Operational flood management includes four activities:
- *Detection* of the likelihood of a flood forming (hydro-meteorology),
- *Forecasting* of future river flow conditions from the hydro-meteorologiral observations,
- *Warning* issued to the appropriate authorities and the public on the extent, severity and timing of the flood, and
- *Response* to the emergency by the public and the authorities.

The post-flood activities (depending upon the severity of the event) may include:

- *Relief* for the immediate needs of those affected by the disaster,
- *Reconstruction* of damaged buildings, infrastructure and flood defences,
- *Recovery and regeneration* of the environment and the economic activities in the flooded area, and
- *Review* of the flood management activities to improve the process and planning for future events in the area affected and more generally, elsewhere.

3.2   The EU best practice document and policy development

An informal meeting of Water Directors of the European Union (EU), Norway, Switzerland and Candidate Countries took place in Denmark Copenhagen, 21-22 November 2002. This meeting considered the recent problems of flooding across Europe and agreed on an initiative in this area. Subsequently, a core group led by the Netherlands and France prepared a "best practice document" on flood prevention, protection and mitigation which was accepted by the Water Directors meeting in Athens in June 2003.

This document restates some guiding principles for flood management across Europe, the key conclusions were the need for a comprehensive set of actions and activities which pick up many of the points from the RIBAMOD principles above. The conclusions are summarised as:

- *"Integrated river basin approach*
- *Public awareness, public participation and insurance*
- *Research, education and exchange of knowledge*
- *Retention of water and non-structural measures*
- *Land use, zoning and risk assessment*
- *Structural measures and their impact*
- *Flood emergency*
- *Prevention of pollution"*

In addition, during their June 2003 meeting in Athens, Water Directors agreed:

- the need for a reinforced political commitment to flood prevention and flood protection,
- integrated river basin management as the tool of choice to address flood prevention and flood protection; and
- EU funding mechanisms as powerful means of promoting investment in flood prevention and protection schemes.

Following this, the European Commission, plans to bring together all the flood-related activities at EU level, linked to an *"Action Programme"* on Flood Risk Management (See Section 5.2 below).

4. ANATOMY AND ANALYSIS OF FLOOD RISKS

4.1 Model of risk analysis and management

The analysis and management of flood risks has several components. The **FLOODsite** research project is built upon a three-fold model of risk – a risk analysis model, a socio-economic model and a flood risk management model. The risk management model is essentially the RIBAMOD principles stated above (Section 3.1) which separates out pre-flood, flood emergency and flood recovery activities. These will not be discussed in detail in this paper as we concentrate here on the analysis rather than the management of the flooding risks. Indeed a thorough understanding of the risks will assist in developing the appropriate management strategies.

One formal approach to risk analysis is the Source-Pathway-Receptor-Consequences (SPRC) model. Broadly speaking, the Source and Pathway determine the probability and nature of the hazard and the Receptor and Consequences determine the impact of flooding on whatever human society values – people, built and environmental assets, cultural heritage etc.

4.2 Source, pathway, receptor and consequences

In the context of river flooding, the *source* of risk is the heavy rainfall, snowmelt, river discharge, storm surge or whatever gives rise to the potential for flooding. A characteristic of flood sources is that there is little or no ability to control the frequency of occurrence of the source. The *pathway* is the route which the flood source takes to the receptor. The pathway may include river channel, floodplain or embankments (if submerged or breached). The performance and characteristics of the pathway can be modified through engineering and maintenance activities. The *receptor* is the asset or person that can experience (usually undesirable) consequences from the flooding. The *consequences* are damages (whether capable of expression in monetary terms or not) and possible benefits from the flooding. The receptors of flooding can be influenced through strategies, policies and information which change the frequency of exposure of the receptors to flooding and the consequences of flooding may be altered by changes in design and load flood protection measures.

In order to identify receptors and understand the potential consequence on them the flooding needs to be characterised through questions. Can the land flood? What area is affected? What causes the flooding? How often do floods occur? How deep is the flooding? How rapidly does the flood rise? How fast does the water flow? How long does the flooding last? Can any warning be given? Answers to these questions will enable the individuals, groups, businesses, infrastructure, environmental and cultural assets to be identified which may be *exposed* to floods together with their *vulnerability* to experience damage or loss.

4.3   Scale of risk analysis

The scale of any assessment to some extent dictates the tools that may be used for the risk analysis. For example the identification and solution of the flooding problems for a particular community or area will need detailed information on levels at which properties flood, the occupation, use and contents of buildings, the routes which flood water take in the area, etc. However for broad appraisals of flood management strategies at a basin or sub-basin scale, such detail can be aggregated or approximated through proxy relationships.

Purnell (2002) and Richardson (2002) report on national scale flood risk assessments for England and Wales, which were undertaken by Government to identify the scale of the assets at risk from flooding and the potential effects of future climate change on the valuation of flood risk. Purnell (2002) indicates that approximately 10% of the UK population lives in an area of flood hazard (1% probability), with an asset base of over €300 Billion. Richardson (2002) indicates that the impact of climate change on river flood damage to the mid 2080's is to double (in current value) the annual damage valuations. Although such figures are approximate they represent key information in justifying national expenditure from taxation on flood risk management. Indeed the scale of the consequences was one of the factors that led to the Foresight exercise on flood management (See Section 5 below).

Hall et al. (2003) describe methods for national-scale flood risk assessment. These aggregate information on a 1 km basis in "flood impact" zones. These zones are delineated by river defences and indicative limits of inundation from national mapping undertaken in the late 1990's and available to the public in the Environment Agency web site. The likelihood of defence failure depends upon both the loading from the source and the properties of the defences taken from the national database of defence location, type and condition. The flood damage was estimated from information of property and commercial flood damage according to depth of inundation, coupled with socio-economic information from the 2001 national census to determine a social flood vulnerability index. Such national or regional scale analyses can be used to develop, compare and assess broad policies and strategies for flood risk management.

For the appraisal of a specific set of flood defence infrastructure, these national scale assessments may indicate that there is a potential case for work to be done. However, an economic appraisal of the proposals will need specific information on the location and cost of the works proposed and the flood damages and impacts they will reduce. Risk assessment can inform the selection of the options for and design of individual scheme works, providing a powerful means of negotiation between stakeholders with conflicting priorities, see for example Shoustra *et al* (2004) and Gendreau and Gilard (1998).

4.4 Valuation of consequences of flooding

Where decisions are made on a cost-benefit appraisal of investments, a risk analysis forms a core component of the appraisal. Thus there is a need as far as possible to devise a common metric for all the consequences of flooding; this presents considerable challenges and difficult choices. Some aspects of flood damage are readily measured in monetary terms including direct costs of inundation damage, clean-up, emergency operations, repairs etc and indirect costs of business closure and transport disruption. However, some human impacts such as distress, anxiety, fear and outrage are more difficult to value in financial terms even though there are available estimates for the "economic" loss of fatalities caused by floods. An important question, particularly in the context of sustainability appraisal is whether reduction of all impacts to monetary terms is desirable or whether multi-criteria analysis will provide a more robust and publicly transparent means of decision making.

4.5. Uncertainty and risk analysis

A further aspect of risk analysis is that the information for the analysis is imperfect and many methods of calculation and models are approximations to real life and this uncertainty gives rise to potential differences between assessment of some factor and its "true" value. Hence there is a lack of "sureness" surrounding the assessment (NRC, 2000) and thus additional chance that, say, some new flood defence infrastructure will not provide the intended standard of protection.

Understanding this uncertainty within our predictions and decisions is at the heart of understanding risk. Within uncertainty it is possible to identify:

- *Knowledge uncertainty* arising from our lack of knowledge of the behaviour or imperfections in our representations of the physical world,
- *Natural variability* arising from the inherent variability of the real world and
- *Decision uncertainty* which reflects the complexity of our social and organisational values and objectives.

However, this classification is not rigid or unique, for example, uncertainty on weather or climate will be taken as "natural variability" within flood risk management but as "knowledge uncertainty" in the context of climate simulation. The **FLOODsite** research project plans to develop a framework for analysing uncertainty in flood risk assessment and management.

A significant challenge for flood risk analysis and management practice is how to represent and communicate best the uncertainty in the assessments. These may be through a probability distribution from Monte-Carlo modelling or from more simple representations as "credible" bounds on some key parameters. The choice of representation of uncertainty may also depend upon the application area and type of decision to be made

(McGahey and Samuels, 2004). Since the subject of uncertainty in climatological, hydrological and hydraulic modelling is an area of active research, it is to be expected that professionals involved in the assessment and management of flood risks will need to review and update their knowledge and practice on this topic. The Precautionary Principle of sustainable development prevents lack of scientific certainty as being a reason for inaction, however, the bounds of uncertainty should inform decisions made.

5. FUTURE PERSPECTIVES

There is no doubt that risk analysis will be a key feature in the management of flooding over the coming decades within Europe. Some important current developments at national and European level are the UK Foresight programme on future flooding, the EU Action Programme on flood risk management and activities in the EU Sixth Framework Programme of Research and Development (FP6).

5.1 Foresight future flooding

Over the period October 2002 to March 2004, the UK Office of Science and Technology investigated drivers, responses and scenarios for flood risk over a timescale of about 100 years to inform public policy and expenditure. The flood risks were analysed at a scale of a 10km grid for four socio-economic scenarios, which were linked to global emissions scenarios and simulations of future climate. The scenarios represent different policy frameworks for the country and were entitled World Markets, National Enterprise, Global Sustainability and Local Stewardship. The Foresight project considered flooding from all causes:- urban storms, river, estuarine and coastal flooding. Drivers of flood risks were identified and ranked under each of these scenarios and the potential flood damages estimated for the 2080s.

Substantial differences emerged between the scenarios with the damage increasing in all scenarios if current policies are maintained. Current annual flood damage is estimated as about €1 Billion, with this rising to over €30 Billion without additional mitigation strategies in the worst scenario (World Markets). Future flood risks depend strongly on assumptions on global emissions of greenhouse gases which provides a clear link between international policy and impacts at the national scale. The report (OST, 2004) poses many questions to policy makers such as:

- Should the increasing levels of flood risk be accepted or actions taken to reduce them?
- How important is managing climate change to the risks faced from flooding and how best can this be achieved?
- How should land be used in balancing the wider economic environmental and social needs against creating a legacy of flood risk?

- What is the balance between societal responses to flood risk and the implementation of bigger structural defences?
- Who should pay for flood defence – the balance between government, developers, the individual and insurance?

5.2   The EU Action PROGRAMME

On 12 July the European Commission issued a communication on flood risk management, flood prevention, protection and mitigation (Document *COM (2004) 472 final*). Within that communication it is proposed that the Member States and the Commission should work together to develop and implement a co-ordinated flood prevention, protection and mitigation action programme. The essential features of this action programme should include:

a) *"improving co-operation and coordination through the development and implementation of flood risk management plans for each river basin and coastal zone where human health, the environment, economic activities or the quality of life can be negatively affected by floods;*

b) *developing and implementing flood risk maps as a tool for planning and communication;*

c) *improving information exchange, sharing of experiences and the co-ordinated development and promotion of best practices;*

d) *developing stronger linkages between the research community and the authorities responsible for water management and flood protection;*

e) *improving co-ordination between the relevant Community policies;*

f) *increasing awareness of flood risks through wider stakeholder participation and more effective communication."*

The development of the Action Programme is one of the targets set for the 2004 Dutch Presidency to the EU. The communication makes specific reference to the research being undertaken at Community level in FP6 as the **FLOODsite** project to contribute to the improvement of integrated flood risk analysis and management methodologies.

5.3   European research activities

The EU has sponsored research on hydrological risks for over a decade, commissioning over 100 projects with a funding of approximately €90 Million (Samuels, 2003) up to and including the Fifth Framework Programme. In FP6 several activities related to flood risks are anticipated including the Integrated Project **FLOODsite**, an Integrated Project on Flood and drought Forecasting (closure of call October 2004) and a Coordination Action, **CRUE**. FLOODsite comprises 35 research tasks in seven themes with significant actions on risk analysis and risk management; the project website contains a summary of the work programme. **CRUE** commences in November 2004 and aims to provide greater cooperation between national research programmes on flooding which together spend in excess of €30 Million per annum in the 25 Member States.

6. DISCUSSION OF THE CHALLENGES IN RISK ANALYSIS FOR FLOOD MANAGEMENT

Patterns of settlement, land use, flood defence and drainage in Europe have a significant legacy from previous generations of interventions in the natural systems of rivers and flood plains. Seeking the path of sustainable development implies that our generation's flood management practices should not place additional burden on future generations. Hence risk analysis for flood management in current and future conditions should be undertaken to inform policy choices, land use planning and public investment in infrastructure. This is implied by the aspirations in the EC Communication on floods and the UK Foresight process indicates a mechanism and approach for risk analysis over a time horizon of about 100 years.

Many disciplines are involved in flood risk analysis and so it is essential for mutual understanding that there is a common language and concepts for the subject.

Risk analyses may need to use multi-criteria decision tools to compare and balance factors that are incommensurate; this should facilitate transparency of decision making and building consensus between stakeholders and beneficiaries. Risk analysis can be used to structure the dialogue between the stakeholders in and beneficiaries of flood management activities. To do this effectively the risk analysis must be documented so that the key principles and results are accessible to a wide readership.

The world is dynamic in terms of both natural (environmental) and social systems (individual aspirations, societal expectations, and political priorities). Thus, in a risk analysis, stationarity must not be assumed in terms of the "sources" which force the flooding system, nor in the vulnerability and consequence on the "receptors" of flooding. Risk analysis can however, provide a framework for making decisions which are robust to the anticipated spread of uncertainty in future conditions through scenario analysis.

Flood risk analysis needs to make use of and integrate many sources of information. Data management systems need to be established to record the source and quality of the data used in the risk assessment, so that refinements can be targeted on those aspects which contribute most to the uncertainty. For example, available data on aspects of the system is often sparse, such as the construction materials and ground conditions for historic embankments.

Flood risk analysis will make substantial use of models of both physical processes and systems and of social responses and impacts. Research is underway in many national and international programmes, which will improve such models and provide integrated assessment frameworks. The use of probability concepts in the design of engineering works coupled with uncertainty estimates in a risk analysis will require a change in practice and understanding amongst professionals. Such advances need to be brought effectively into flood risk management practice though programmes of education, training and professional development for the practitioners involved as well as through appropriate software applications.

ACKNOWLEDGEMENTS

I acknowledge the support of the NATO Advanced Workshop in funding my presentation of the paper, and the **FLOODsite** project (contract GOCE-CT2004-505420) for its preparation as part of the project networking and dissemination activities. The opinions expressed in this paper are personal and do not necessarily reflect those of HR Wallingford or the EC Research Directorate General.

References

Belz J.U., Busch N., Engel H. and Gasber G. (2001) Comparison of River Training Measures in the Rhine-Catchment and their Effects On Flood Behaviour. *ICE Proceedings, Water & Maritime Eng*, 148, 123-132, Issue WM3, Sept.

Berz G. (2000) Flood Disasters: lessons from the past-worries for the future. *ICE Proceedings, Water & Maritime Eng*, 142, 3-8, Issue WM1, March.

Bronstert A., Ghazi A., Hladny J., Kundzewicz Z. and Menzel L. (1999) The Odra / Oder Flood in Summer 1997. *Proceedings of the RIBAMOD European Expert Meeting on the Oder Flood 1997*, EC Directorate Science, Research and Development, Brussels, ISBN 92-828-6073-6.

DEFRA (2002) *Risk Performance and Uncertainty in Flood and Coastal Defence*. Report FD2302/TR1, Joint Department for Environment and Rural Affairs and the Environment Agency Research Programme, Available from the R&D programme results web portal: www.environment-agency.gov.uk/subjects/flood/351291/211195/?version=1&lang=_e.

DEFRA (2004) *Making space for water – developing a new Government strategy for flood and coastal erosion risk management in England*. Consultation document available at the web address: www.defra.gov.uk/corporate/current.htm.

Gendreau N. and Gillard O. (1998) Structural and non-structural implementations – Choice's arguments provided by inondabilité method. In Casale R, Pedroli G.B. and Samuels P. (Eds.), *RIBAMOD River basin modelling management and flood mitigation Concerted Action, Proceedings of the First Workshop on Current Policy and Practice*, EUR 18019 EN, ISBN 92-828-2002-5.

Hall J.W., Dawson R.J., Sayers P.B., Rosu C., Chatterton J.B. and Deakin R. (2003) A methodology for national-scale flood risk assessment. *ICE Proceedings, Water & Maritime Eng*, 156, 235-247, Issue WM3, September.

Jorissen R.E. (1998) Safety, risk and flood protection, *RIBAMOD River basin modelling management and flood mitigation Concerted Action, Proceedings of the First Workshop on Current Policy and Practice*, Eds Casale R, Pedroli G B & Samuels P, EC Research Directorate General, EUR 18019 EN, ISBN 92-828-2002-5.

Klaassen D.C.M. (1998) Flooding risks of floodplain areas of the Netherlands, *RIBAMOD River basin modelling management and flood mitigation Concerted Action, Proceedings of the First Workshop on Current Policy and Practice*, Eds Casale R, Pedroli G B & Samuels P, EC Research Directorate General, EUR 18019 EN, ISBN 92-828-2002-5.

Kundzewicz Z. and Samuels P.G. (1998) *RIBAMOD River basin modelling management and flood mitigation Concerted Action. Proceedings of the Workshop and Second Expert*

Meeting on Integrated Systems for Real Time Flood forecasting and Warning, Published by DG XII, European Commission, Luxembourg, ISBN 92-828-6074-4.

McGahey C. and Samuels P.G. (2004) A practical approach to uncertainty in conveyance estimation. *39th DEFRA Flood and Coastal Management Conference*, University of York, June 2004, Proceedings to be published by DEFRA, London.

NCR (2002) *IRMA-SPONGE: Towards sustainable flood risk management in the Rhine and Meuse basins*, Netherlands Centre for River Studies, Delft, Publication 18-2002, ISSN 1568-234X.

NRC (2000) *Risk analysis and uncertainty in flood damage reduction studies.* National Research Council, National Academy Press, Washington, ISBN 0-309-07136-4.

OST (2004) *Future Flooding – Executive Summary.* Office of Science and Technology, Victoria Street, London, SW1H 0ET, available at web site www.foresight.gov.uk.

Purnell R. G. (1993) Flood defence organisation and management. *Proceedings of the British-Hungarian Workshop on Flood Defence*, VITUKI, Budapest ISBN 9635111134.

Purnell R.G. (2002) Flood Risk – A Government Perspective, *ICE Proceedings, Civil Engineering*, 150, 10-14, Special Issue 1, May.

Richardson D. (2002) Flood Risk – the impact of climate change, *ICE Proceedings, Civil Engineering*, 150, 22-24, Special Issue 1, May.

Samuels P.G. (2003) Flood risk and flood forecasting – the state-of-the-art in EU research.. *Proceedings of EU-MEDIN conference on disaster research - The Road to Harmonisation*, Thessaloniki, *to be published by EC Research Directorate General.*

Shoustra F., Mockett I., Van Gelder P. and Simm J. (2004) A new risk-based design approach for hydraulic engineering, *Journal of Risk Research*, 7 (6), 581-597.

Wang Z.Y. and Plate E. J. (2002) Recent Flood Disasters in China., *ICE Proceedings, Water & Maritime Eng*, 154, 177-188, Issue WM3, September.

White W.R. (2001) Water in Rivers: Flooding. *ICE Proceedings, Water & Maritime Eng*, 148, 107-188, Issue WM2, June.

Zorkóczy Z. (1993) Flood defence organisation and management in Hungary. *Proceedings of the British-Hungarian Workshop on Flood Defence*, VITUKI, Budapest ISBN 963 511 113 4.

Chapter 3

HYDRO-INFORMATICS' TOOL IN PRACTICE
"BEFORE – DURING – AFTER" AUGUST 2002 FLOOD
ON VLTAVA AND ELBE CATCHMENTS – CZECH
REPUBLIC

JAN ŠPATKA, PETR JIŘINEC, PETR SKLENÁŘ,
EVŽEN ZEMAN
DHI Hydroinform a.s., Prague, Czech Republic

Keywords: Flood, mathematical modelling, flood zoning, flood protection, flood protection
measures.

1. INTRODUCTION

For the Czech Republic, floods represents the highest direct risk stemming
from natural disasters and they can be a casual factor of serious crises,
which can be associated with high material property damages, ecological
and cultural damages and also losses of human lives in the areas affected
by floods. The flood events have their genesis almost exclusively inside
the borders of the country and are provoked by snowmelt in the period
from December to April, and by precipitation in the period from May to
August.

The precautionary principle and principles of systematic preventive
measures were highly neglected because the last floods with fatal
consequences occurred in the Czech territory at the end of 19th century.
The above facts were fully proven during catastrophic floods which
occurred in summer 1997, 1998 but primarily in 2002 (see Figure 1).
Market driven economy also did not recognise a priority in flood
protection in late nineties, so it was possible to recognise a rather low level
of awareness of the flood risks and certain stagnation of activities focused
on the development of preventive flood measures systems in last decades
until 1997 a severe and regional flood stroke the large area of the Morava
River basin. Based on the detailed assessment results of the catastrophic
flood in 1997 and experience gained from rehabilitation activities, the
Czech Government has assigned a task to prepare a Flood Protection
Strategy as a basis for systematic approach in this area and a basic material
for preparation of necessary measures.

J. Schanze et al. (eds.), Flood Risk Management:
Hazards, Vulnerability and Mitigation Measures, 35–46.
© 2006 Springer.

36

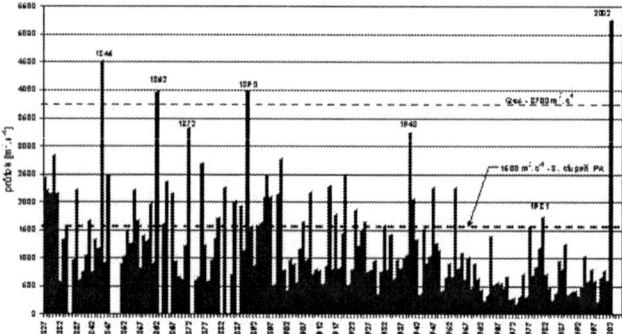

Povodně na Vltavě v Praze

*Figure 1. A comparison of floods which occurred in Prague within last 175 years –
prepared by CHMI – 2002*

New concept of flood prevention was defined by the newly defined
Flood Protection Strategy. The most important highlights are:

- The flood management structures have to be assessed in conceptual
  way (the impact of the structure has to be assessed not only locally
  but also generally in the whole river basin), only mathematical models
  provide the relevant answers on impacts,
- Design floods are based on historical floods including their dynamic
  characteristics and the impact of the newly designed structures has to
  be compared with the current situation – with and without the
  structure itself,
- There has to be emphasized an interest to enlarge the natural flood
  plain retention capacity along the river and any restriction of flood
  retention volume should be compensated by other complementarily
  designed measures,
- Environmental aspects and hazards have to be assessed,
- Best forestry and agricultural practice has to be taken into account in
  order to increase the retention capacity in upper parts of the basin,
- Moveable-temporary constructions are very important supplementary
  activities in flood mitigation effort,
- Sediment transport phenomena has been ignored in the past periods
  for many times and the designed structures has failed to operate
  accordingly due to this phenomena,
- Urban drainage systems (primarily sewers and local streams) have
  direct impact on functionality of the flood mitigation structures and
  the assessment of mutual relation between two systems has to be
  performed in order to get a platform for operational rules and actual
  relevance of the scheme (pumping of gravity driven sewer system of
  the city with 1M inhabitants over flood water levels might create a
  crucial sanitation problem for longer flooding period in comparison
  with leakage from dikes),
- All newly proposed/designed structures have to be assessed also for
  the impacts of consequences of discharge considerably higher than the

design flood. This is very important for flood management planning (If ... , then ...scenarios)

- All newly designed structures have to be equipped with emergency release scheme in case of higher discharge (or water levels) than designed ones.

An example of Prague and Elbe, described in this paper below, was taken to provide an overview how the strategy is reflected in reality.

2. FLOOD MODEL OF ELBE RIVER "BEFORE" THE 8/2002 FLOOD EVENT

After severe flooding in 1998, the Elbe River Water Authority initiated the development of an advanced mathematical model complex for use in the flood prevention efforts. The models have been going to cover the 110 km of the River Elbe between the confluence with the River Vltava and the border with Germany. Both 1D (one-dimensional) and 2D models have been established and linked to GIS (Geographical Information Systems) for flood mapping acquisition.

Detailed 2D model and is described in detail in the computational network of the MIKE 21C model with dimensions of 11 032 x 342 points. The computational network was concentrated in the area of constructions (bridges, proposed riverbed drops) to a distance between points of 3 – 5 m (in lengthways and transverse direction), whilst in places of relatively flat flood plain the distance between points is 7 – 10 m.

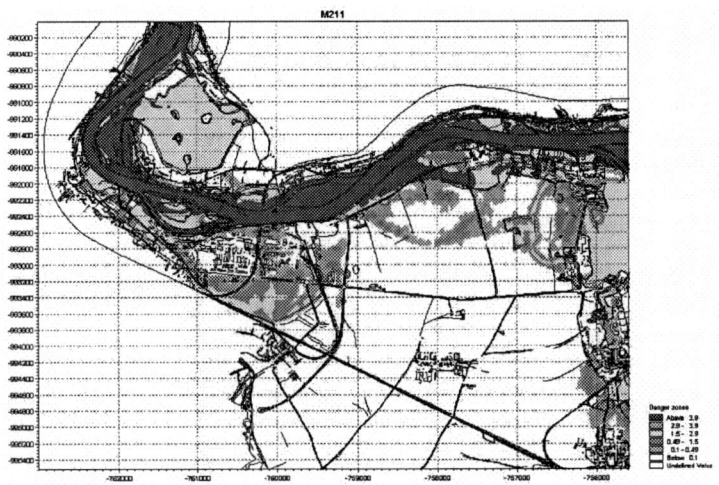

Figure 2. Example of danger zones on Elbe River

Based on the maps showing the areas flooded, the water depths, flow velocities, etc, it would be possible to improve the flood mitigation measures along the Elbe River. These maps would also support the authorities in order to regulate the urban development in flood prone areas. For instance it would be possible to guarantee that chemical plants and other industries with high risk of dangerous pollution have been located well outside the zones where flooding can occur.

In areas needing flood protection there has sometimes been new development, due to the fact that for a long period no severe floods had been recorded and human memory had faded. Lack of awareness of flood risks has led to risky use of land in flood plains.

The planned simulations of the Q20 year and Q100 year floods were completed in June 2002, e.g. before the August flood disaster. However, the results were not yet reported, and the relevant Czech authorities had not had any chance to implement improvements in their flood management procedures on the basis of this important new source of information. Hence, only the swift actions by the staff of DHI Hydroinform in combination with equally swift and courageous decisions made by the managers in charge of the rescue operations made it possible to use the models during the disaster.

3. USING THE FLOOD MODEL OF ELBE RIVER "DURING" THE 8/2002 FLOOD EVENT

Despite the fact, that this system was not completed on set of the even worse flood disaster in August 2002, the models nevertheless played a crucial role during the disaster. Unlike the authorities in the neighboring countries, the Czech authorities had access to accurate prediction of flood elevations and made very efficient use of the information in the difficult rescue operations.

Sunday evening, August 11th, 2002 – two days before the peak of flooding of the Vltava in Prague, upstream from the Elbe area, it was evident that the disaster would also affect the areas covered by the Elbe flood models. An agreement was then quickly made between a model owner, the Elbe River Basin Water Authority, and DHI Hydroinform to make the model available for operative forecasting of flooded areas and for support of the flood councils and other state administrations.

First of all, maps of already modeled flood areas (for Q20 and Q100) were copied and distributed. However, it soon became evident that the flood would be more severe than a Q100 year event, which corresponds to a peak discharge of around 4.000 m3/s. Consequently, new simulations had to be executed based on CHMI's forecasted discharges which were in the range from 4.500 to 5.500 m3/s. Due to the urgency of the situation, the faster 1D+ (MIKE 11) model was used for these simulations.

Modeled peak water levels for predicted discharges were drawn onto longitudinal profiles of the river and sent via e-mail to the responsible authorities. For the largest predicted discharge of 5.500 m3/s maps of flood lines and maps of water depths were generated and also distributed in the electronic form via e-mail and FTP as well as in a paper form as A3 map sheets for operational use. These 39 maps were distributed by the Elbe

River Basin Water Authority and Operation Dispatching of Integrated Rescue System via helicopters.

Thanks to the availability of these maps, the endangered areas were known and the evacuation of people and dangerous substances could be coordinated and managed.

4. FLOOD MODEL OF ELBE RIVER "AFTER" THE 8/2002 FLOOD EVENT

Conceptual design of flood protection measures (dikes, walls, etc.) was proposed based on computed hydraulic characteristics and impact of proposed flood protection measures on flow characteristics was evaluated by new calculation.

For example, quite new design of "West bridge near Litomerice town" (level of the road in entire flood plain may not be flooded in the time of high water as in August 2002) was evaluated by the model. Impact of the bridge and roads on high causeways on flow characteristics was mineralized having used results of many computed variants.

Not only high flow, but also low flow and sediment transport phenomena were investigated by this model.

It has long been necessary to improve the shipping conditions on the 40 km long Elbe reach between Strekov and Hrensko. On the basis of previously conducted studies and analyses, the optimum conception for attaining this target has been assessed to be one which envisages the construction of two new barrages Male Brezno and Prostredni Zleb, and the implementation of partial channel dredging in limited sections of the waterway. Building of both barrages including needed river bottom dredging might change sediment transport regime of Labe River. For this reason future sediment transport and morphological changes were studied after construction under flood condition.

Mathematical model has described future sediment transport regime of Labe reach down Strekov after construction of designed barrages during and after flood events. In several relatively short sections of the channel (hundreds of metres up to individual kilometres long) scours were formed with the subsequent nearby deposition (see Figure 3).

The natural helical flow in the elbows has a conspicuous effect upon these processes. In all river reaches no continuous sediment transport was observed, only local morphological changes appeared. The bed sediment was never carried away for a longer distance. Remarkable morphological changes are linked to the corridor sections proposed for dredging and likewise are connected with high values of vertical velocities and also with the transverse velocity values, which were ascertained in previous research in this locality. Simulation results demarcated areas for necessary fairway dredging and can be used as recommendations for optimum realization of deepening of river channel.

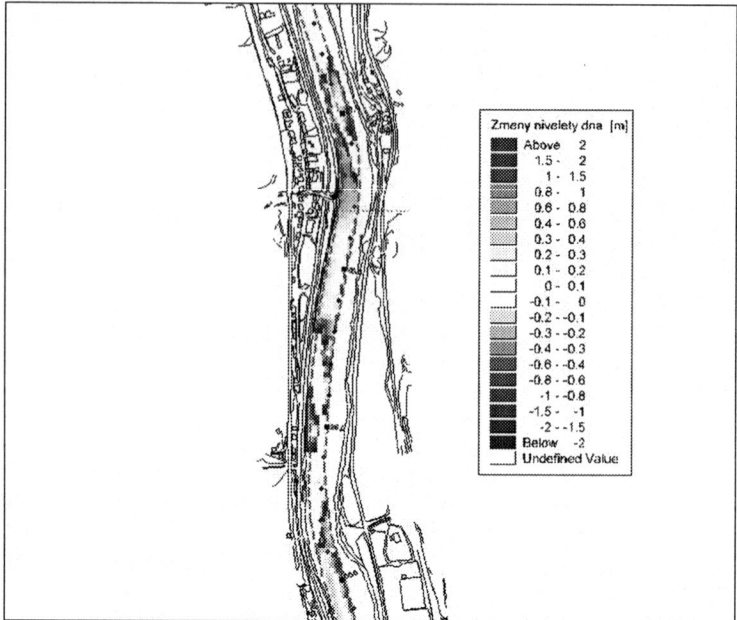

Figure 3. Example of morphological changes due to sediment transport phenomena

5. FLOOD MODEL OF PRAGUE "BEFORE" THE 8/2002 FLOOD EVENT

In mid of 90´s, the City Council of Prague and the Vltava River Basin Authority decided to improve the preparedness of Prague for floods after enormous floods that occurred in Europe in mid-90´s. During the period of preparation and feasibility of the general flood protection scheme for Prague several alternatives were considered including "action 0": doing nothing.

After some decision-making process, the mathematical modeling tools were finally selected for an assessment of the flood protection strategy and derived measures for the Capital city of Prague and its close neighborhood.

The first comprehensive study, finished in 1997, had to be based on rather complicated hydrological data set. Though there was unique historical data available, there were many changes in urbanized areas in the basins as well as man-made changes in the river beds themselves. The study included hydrodynamic 1D+ mathematical model of Prague and also the phase of rainfall-runoff modeling.

The modeling effort was among others focused on objective assessment of the myth, that the cascade protects the city from catastrophic floods. The reality is that the volume of all the Vltava cascade reservoirs represents indeed only the fraction of the total volume of the 1890 flood wave. This means that the cascade is able to eliminate the impact of minor floods, but has only slight influence on the damaging flood events.

The 1D hydrodynamic model itself covered 33 km long river reach, limited by official borders of City Prague. The global model set-up of the Vltava basin above Prague was applied in order to obtain transformed flood waves in the Vltava River based on operational rules of all large reservoirs. The operational rules were set-up by the Vltava Water Basin Authority dispatchers, which were present during the simulation and influenced the simulation similar to reality.

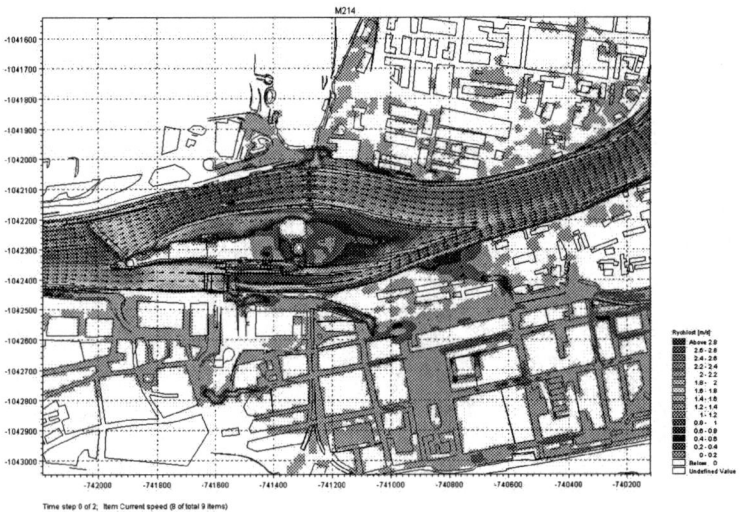

Figure 4. An example of the velocity fields – flood mapping in Prague

The results, outputs and model setups were supposed to be used in several areas:

- Flood mapping for town planning schemes
- Rescue operation plans in emergency situation when Prague is flooded
- Evaluation of new flood mitigation plans and their impacts on overall flood condition
- Evaluation of operational rules of the Vltava Water Basin Authority under the flood condition
- Preparation of mapping outputs for GIS layers supervised by the city council of Prague and similar data processing activity for the Vltava Water Basin Authority
- Support for detail city planning activities

All these aspects were utilized by the city council authorities. The flood mapping results were accepted and by a comprehensive comparisons with the former flood contours several places were discovered, where the currently valid map had performed a low accuracy. The weakest points of flood defense were specified and suggestions for improvements were assessed. The 2002 flood proved, that most of the conclusions based on this study were correct and in some cases provided the crucial margin in decision making process of the local authorities – the best example are the mobile barriers which saved the Old Town in August 2002.

Figure 5. Charles Bridge in Prague on horizon – Forefront- non-permanent flood protection anchored to the pedestrian path with the reinforced foundation platform – the photo was taken during the flood 8/2002 in Prague

After the success of the first 1D Prague Flood Model, the cooperation with the Vltava Water Basin Authority and the City council authorities has been constantly at very high level. When the DHI developed a new tool for 2D modelling – MIKE 21 C, the decision to update the Prague Flood Model came almost naturally.

The grid of a whole model has 1 081 348 computational nodes, when the spatial resolution of a grid varied between 2-3 meters, necessary for describing the Old Town area with narrow and winding streets, and 5-9 meters in the floodplain, less important areas.

The results of a new 2D flood model of Prague showed some interesting moments, which could not be recognized in the original 1D model – the flow direction changed rather differently from the original estimation, which had the significant influence on overall flood behaviour of these areas. The flood 8/2002 was definitely far higher than the 1890 flood, used in the framework of all the flood models so far. According to the first estimates, the peak discharge of 2002 flood reached 5300 m3/s, contrary to 4030 m3/s of the peak discharge of the design flood - modified flood wave from the year 1890. As it was impossible to carry out proper calibration of the models before 2002 because of lack of relevant water marks and already mentioned river morphology and all basin changes, all the calibration factors - bed resistance coefficients - were set at the upper part of their ranges for safety reasons. The first part of the flood protection measures were built with rather high safety margin, which saved the Old Town from flooding. But it was tight – only a few centimetres were left out of the original 50 cm margin.

6. FLOOD MODEL OF PRAGUE "DURING" THE 8/2002 FLOOD EVENT

Nevertheless the benefit of the 2D approach was un-doubtable and it was obvious, that it is only a question of time, when the hardware development enables the fully dynamic simulations to be performed without time constraints. All the generated maps were used during the Flood 2002 event. Moreover, mathematical models presented here were used operatively for support in the Flood protection council of Prague. According to the CHMI forecast, the actual flood maps were generated and delivered to the Flood management committee. Rescue workers used already available outputs of the flood maps with clear indications of depths, water levels and in many cases also velocity fields under the discharge, which were usually lower than the catastrophic ones in August 2002. But the maps gave the rescue workers certainty and overview so desperately needed in such cases of crisis.

Meteorological situation, which caused the flood event in August 2002 could be shortly described as a rare condition, where two rather well-developed frontal borders had remained unusually long time above the Czech territory.

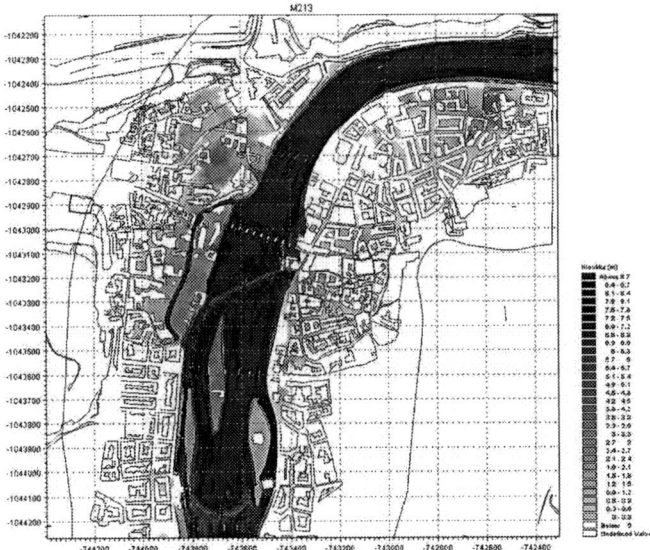

Figure 6. An example of the depths mapping in Prague

The retention capacity of the reservoirs of the Vltava cascade was kept according to the valid ruling curves, but the dispatchers were helpless against the volume of water rushing down the basin. However thanks to available limited retention volume in the Vltava cascade the dispatchers of the Vltava River Basin Authority were able to keep the outflow of regulated cascade at the reasonable level of discharge for a while, which enabled Prague authorities to anchor all boats in the inland ports and to put up the moveable flood protection sections in the central area of Prague. This fact also assisted in the way that the other flood protection measures were in majority executed and evacuation in Prague could take place without chaotic patterns. However due to high volumes coming from all rivers into the cascade the outflow from the cascade became unmanageable soon, and it caused never recorded flood even in lower parts of the Vltava river reach.

It was clearly indicated that the monitoring network was underinvested within the last century. Most of the hydro-posts were destroyed, some were not able to transmit or even record under the very high water level condition during the flood 8/2002.

A meteorological forecast was available and quite accurate. Hydrological forecast was not available fully in the later stages, because most of the models were out of range and on-line data were, later during the event, hardly available, because the posts were either destroyed or blind. Rating curves were not accurate enough or did not exist in the upper area for very high discharges. Regardless the shortages and problems it is possible to state that the basic information was available thanks mainly to modern hydro-informatics tools and mathematical models and the media (primarily TV) spread the information quite well.

7. FLOOD MODEL OF PRAGUE "AFTER" THE 8/2002 FLOOD EVENT

Very natural and the only logical step after the flood was to calibrate the model for the peak discharge which occurs during the flood 8/2002. During the flood the reliable flood water marks were fixed and recorded, and the authors of the model had the perfect overview of everything what happened in particular phases of the event. Thanks to that, were possible to polish model calibration a apply recalibrated model on further Prague flood protection studies.

Because the Prague Flood Protection Project has become an issue, the first basic step of the study was to assess different scenarios of potential changes examining the influence of the already finished flood protection in some various states of completion and modifications. Other scenarios were used to evaluation of proposed line of the flood protection reducing the flood plain and different variant of central waste water treatment plant protection.

For example, one of these scenarios assumed that the central waste water treatment plant (CWWTP) is fully protected, the second one assumed the CWWTP is completely removed and another scenario assumed that the CWWTP would be arched over by the flat roof at the elevation of current dikes.

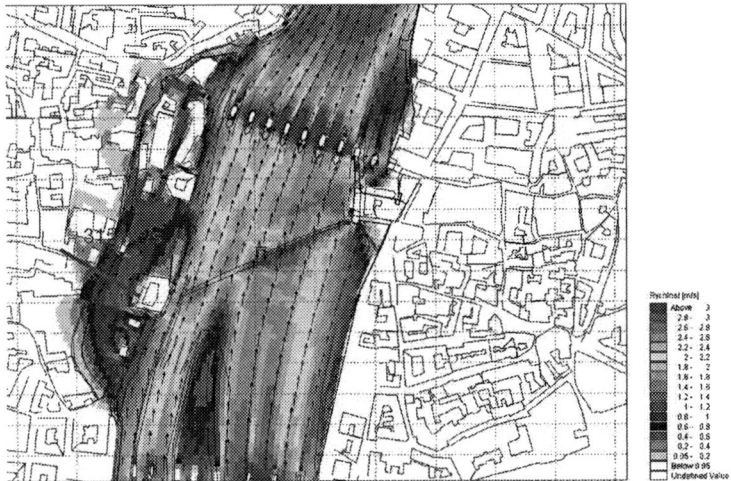

Figure 7. The close up of flowing through Kampa park – flow velocities and the discharge fraction

The results of another simulations successfully proved, that the Prague Flood Protection will not considerably affect the flow conditions during high floods, as it will cut off mainly only the passive parts of the flood plains.

Figure 8. Moveable construction as completed prior flood in August 2002 – phase 0001 – designed by Hydroprojekt a.s. – total cost 90M CZK.

The last simulation proved the expected fact, that the Kampa protection – part called "P0002" almost would not affect the situation of the Charles Bridge at all. The discharge through the Kampa Park is only 300 m3/s out of 5300 m3/s the total, so the changes are fractional.

The described results of simulations carried out were accepted by all involved authorities and 2D Prague Flood Model has become an integral part of the design process of the Prague Flood Protection. As the next task the evaluation of the final designed form of Prague Flood Protection was carried out at the beginning of 2004, and when the final 2002 hydrographs are evaluated, the fully dynamic simulation of the 2002 flood wave will be executed.

8. CONCLUSIONS

Finally the 2002 flood fully proved the usefulness of mathematical models for flood protection at any stage, for the flood measures design process, for improving the flood forecasting, for producing easily understood maps and charts and for improving the rescue activities. Mathematical models now have become an integral part of Flood Protection scheme in the Czech Republic.

References

CHMI (2002) Preliminary report on hydrometeorological situation of the Flood in August 2002. 2nd preliminary version of the report August 29, CHMI, Prague.

DHI Denmark (2000) Flood Management in the Czech Republic (1998 - 1999) – Final Report. DHI, DHI Hydroinform, Morava Water Board Brno.

DHI Denmark (2002) Flood Management in the Czech Republic Phase II (2000 - 2001) – Final Report. DHI, DHI Hydroinform, Morava Water Board Brno.

DHI Hydroinform a.s. (2001) 2D Prague Flood Model. Report of study, Prague March 2001.

DHI Hydroinform a.s. (2003) 2D Prague Flood Model, version 2002. Report of study, Prague, January 2003.

Hydroinform a.s. (1997) Prague Flood model – Comprehensive study. Prague.

Kubat J. (1999) 1997/1998 Floods in the Czech Republic: Hydrological evaluation. Marsalek J., Watt W.E., Zeman E and Sieker F. (eds.) Flood Issues in Contemporary Water Management. *NATO Science Series, Series 2. Environmental Security* 71, 25-39, Kluwer Academic Publishers.

Patera A. (1999) How floods influence the development of Prague city. Patera A. (ed.) The Vltava River Basin Authority, Prague.

Chapter 4

EXPERIENCES IN APPLICATION OF HEC-RAS MODEL UNDER CIRCUMSTANCES OF FLOOD WAVES

SANDOR KOVACS[1], ATTILA KISS[2] AND
JANOS SZEKERES[3]
[1]*Middle Tisza-valley District Water Authority, Szolnok, Hungary,*
[2]*Körös-valley District Water Authority, Gyula, Hungary,*
[3]*Water Resources Research Centre, Budapest, Hungary*

Keywords: Hydrodynamic modelling, HEC-RAS, floodplain rearrangements, reservoirs.

1. INTRODUCTION

The very varied flow regime of River Tisza is characterized by repeated floods with gradually increasing and sometimes very high peak water levels and by frequent, long-lasting low flows. The main causes of these characteristics are: the shape of drainage basin; the discharge of tributary streams related to the one of the main channel; the high sediment transport; the low slope of the low-land reach and the trained low-flow and high-flow river channel. The present length of R. Tisza is 945.8 km and its drainage basin is 157 200 km^2. According to discharge measurements of the recent decades, the maximum discharge of Tisza River at section Tivadar (705.7 rkm) exceeds 4 000 m^3/s, in the middle reach between Szolnok and Kisköre: 2 600-2 900 m^3/s and in the lower reach, at section Szeged (173.6 rkm) above 4 000 m^3/s.

Before the comprehensive river training, River Tisza has been an intensively meandering, low-slope stream. Due to training measures carried out in the XIX century, the length of the river has decreased by 30%, its planform has changed and its slope became somewhat steeper. In the Hungarian Great Plain, the slope of the river varies between 1 and 6 cm/km. Flood protecting dikes were constructed along the river. Presently, the levees reach from Técső (887 rkm) till the confluence with River Danube. Between these dikes the water and sediment transport take place in a much narrower compound channel than the original floodplain was.

J. Schanze et al. (eds.), Flood Risk Management:
Hazards, Vulnerability and Mitigation Measures, 47–58.
© 2006 *Springer.*

Similarly, the river channel in the low-land reaches of the tributaries have also been trained and provided with several km long flood protecting levees. The width between the dikes varies from 350 and 6 000 m.

Between 1998 and 2001, unprecedented high flood waves had run off on River Tisza. In the upper and middle reaches the peak water levels have surpassed the previously observed maxima by 1.3-1.4 m. These new maximum values initiated the review of training principles on River Tisza.

The "Further Development of the Vásárhelyi-Project" aiming at the flood-flow regulation of R. Tisza and at the regional developments, tries to achieve one of its main goals: to decrease the maximal flood level by 1 m. It is to be accomplished by improving the water conveyance capacity of the flood river channel (to re-establish the pre-1970 conditions) and by establishing emergency flood reservoirs on the Hungarian floodplain (Figure 1).

Figure 1. Plan of the Tisza River

2.   HYDROMETEOROLOGICAL FACTORS

The unfavourable combination of hydrometeorological factors, the coincidence of flood waves of Tisza R. and the ones of its tributaries basically determine the values of maximum water levels. In the followings, it is investigated whether the hydrometeorological phenomena occurred in the years 1999/2000 may be regarded as exceptional.

The flood wave of April 2000 was preceded by a period of long-lasting and high precipitation. In the mountains, the accumulation of snow has started already at the beginning of the previous November. The water content of the snow layer has gradually increased due to smaller warm spells and repeated snowfalls. Despite of the fact, that till the end of March the water content of the snow layer has not reached the value of 1999, in some places unprecedented high snow depths and densities have been recorded at some regions. The quantity of the accumulated snow was high, however, it still could not be considered exceptional. (In the years 1970,

1985 and 1987, the terrain was covered by a snow layer of similar quantity.)

In the upper reach of River Tisza, runoff hydrographs of 3-4 local peaks frequently occur, producing gauge heights between 400 and 600 cm. These used to superimpose in the middle reach of Tisza R., running off in the form of one significant high flood wave. In the upper reach, unusual phenomena in 2000 were the closeness and size of the individual flood waves. Induced by a continuous precipitation activity between 8th of March and 8th of April, three flood waves started to run off the sub-catchments, one after another and increasing in size. The first one had a peak gauge height at Vásárosnamény (684.4 rkm) H_{max}= 769 cm, filling up to flood plains on the middle reach of Tisza River. After a short recess period, the second flood wave has arrived to the same section with H_{max}=806 cm which could have produced at the section Szolnok (334.6) a gauge height of 900 cm. Thereafter a third flood wave started to run off, exceeding the previous two and reached Vásárosnamény with H_{max}=882 cm.

Three flood waves of this size have never occurred yet at Vásárosnamény within one month. Simultaneously with the flood of the upper reach, flood waves have started on each of the tributaries of River Tisza. The last group of flood waves has reached its peak water levels at the state border sections almost on the day (6-7 April). Despite the flood waves of the tributaries have not produced new maximum gauge heights, with the exception of the Fekete-Körös River, their great water volumes filled up the channel in the middle reach of Tisza R. to a very high extent. The third flood wave of Tisza River has running off in that partially filled-up channel.

Is it possible that a flood wave exceeding the April 2000 one could occur in the middle reach of Tisza River? The answer is definitely yes. There have been and probably will be again weather conditions even more unfavourable than were in 2000. It should be considered that how big flood wave could have occurred if the precipitation of 2001 fallen onto a snow cover accumulated in 1999? From meteorological point of view, this condition can not be precluded. One of the historical design floods from hydrometeorological viewpoint was the one occurred in 1888, during which the gauge height of Tisza River at Vásárosnamény has exceeded 790 cm for 16 consecutive days. Simultaneously, the water level in the tributaries was exceptionally high. In the recent years, the duration of gauge heights above 800 cm at Vásárosnamény was never longer than 3.5 days. In 1888 the maximum gauge height at this section was 900 cm while the flood protecting levees on the tributaries have been very severely damaged.

Figure 2 shows the time series of gauge heights at Vásárosnamény and Szolnok in 1888 and in 2000. The significant differences in flood wave volumes and water levels can clearly be seen.

In 1999 and 2000 flood waves exceeding the design flood water levels and having long durations have occurred in the middle reach of Tisza River. On the other hand, in 1998 and 2001, exceptionally high flood waves have been generated in the upper reaches. The autumn flood wave in 1998 was entirely due to the high amount of precipitation. In 1998

50

between 25-30 October and 1-5 November, the drainage basin of the Upper Tisza R. has received 136.9 mm precipitation as areal average. The flood wave in March 2001 was partially generated by the melt of 2 km3 snow pack covering the basin, however, the main cause of it was the heavy rainfall (124.5 mm in the period of 3-5 March). The meteorological situation was similar to the one in the autumn of 1888. Both flood waves have produced significant but no decisive rises of water level.

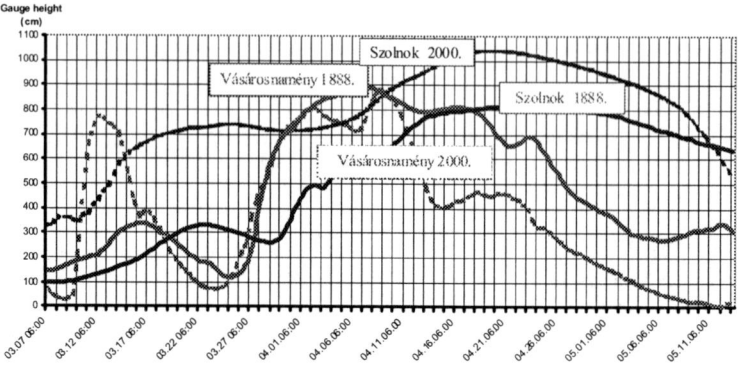

Figure 2. Hydrographs of the flood waves in 1888 and 2000

3. DISCHRGE CARRYING CAPACITY OF THE RIVER CHANNEL

The flow carrying capacity of a river channel can be expressed most accurately by the analysis of discharges. This statement holds on the middle reach of Tisza River having low slopes of channel and water level, despite the backwater effects of Hármas-Körös and Maros Rivers and partially that of Danube R. These influence the runoff of flood waves and the shape of flood discharge loops as well.

One important problem of discharge analyses was whether the results of the historical discharge measurements (from the years 1895 and 1932) should be accepted and involved? It will turn out from the followings that these measurements by no means should be disregarded. The runoff phenomena occurring in the regulated riverbed in the last 100 years can be explained and understood by processing and analysing the mentioned data only.

In Figure 3 the results of discharge measurements carried out during the flood waves of extraordinary size in the years 1895, 1932, 1970, 1979, 1999 and 2000 are plotted. It has to be remarked that in 1895 and 1932 the discharge measurements were performed at Tiszapüspöki and at the riverbed narrowing Vezseny, instead of the section Szolnok. The gauge heights, however has been correlated to the staff gauge of Szolnok.

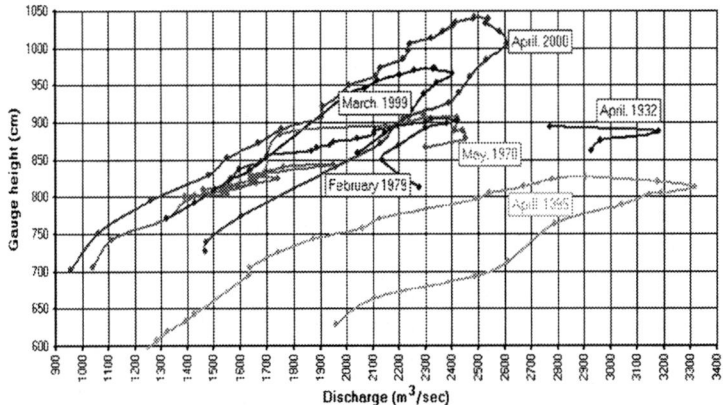

Figure 3. Discharge loops at Tisza-Szolnok

The looped flood discharge rating curves are differing remarkably from each other, however, in some cases also similarities can be detected. The maximum discharges measured in 1895 (H_{max}=827 cm) and in 1932 (H_{max}=894 cm) were found similar and both have exceeded 3 100 m³/s. (In 1932 the discharge was measured in the vicinity of the maximum water level only). The maximum discharge in 1932 was smaller than that in 1895 by 140 m³/s, however, the peak gauge height was higher by 67 cm. During the 32 years long period between 1895 and 1932, the planform of the flood protecting levees has not changed significantly. The rise of water levels can be best explained by the alterations of the discharge carrying capacity of the river channel. Unfortunately, during the flood wave in 1919 there was no possibility to measure discharges.

The gauge height and discharge data measured in 1970 (H_{max}=909 cm) and in 1979 (H_{max}=904 cm) were very similar. Discharges above 2400 m³/s belonged to gauge heights slightly exceeding 900 cm. Between 1932 and 1970, 38 years have passed, almost the same as between 1895 and 1932. The flood waves in 1970 and 1979 produced peak gauge heights slightly above the 1932 maximum only, however, their peak discharges were lower by 400-500 m³/s.

During the runoff of recent flood waves, significantly higher maximum water levels have occurred than in the past. In 1999 (H_{max}=974 cm) similar peak discharges were measured as in 1970 and 1979, however, the maximum gauge height was higher by 65 cm. The flood wave in 2000 (H_{max}=1041 cm) produced a maximum discharge of 2600 m³/s only. The steepness of loops of the discharge rating curves has gradually increased during the past decades. It has to be remarked that presently at the section Szolnok already 3.5 m³/s increase is sufficient for 1 cm water level rise.

The prediction of effects of emergency flood reservoirs and improving the flood conveying capacity was based on scientific investigations. The most important part thereof was the numerical hydraulic computations including 1D hydrodynamic modelling. The paper summarizes the hydrodynamic carried out using the *HEC-RAS* software.

4. HYDRODYNAMIC MODELLING

4.1 Stream network

The database of the model presently comprises the 740 km long Tiszabecs-Titel reach of R. Tisza, including its 8 tributaries and two side arms as concentrated contributions. In this way, the total length of streams involved in the computations is more than 1500 km.

4.2 Data of cross-sections

The HEC-RAS model (like other 1D models) is capable to take cross-sections given by the usual distance – elevation data-pairs and to process GIS-based (geodetic information system) databases. The cross-sections in the Kisköre – southern state border reach were established using a digital terrain model. The latter was supplemented by local geodetic surveys on certain parts of the floodplain: overgrown by dense vegetation or along the natural levees. The river reach between the northern state border and Kisköre was modelled mainly using the survey data obtained after 1999. Concerning the completion of the river system, a valuable contribution was given by the Serbian water management authorities providing us with the cross-sections for the reach of R. Tisza between the southern state border and the Tisza-Danube confluence. The river system of Tisza and its tributaries was approximated by more than 1500 cross-sections.

4.3 Hydrotechnical structures

The HEC-RAS model used for the detailed description of the entire stream system enabled the consideration of effects caused by various hydrotechnical structures like bridges, river barrages, culverts, weirs, sluices, bottom sills, side-weirs and sluices, static reservoirs, pumping water intakes or water supply canals. At the present, 82 bridges have been incorporated in the model.

4.4 Roughness (smoothness) coefficient

In the course of the model calibration, the roughness (smoothness) coefficients shown in Table 1 were applied for taking into consideration of the agricultural and vegetation condition in the floodplain and for calculation of the water conveying capacity of the high-flow channel. The listed categories were determined from aerial photos, orthophotos and by field reconnaissance. The roughness coefficients were changed crosswise, according to the floodplain agricultural/vegetation data. The applied co-efficients correspond to the ones suggested by Chow (1959) and used also by the model HEC-RAS. The ranges of roughness coefficients belonging to certain categories might overlap each other, since e.g. it is impossible to clearly distinguish the categories "scarce scrub" and "dense scrub".

Table 1. Smoothness coefficients

No.	Type	n $(s/m^{1/3})$		k $(m^{1/3}/s)$	
		min	max	min	max
0	Main River bed	0.150	0.017	6.67	58.8
1	Pasture	0.050	0.025	20.00	40.0
2	Plough-land	0.050	0.020	20.00	50.0
3	Scarce scrub	0.080	0.035	12.5	28.6
4	Dense scrub	0.160	0.040	6.25	25.0
5	Forest	0.120	0.030	8.33	33.3
6	Forest with brush	0.200	0.080	5.00	12.5
7	Coarse gravel	0.070	0.030	14.3	33.3

Having determined the basic data and the categories of roughness coefficients, the procedure of model calibration were made in the following steps:

Roughness coefficients of the main channel were determined for flood waves approaching to but not surpassing the first alertness grade of flood protection. By proper selection of the smoothness coefficients we succeed to model the water levels in the main channel with an appropriate accuracy. It was accepted, however, that this procedure might provide greater errors in the range of low flows.

In the flood plain equal-smoothness stripes were indicated, perpendicular to the main channel, for which the mean smoothness-coefficient values of the above categories were selected. By selecting these stripes, the roughness conditions of the terrain between two neighbouring stripes were taken into consideration using a good engineering judgement. It means that the widths of the stripes were selected to be equal with the average width of the terrain between them.

The calibration of models for flood conditions was made by changing the smoothness coefficients, generally for the whole basin of Tisza River, by keeping them within the smoothness ranges indicated in the table (Figure 4).

The advantage of this calibration method that the selection of smoothness coefficients to be altered for planning future conditions is simple and corresponds to the physical conditions of the river systems. Consequently, the steps of modelling a so-called "hydraulic corridor":

- Determine the modified width of the flood-plain stripes (by changing the existing widths or by selecting new stripes),
- Give smoothness coefficients for the modified stripes according to the categories (e.g. arable land, etc.) determined during the calibration.
- Basic hydrological data and boundary conditions

In the first phase of computation, the hydrological database consisted of the hourly water level elevation (Z) and flow discharge (Q) data of flood waves occurred between 1998 and 2000. For the calibration and validation of the model, the database was supplemented by the time series of hourly gauge height data read on nearly 50 staff and dike watch-house gauges.

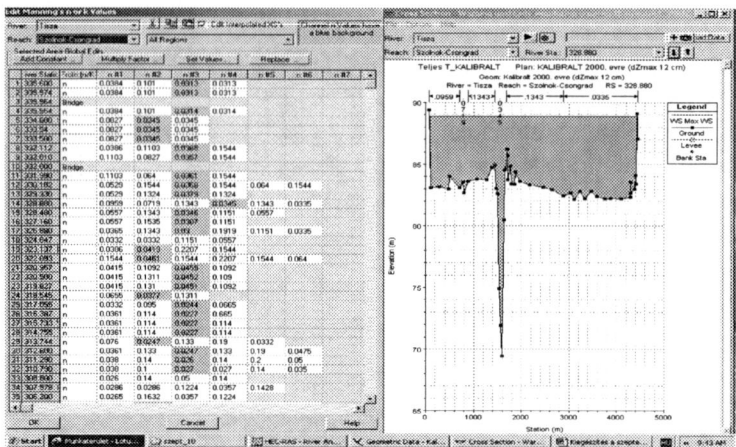

Figure 4. Detailed representation of roughness coefficients in the cross-section

4.5 Calibration

The stepwise calibration of the model has been started using shorter river reaches. By joining up the individual reaches, the complete model of Tisza River between Tiszabecs and Titel was developed. The result of calibration is shown in Figure 5.

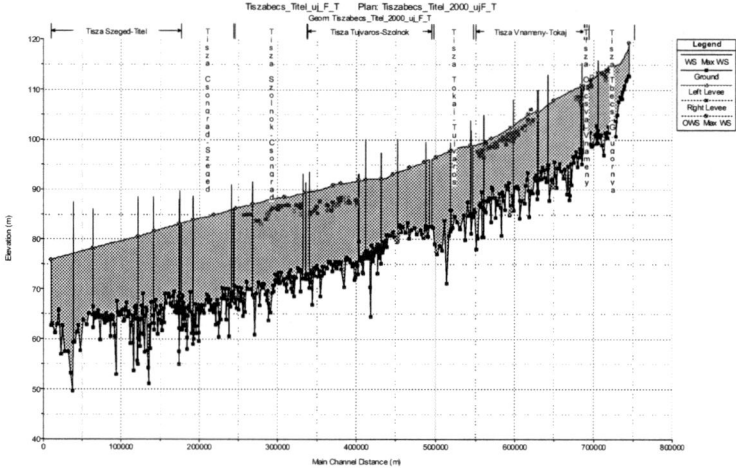

Figure 5. Results of calibration in the river reach Tiszabecs-Titel

The solid line (on top of a blue colour area) is the enveloping curve of the calculated water levels, i.e. the longitudinal section of the local flood crest elevations. The red points depict the maximum water levels observed in 2000 at the staff and the dike watch-house gauges. The pink and dark blue points under the water surface show the elevation of the obstructions (e. g. summer levees) on the left and right floodplain. Along the Tiszabecs – Titel River reach the absolute differences between the observed and calculated maximum flood water levels varied between 0 and 10 cm, thus regarded as a very good result.

4.6   Verification

A calibrated model has to prove that it can be successfully applied not only in the case (hydrological event) for which it was calibrated but in case of any given flood wave. Therefore, the verification was made by running an event, independent from the former ones and leaving the previously determined parameters unchanged. The flood waves applied for the calibration (from the year 2000) and for the verification (from the year 1999) occurred in the same season, thus, the vegetation conditions can be regarded the same. The only difference was the amount of water retained in sections of the flood plain.

5.   STUDY OF EFFECTS INFLUENCING FLOOD RUNOFF

5.1   Investigation of human interventions in flood plains

It was repeatedly emphasized at different forums that the basic aim of floodplain re-arrangement is the possible restoration of the flood conveying capacity of R. Tisza as it was before 1970.

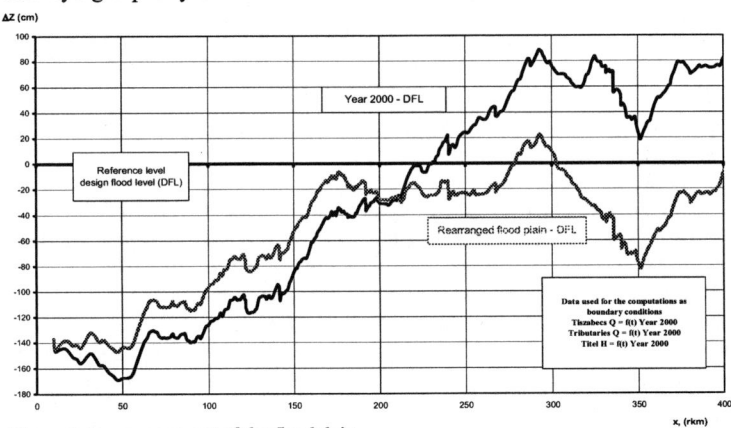

Figure 6. Rearrangement of the floodplain

56

In the reach of Tisza River between Kisköre and the southern state border the following modifications are planned in the floodplains: clearing the vegetation from an about 600 m wide strip for enabling flood conveyance, removal of summer dikes, digging deeper ditches following the drift line of floods in order to accelerate flood runoff, displacement of flood protecting levees wherever necessary. Similar interventions are planned also in the region of Tivadar. Figure 6 shows the planned rearranged floodplain in a cross-section of the Vezseny bend of R. Tisza.

The results of numerical modelling are to be seen in Figures 7. and 8. An important part of the study is the comparison of water levels occurring due to the planned interventions with the design flood level (DFL).

The maximum flood water levels observed during the flood of 2000 have exceeded DFL above a 230 rkm until 544 rkm. In certain sections the flood levels were above DFL by 80 cm. It can be seen that in case of carrying out the planned floodplain alterations, the enveloping curve of the local maximum water levels still exceed DFL between rkm 278 and 302. It has to be mentioned that the flood level changes shown in Figure 7 and 8 would occur only if the planned interventions in the floodplains would be carried out along the whole river reach investigated and within a relatively short time interval.

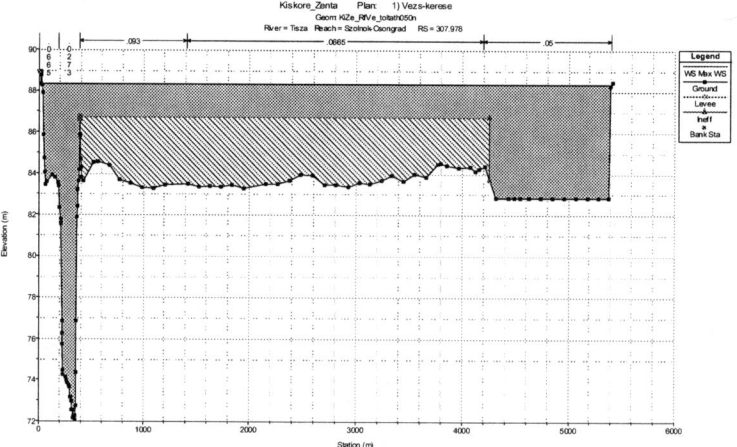

Figure 7. Effects of interventions in the floodplain in the reach between Kisköre – mouth at the River Danube

It has been stated that the natural drop of flow conveyance capacity of the main channel and the floodplain of River Tisza is very significant. Interventions applied on short river reaches could improve the local situations only temporarily. Following the investigations concerning the floodplain alterations, the effects of floodplain reservoirs were modelled with and without interventions in the floodplain. In the first phase of the project "Further development of the Vasarhelyi concept", six reservoir sites have been selected. The model study was made using historical and so-called "generated" flood waves. The number of modelled variants was more than 300.

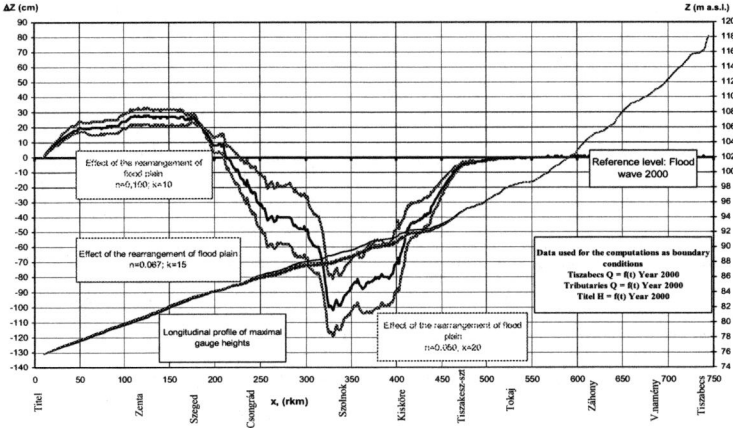

Figure 8. Effect of floodplain rearrangements related to the year 2000 flood level

5.2 Study of the six reservoirs selected in the 1st phase of development with and without interventions in the floodplain (using past design conditions)

The effects of the interventions in the floodplain and those of the reservoirs were investigated in case of flood waves of different character. This paper presents the model runs using the flood waves in the years 1998 and 2000. As it was mentioned above, an exceptional high flood wave has occurred in 1998 on the upper reach of Tisza River, while in 2000 the design flood water levels have been exceeded on the middle reach. Correspondingly, the effects of floodplain interventions and reservoirs were different due to the diverse character of flood waves.

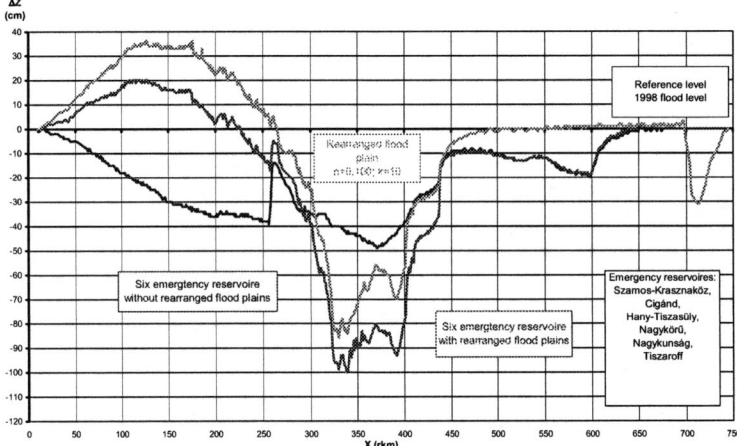

Figure 9. Effect of the six reservoirs determined by the governmental decision without and with floodplain rearrangement in case of the 1998 flood wave

58

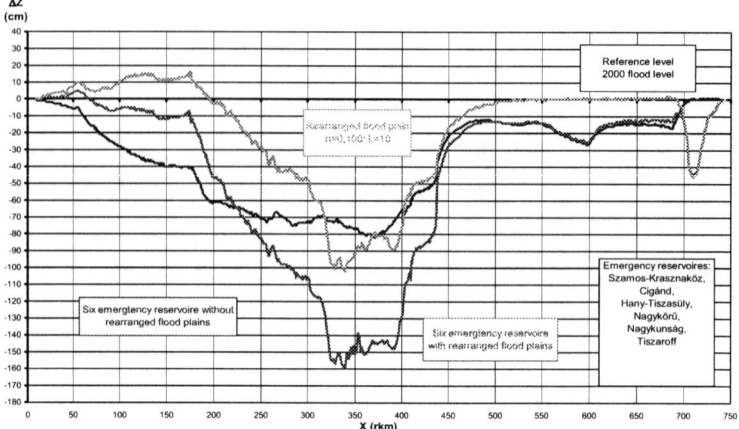

Figure 10. Effects of the six reservoirs determined by the governmental decision without and with floodplain rearrangement in case of the 2000 flood wave

6. CONCLUSION

The model enabled us to run past and recent floodwaves and to analyse the effects of various interventions and those of the reservoirs. Several hundred model runs were carried out investigating different flood situations and design alternatives. It is envisaged that in the coming years and decades the discharge carrying capacity of River Tisza and its tributaries will further change, due to various interventions in the floopplain or in case of their absence due to natural processes. Consequently, the present calibration of the model should also be changed. And the water level decreasing effects of the reservoir operation will differ from the results of the present modelling. The parameters of the model should be updated considering the effects of the new interventions and the results of the new river channel surveys.

In order to achieve this goal, the database should be continuously revised and maintained. The HEC-RAS model, connected to the hydrographic information system and to the gauge height and discharge forecasts, could form a base of a flood-time operating system of Tisza River.

References

Budapest Technical University (2003) Establishing the scientific background of flood peak decreasing by floodplain reservoirs and floodplain intervention. Improving the Vasarhelyi Project (1st phase), 1st Progress Report, September 2003, Budapest, (in Hungarian).

Chow V. T. (1959) *Open Channel Hydraulics*. McGraw-Hill, New York, USA.

Chapter 5

STUDY OF POSSIBLE FLOODING IN THE CHEMICAL FACTORY AS A BASIS FOR FLOOD MANAGEMENT PLAN IMPROVEMENT

MIROSLAV LUKAC, MAREK COMAJ[1], MARTIN MISIK[2], JAN SPATKA AND MARCELA SVOBODA[3]
[1]*Water Research Institute, Bratislava,* [2]*DHI Slovakia, Bratislava, Slovakia,* [3]*DHI Hydroinform, Prague, Czech Republic*

Keywords: Hydrodynamic modelling, MIKE 11, MIKE 21, flood mapping, flood management.

1. INTRODUCTION

The Slovnaft refinery complex is one of the key elements in the Slovak economy. It is situated in Bratislava, the capital of Slovakia, close to the Danube river. Catastrophic floods, which occurred in the central Europe in the summer of 2002, emphasized urgent need for revision of flood management plan in this important factory. The peak discharge of Danube in Bratislava was close to Q_{100}. The flood consequences in Bratislava were not so dramatic as in the Czech Republic, however the weak elements of flood control in the city were discovered. The refinery, area of which is about 5.2 km^2, is situated in a very flat territory. A lot of refinery objects and technological equipment are vulnerable in the case of flooding, which could occur as a consequence of various crisis scenarios. Overtopping of flood protection dykes or flood protection lines in the city or in the refinery surrounding area, as well as dyke breaks represent the most probable flooding scenarios. The flood management plan of the refinery was out of date, not taking into account dynamic character of possible flooding. Therefore, the refinery representatives turned to Water Research Institute (WRI) experts in the end of 2002 with a request to elaborate study of possible flooding. The cooperating organizations were - DHI Hydroinform, Prague, DHI Slovakia, Bratislava and Slovak Water Management Enterprise, Danube River Basin Authority in Bratislava. It was decided to apply hydroinformatics software tools of the DHI Water & Environment, Denmark in the frame of study, in order to represent properly dynamic effects of the flooding phenomenon.

J. Schanze et al. (eds.), Flood Risk Management:
Hazards, Vulnerability and Mitigation Measures, 59–67.
© 2006 *Springer.*

2. MODELLING STUDY OF POSSIBLE FLOODING IN THE REFINERY

2.1. Conception of study

The study of possible flooding in the Slovnaft refinery (Misik et al. 2003) was based at hydrodynamic models simulations. It was divided into following stages:

- Setup of one-dimensional (1D) hydrodynamic model of the Danube river, its calibration and validation,
- Setup of enlarged 1D model, which enabled simulation of flooding in the city intravilan and definition of boundary conditions for the crisis scenarios,
- Visualization of 1D model results in the refinery surrounding, using digital terrain model (DTM),
- Setup of detailed two-dimensional (2D) hydrodynamic model of the refinery territory,
- Simulations of selected crisis scenarios, using about mentioned hydrodynamic models.

MIKE 11 (1D) and MIKE 21 (2D), software tools of the DHI Water & Environment, were applied in the frame of the study.

2.2. 1D Hydrodynamic modelling

Setup of 1D model of the Danube river represented the first step of the modelling study. The catchment area of Danube river in the gauging station Bratislava is 131,329 km^2. The records from this station are available since 1876 and the first discharge measurements of the Danube on the Slovak territory were made in 1882 (Svoboda et al. 2000). Discharge in Bratislava ranged from 580 $m^3.s^{-1}$ (January 1909) to 10,400 $m^3.s^{-1}$ (July 1954), in the observation period between 1901 and 2004.

Hydrodynamic 1D model of the Danube river channel covered the reach between gauging station Danube - Devin (river chainage km 1879.780) and the Cunovo weir (km 1851.750). The channel topography was schematized with 84 cross-sections, which were surveyed in 2002. It was needed to schematize also the Little Danube branch, which is neighbouring the nothern part of the refinery area. The flood protection dykes are situated almost all along the modelled reaches of the rivers, but it was proven by the previous study (Misik and Comaj 2003), that several sections of dykes or flood protection lines are not safe against the floods with return period equal or lower than 100 years.

The model boundary conditions were defined as follows:

- upstream boundary in the gauging station Danube - Devin, discharge time serie $Q = f(t)$,
- downstream boundary at the Cunovo weir, water level H in the reservoir,

• downstream boundary at the Little Danube branch, rating curve of the river channel H = f(Q).

The model of the Danube river channel was calibrated and verified using reliable data from the recent floods, which occurred in 2002. The peak discharge of the first one (March) reached around 8,500 $m^3.s^{-1}$. The second flood (August) was larger, with the peak close to 10,400 $m^3.s^{-1}$. Besides the hydrographs in the gauging stations, the longitudinal profiles of the water level in the modelled reach were measured during the floods. These data sets represent valuable material for the model calibration and validation. The roughness coefficient was the main calibration parameter. Satisfactory differences (up to 10 cm) between measured and computed values were achieved.

Based at the results of simulations with basic 1D model, it was possible to evaluate level of safety of flood protection dykes and flood protection lines. The safety of left sided dyke, situated along the refinery, is at the desired level. The critical sections from the viewpoint of flood control were identified in the Bratislava intravilan. Several flood protection lines could be overtopped at the discharges close to or higher than Q_{100}.

The basic model of the Danube river channel was later enlarged in the city intravilan. Based at the DTM, 60 additional model branches were defined in the city and in the refinery surrounding, in order to schematize possible flow in the model area during flood. Additional branches were interconnected mutually, as well as with the Danube river branch, using the link channels.

To prepare boundary conditions for detailed 2D model was the main purpose of enlarged 1D model. Simulations were performed for selected crisis scenarios, described in the section 2.3. It was not possible to calibrate reliably the enlarged model, because of not existing water marks in the enlarged model area. During the time of observations, no floods occurred which would flow through the city intravilan. There were some historical floods which flooded the city center – in 1501, 1526, 1721, 1809, 1850, 1899 (Svoboda et al. 2000), but the nowadays hydraulic conditions are different. In the history, the flood control measures either did not exist, or were not sufficient.

2.3   Definition of crisis scenarios

The selection of crisis scenarios was based at the results of river model simulations and experiences from the previous significant floods at the Slovak section of the Danube river.

Two basic scenarios were studied, using 1D model:
1. overtopping of flood protection lines and water flow through the city towards the refinery,
2. breaking of flood protection dykes.

In the simulations of crisis scenarios, two different flood waves were used as an upstream boundary condition. The first one was the flood from

62

August 2002 and the second one reconstructed hydrgraph of historical flood from 1501 (Svoboda et al. 2000). The peak of 1501 flood was estimated to around 14,000 $m^3.s^{-1}$ in Vienna. The comparison of both hydrographs is given at the figure 1.

The flood protection lines could be overtopped at several locations in the case of catastrophic flood. The water starts to flow through the city at the discharge of about 11,000 $m^3.s^{-1}$. The total peak discharge in the city resulting from the overtopping of flood protection lines was 215 $m^3.s^{-1}$. The flood protection lines were overflown during 160 hours, in the case of catastrophic flood. The extent of flooded city area could be rather high, because of flat terrain. At the same time, flooding of large area will also result in the flooding routing. The peak flooding discharge decreased to 55 $m^3.s^{-1}$ in the model branches situated close to the refinery. This water would be drained by the Little Danube. Its channel has sufficient capacity to accommodate also this excessive volume of water, without overtopping of flood protection dykes. It was concluded from this simulation, that overtopping of flood protection lines in the upstream situated city should not result in the flooding of refinery area.

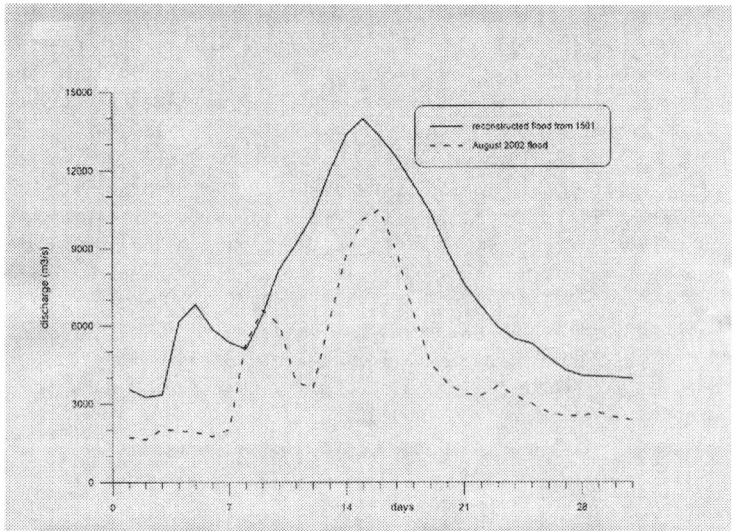

Figure 1. Comparison of flood hydrographs from 1501 (reconstructed) and 2002

Breaking of flood protection dykes represents more dangerous situation. During the 1965 flood a Danube dyke was broken at two sites, around 70 and 115 km downstream from Bratislava. The breaches resulted in the flooding of very large area (around 52,000 hectares, total volume of flooding around 1.200 mil. m^3, 50,000 inhabitants evacuated) in the Slovak part of the Danube lowland and extensive damages. Significant seepage through the dykes occurred also during the August 2002 flood, in the location situated very close to the Slovnaft refinery.

A dam break module of MIKE 11 was applied for the simulation of dyke breaks. The parameters of dyke breaches and their locations are questionable. In the simulations, the parameters of breaches were inspired by the 1965 flood. Finally, three scenarios of dyke breaks were selected for a detailed 2D model simulations of refinery flooding. The summary of the main parameters of individual scenarios is given in the Table 1.

Table 1. Basic data of flooding scenarios

Scenario-no.	Location of breach (river km)	Peak discharge in the Danube $(m^3.s^{-1})$	Breach width (m)	Peak discharge through the breach $(m^3.s^{-1})$	Flooding volume (mil.. m^3)	Flooding duration in the breach site (hours)
1	1864.4	14,000	50	290	121	220
2	1862.5	14,000	100	360	175	310
3	1864.4	10,390	50	125	28	110

The results of simulations with the enlarged 1D model were visualized, using the Flood mapping module of MIKE 11. Resulting water levels in the model area were transformed into the DTM and the maps of flooding depth prepared. The examples of Flood mapping visualization are given in the figures 2 and 3.

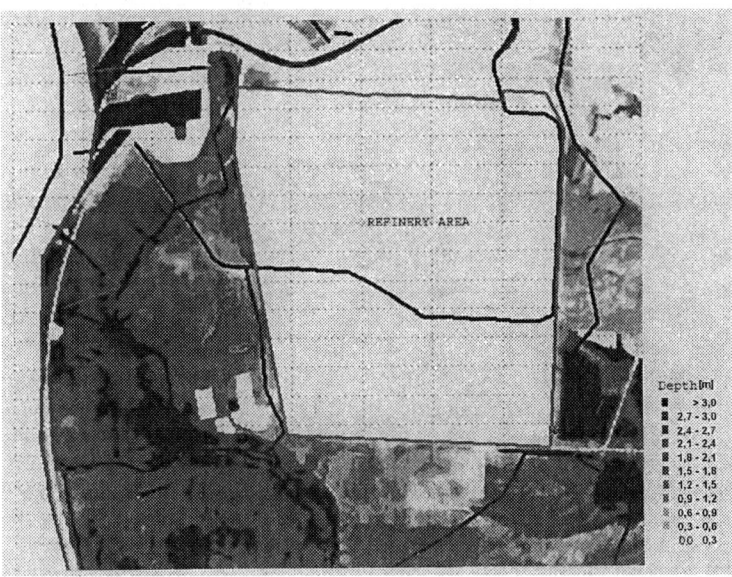

Figure 2. Flood mapping in the refinery surrounding - scenario no.1

64

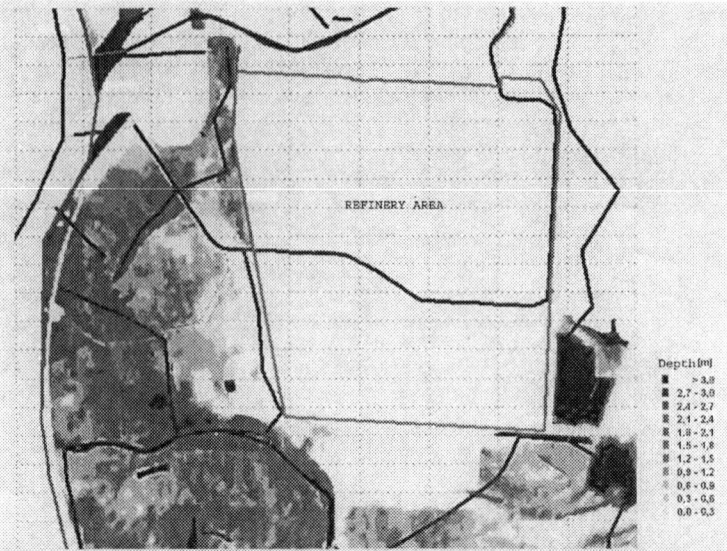

Figure 3. Flood mapping in the refinery surrounding - scenario no.3

2.4.   2D Hydrodynamic modelling

Detailed 2D hydrodynamic model of the whole refinery area was set up, in order to simulate possible flooding. The computational grid of model was based at the DTM of the refinery area. Contours of buildings, internal small dykes, roads, access ramps, fences, tanks and other refinery objects were digitized and transformed to the model grid. The final computational grid was orthogonal with 3x3 m density, containing around 650,000 computational points. The roughness conditions in the model area were schematized, based at classification of land use and vegetation cover.

There are several locations, through which the water can flow into the refinery area - entrances, gates, railway gateways, inflow/outflow canal of water used for industrial purposes. The concrete fence, bordering refinery area was assumed to be watertight. Boundary conditions for 2D model were defined in these locations, based at the results of 1D modelling and Flood mapping. Boundary conditions were defined as the time series of water depth h = f(t) at the boundaries.

Three scenarios summarized in the table 1 were simulated. Simulations were very time consuming, with respect to the extent of computational grid. It was needed to apply very short time step (less than 1 second at the beginning, later not more than 5 seconds) in the simulations, because of numerical stability problems. Despite the usage of powerful computers, computational time was about 2,5-times longer than the real time of simulated scenarios.

The results of 2D modelling were displayed graphically in the form of flooding maps, with the output values of water level, flooding depth, flow velocity magnitude and vectors. These maps were printed for the refinery area as a whole, as well as in six more detailed segments. The example is given at the figure 4. Duration of simulated scenarios was up to 200 hours, in order to investigate not only the flooding of area, but also the drying progress. The results of simulations were stored every three hours.

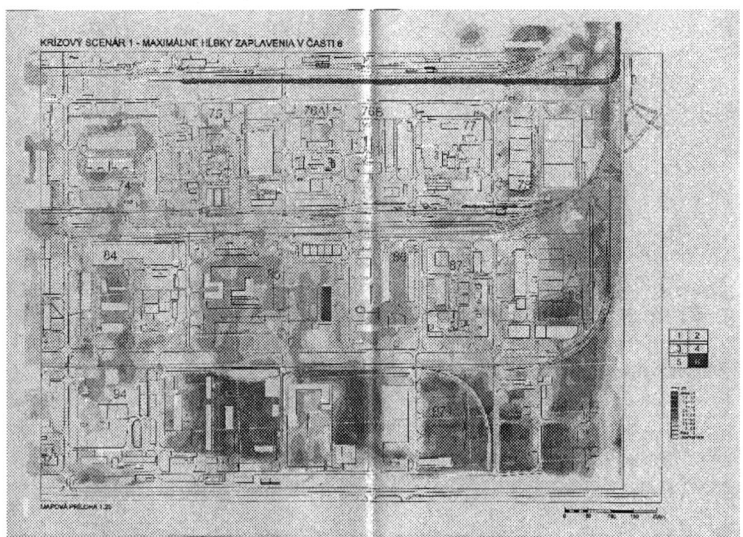

Figure 4. Flooding map for the part of refinery area - maximum depth

Intake canal of the industrial water (which is pumped from the Danube) represents the most critical element from the viewpoint of refinery flooding. Its capacity is only around 6 $m^3.s^{-1}$, which is not sufficient for the draining of flooding water. Water would fill the canal and start to spread in the refinery area 5 hours after the Danube dyke break in the case of the most unfavourable scenario no.1. Situation would not be so critical for the other two scenarios. Refinery flooding would start 37 hours (scenario no.2) or 30 hours (scenario no.3) after the dyke break.

Culmination of flooding would occur 32 hours after the dyke break, in the case of the most unfavourable scenario and it would last around 30 hours. Flooding water would spread almost in the whole refinery area. The average flooding depth reached 0,5-0,8 m, maximum values around 1,4 m. Water would also flow out of the refinery area, through the outflow canal and several gates. Maximum flow velocities ranged in the interval 0,6-0,7 $m.s^{-1}$ in the prevailing paths of water flow, but the water was stagnant in the majority of refinery area during culmination.

Dyke breach for scenario no.2 was situated more downstream, comparing with the scenario no.1. It resulted in substantially smaller and time delayed flooding of the area. The maximum flooding depth reached around 1 m. Dynamics of flooding for the scenario no.3 was similar to this for scenario no.1. The basic difference was represented with lower breach discharge and flooding volume (see Tab. 1). The majority of refinery area would be flooded. Flooding culmination would be reached around 90 hours after the dyke break. The maximum flooding depth was 1.4 m in this case.

It was not possible to calibrate 2D model using measured parameters of flooding (observed water levels, flow velocities, extent of flooding), because the refinery area in the present state have never been flooded. Therefore, the results should be interpreted with care. Based at the results of 1D and 2D modelling, final conclusions and recommendations were formulated, which were transformed into the updated flood management plan of the Slovnaft refinery.

3.   DISCUSSION AND CONCLUSIONS

Advanced modelling techniques were applied in the study of possible flooding of the Slovnaft refinery. Selection of final crisis scenarios was based at careful evaluation of hydraulic conditions in the area of interest. Dynamic mechanism of flooding was analyzed for a different situations - dyke breaks and overtopping of flood protection dykes or lines.

Scenario no.1 represents the most unfavourable case - dyke break situated close to the refinery at the occurrence of catastrophic flood, the return period of which is far behind 100 years. The probability of such an event is very low, but the experiences from recent floods in the region warn. The probability of the occurrence of situation similar to this, simulated in the scenario no. 3, is higher. On the other hand, its con-sequences should not be so dramatic, if proper flood control measures will be applied in time.

Several factors influence reliability and precision of results - level of schematization of the physical phenomena with the model, inaccuracies of input data (river channel topography, DTM of flooded terrain), uncertainties of hydrological data and dyke breaks parameters, etc. The best available data sets were used and the client was informed about the precision and uncertainties of the results.

In general, results of simulations provided very useful information about the dynamics of possible flooding. Flooding maps of water level, water depth and flow velocity in the model area pointed at the zones in the refinery, which are the most vulnerable. It has to be emphasized, that simulations were performed with an assumption, that no flood control measures would be applied. The real situation will be different, if there will be enough time for the application of effective flood control measures. Results provided basis for a decision makers about the priorities in the flood control of the refinery. The study conclusions and recommendations have already been reflected in the updated flood management plan of the Slovnaft refinery.

References

Misik M. and Comaj M. (2003) Computation of the Danube water level in the reach Devin - Cunovo at the discharges Q_{100} and Q_{1000}. Final report, Water Research Institute, Bratislava, (in Slovak).

Misik M., Lukac M. and Comaj M. (2003) Modelling study of Slovnaft flooding. Final report, Water Research Institute, Bratislava, (in Slovak).

Svoboda A., Pekarova P. and Miklanek P. (2000) Flood hydrology of Danube between Devin and Nagymaros. National report 2000 of the IHP UNESCO Project 4.1 International Water Systems. Institute of Hydrology SAS, Bratislava.

Chapter 6

FIOOD RISK MAPPING AND MITIGATION IN
CENTRAL ITALY

STEFANO PAGLIARA

Dip.to di Ingegneria Civile - University of Pisa, Italy

Keywords: Inundation, flood risk, risk mapping, mitigation measures, detention basins.

1. INTRODUCTION

"Flooding of urban land is a frequent, natural or manmade disaster in large or small scale" (Yen, 1995). Flood inundation constitutes a complex problem due to the interaction between flow and infrastructures present in the plain. In this study, by means of a two-dimensional unsteady flow model that implements the complete De Saint Venant equations, different conditions of inundated plain are analysed. Floodplain analysis is needed in many practical situations, mainly because is in lowlands that human activities are concentrated. The study of the inundated plain requires at the same time quite of effort in terms of modelling, computational resources and data acquisition. The need for a two-dimensional model and the presence of a great number of different structures in the inundated zone make the problem complex. The effect of the plain topography, the presence of irrigation and/or natural channels, of roads, highways, railways, low dikes and culverts affect the pattern of the flooding flow and can change it considerably.

2. MATERIAL AND METHODS

Nowadays is a necessity to prepare flood maps for civil protection or planning purposes. In Italy and particularly in Tuscany, a lot of rivers have been studied and different kinds of flood maps have been proposed. Aim of this chapter is to give an overview of the work that so far has been done.

The mathematical models used are all 2D models. One of the most suitable models is the one developed at the Public Works Research Institute (Yoshimoto et Al. 1992). In order to simulate the flow in the

69

J. Schanze et al. (eds.), Flood Risk Management:
Hazards, Vulnerability and Mitigation Measures, 69–76.
© 2006 *Springer.*

floodplain, the model uses the fully dynamic two-dimensional unsteady flow equations consisting of:

continuity equation,

$$\frac{\partial h}{\partial t} + \frac{\partial M}{\partial x} + \frac{\partial N}{\partial y} = 0$$

dynamic equation in the x-direction,

$$\frac{\partial M}{\partial t} + \frac{\partial (uM)}{\partial x} + \frac{\partial (vM)}{\partial y} + gh\frac{\partial H}{\partial x} + \frac{1}{\rho}\tau_x = 0$$

and a dynamic equation in the y-direction,

$$\frac{\partial N}{\partial t} + \frac{\partial (uN)}{\partial x} + \frac{\partial (vN)}{\partial y} + gh\frac{\partial H}{\partial y} + \frac{1}{\rho}\tau_y = 0$$

with:

$$\tau_x = \frac{\rho g n^2 u\sqrt{u^2 + v^2}}{h^{1/3}}$$

$$\tau_y = \frac{\rho g n^2 v\sqrt{u^2 + v^2}}{h^{1/3}}$$

where, g = gravitational acceleration; ρ = density of water; M = u·h = flux (flow rate per unit width) in the x-direction; N = v·h = flux in the y-direction; h = water depth; H = water surface elevation; x, y = spatial coordinates; t = temporal coordinate; τ_x = shear stress at the bottom in the x-direction; τ_y = shear stress at the bottom in the y-direction.

The numerical method used to solve the system of partial differential equations uses an explicit staggered finite difference scheme (Iwasa et al. 1980).

3. FLOOD EXAMPLES

a) Simulation of the flood of the Arno river in Florence (1966)

The flood that occurred on 3-4 November 1966 in the Arno river (Tuscany – Italy) has been, for this area, one of the most devastating floods and the one that has been considered as the target for all the measures of risk mitigation.

The city of Florence was flooded and a great number of damage occurred. Figure 1 shows the beginning of the flood and, superimposed, the grid used for the mathematical model.

Figure 1. Beginning of the big flood of the Arno river in Florence (1966)

Figure 2 shows a mathematical reconstruction of the flow field in terms of velocity vectors at a certain moment after the starting of the flood process. The figure shows the different flow direction in the streets surrounding the main railway station (S. Maria Novella); of particular interest is the flow pattern in the square in front of the railway station.

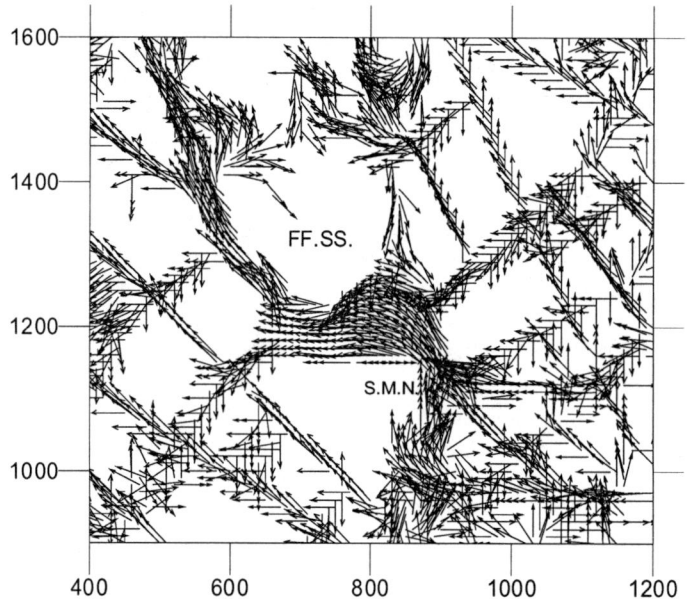

Figure 2. Particular of the flux field (m2/s) close to the central railway station in a certain instant during the flood

72

For this case the result has been given both in term of velocity fields or flux (discharge for unit width) vector maps and envelope of water depths.

b) Complex minor river system: the case of the city of Follonica (Tuscany)

This case is that of a minor system that has been reduced causing floods in an important urban area. Floods are due to the inefficiency of the channel and to the presence of bridges.

Flood areas are reported in Figures 3, 4 and 5. Figure 3 shows the actual situation with big urbanized zones that suffer the inundation; Figure 4 shows the same flood during part of the river sistemation works (i.e. building of new bridges and cross-section enlargement for a tributary); Figure 5 shows the final situation with the building of detention basins in the upstream part of the river basin. The maps clearly indicate the area in which the hydraulic risk is high and how this risk changes due to the building of the mitigation measures.

Figure 3. Flood inundation for the actual conditions (counter lines show different water depth values)

Figure 4. Flood inundation areas in the case of sistemation of a tributary and rimotion of bridges (counter lines show different water depth values)

Figure 5. Flood inundation areas in the case of complete sistemation by means of detention basins. The 200 years flood is contained in the main channel

c) Versilia river floods

Versilia river is located in Italy, in northeast Tuscany. At the end of the last century, 5 floods occurred in the basin causing a lot of damage also

74

in terms of loss of human lives. The mathematical analysis of the flow in the floodplain has been carried out by means of the 2D unsteady flow mathematical model described before accounting for the infrastructures present in the floodplain. Particular stress has been put on the significant changes in the flood flow routing due to urbanisation and construction of the highway and roads.

The biggest floods have occurred on: Sept. 27, 1774, Sept. 25, 1885, July 11, 1902 and June 19, 1996. The last one has caused high damages and the loss of several human lives due, also, to the high level of urbanization reached during this century.

The main purpose of this work is the investigation on the historical levee-break and on the consequent inundation that occurred and the analysis of the dynamic of the flow due to the presence of urbanisation and infrastructures present in the plain.

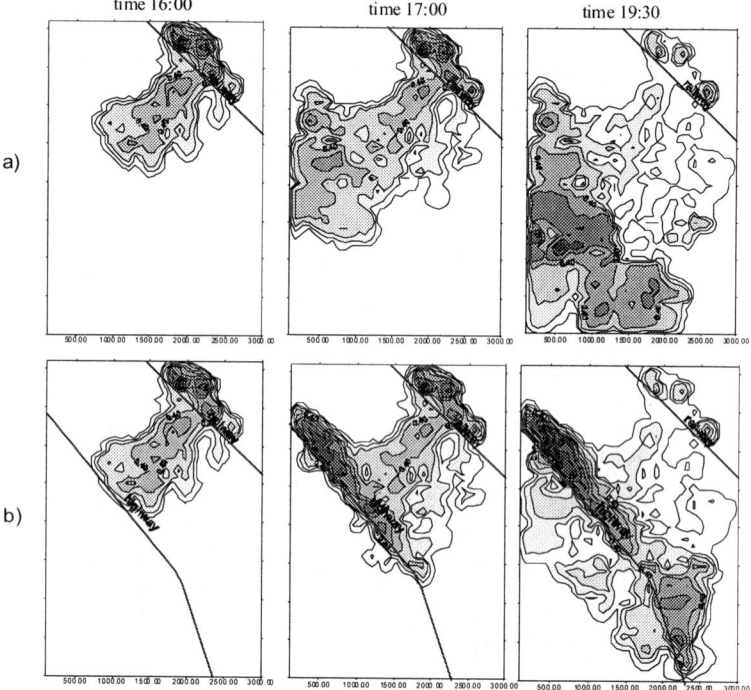

Figure 6. Flood occurred in the Versilia river on June 1996; the flooding started on June 19 at 15:15; are reported the inundated areas at different times a) without the highway built in 1970's b) with the highway

The railway and the highway run in a direction parallel to the coastline; these infrastructures with their banks deeply alter the dynamic of the flood. The peak flow upstream of the break was bigger than 500 m^3/s. A peak of more than 200 m^3/s enters in the floodplain. The total volume of water in the plain has been estimated in 3 millions of cubic meter.

It is noteworthy to remark that the highway has been built in the seventies. A look to the map of figure 6 shows how dramatically this fact has changed the hydraulic of the floods. It shows the comparison between the case of plain without and with the highway bank at different hours during the mathematical simulation of the flood of June 1996. The effect of the bank is evident as well as the presence of the railway in the N-E part of the simulated plan. In Figure 6 the counter lines shows different water depth values.

d) Serchio river: hypothetical flood

This a case in which, for civil protection purposes were carried out a series of simulation in order to obtain a map of flooded areas due to a levee break; the levee break was localized on a very dangerous point of the river in which is present a 90 degree bend and a levee more than 6 meters high. Figure 7 shows the hydrograph upstream and downstream of the levee break for a total volume of about 28 millions of cubic meters of inundation.

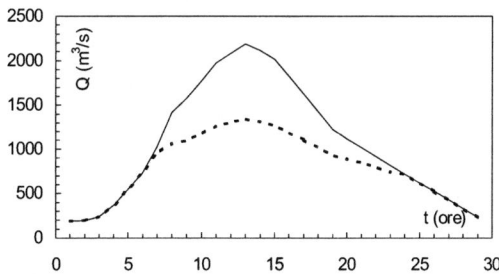

Figure 7. Flood hydrograph in the river upstream (continuous line) and downstream of the levee break (dotted line)

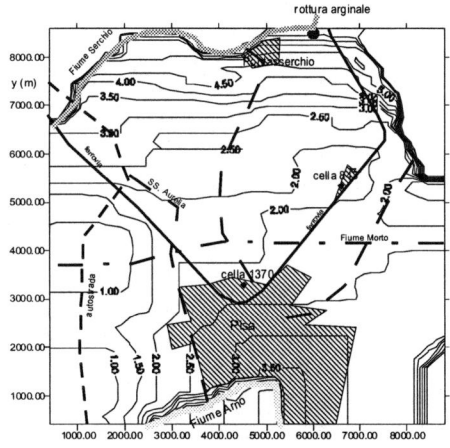

Figure 8. Planimetric view of the studied zone for the Serchio river

76

Figure 8 shows a planimetric view of the floodplain. In the upper part is located the levee break while in the low part is located the city of Pisa. Figure 9 shows the result of two simulations: the first one considering the floodplain without the infrastructures (streets, railways, highways, minor levee systems) and the other one with all the infrastructures that are present in the plain.

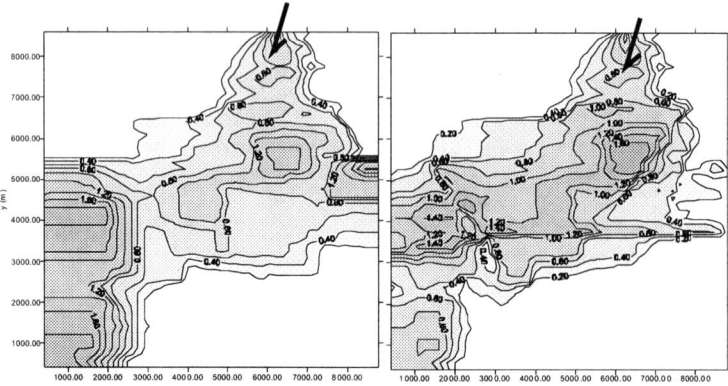

Figure 9. Flooded area for the Serchio river (on the left hypothetical situation without infrastructures; on the right the real situation with all the infrastructures)

It is evident the change in the flow pattern due to the presence of obstacles like infrastructures.

4. CONCLUSIONS

The paper show several applications of flooding models for planning and civil protection purposes. Maps showing water depth and flow velocity represent an useful tool for this kind of applications.

The effect of infrastructures in changing the flood pattern is also highlighted.

References

Iwasa Y., Inoue K. and Mizutori M. (1980) Hydraulic analysis of overland flows by means of numerical method. Annuals, Disaster Prevention Research Institute, Kyoto Univ., 305-317 (in Japanese).

Pagliara S. and Suetsugi T. (1997) Floodplain analysis by two-dimensional model. Technical Memorandum of PWRI No. 3520, Tsukuba, Japan.

Pagliara S. (2001) Hydraulic Modelling and Historical Inundation assessment for the Versilia River. Glade T. et al. (eds.) *The use of historical data in natural Hazard Assessment*. Kluwer Academic Publishers, The Netherland, 141-150.

Yen B.C. (1995) Urban flood hazard and its mitigation. Cheng F.Y. and Sheu M.S. (eds.) *Urban disaster mitigation*. Elsevier Science Ltd.

Chapter 7

ON THE CHOICE OF SPATIAL INTERPOLATION METHOD FOR THE ESTIMATION OF 1-TO 5- DAY BASIN AVERAGE DESIGN PRECIPITATION

SILVIA KOHNOVÁ[1], KAMILA HLAVCOVA[1], JAN SZOLGAY[1] AND JURAJ PARAJKA[2]

[1]*Department of Land and Water Resources Management Faculty of Civil Engineering, Slovak University of Technology, Bratislava, Slovakia,*
[2]*Institute of Hydraulics, Hydrology and Water Res. Management, Vienna University of Technology, Vienna, Austria*

Keywords: Precipitation, rainfall analysis, interpolation methods, frequency curves.

1. Introduction

Spatially distributed measurements of precipitation have gained renewed interest in connection with the spread of distributed hydrological modeling and the increased use of remotely sensed data for a number of tasks such as land use and climate-change impact studies, determination of water budgets at different temporal and spatial scales, etc. Point precipitation measurements are usually only available from a limited number of meteorological stations and spatial estimates of precipitation fields for the described tasks are obtained by interpolation or geo-statistical techniques. Mountainous environments represent a special challenge to spatial data analysis and interpolation because the measured data are sparse, often restricted to lower elevations while the temporal and spatial variability in precipitation can be substantial.

While results on the influence of the choice of interpolation techniques for rainfall field estimation on the quality of runoff simulations by mathematical modeling are numerous, less is known about how the choice of the interpolation method could influence the uncertainty of estimation design values of basin average precipitation. N-year areal average precipitation depth values are increasingly being used in water resources studies and flood frequency analysis accomplished both by regional and simulation methods. In the paper a case study is presented which attempts to shed light in to the problem. First grid maps of the areal

77

J. Schanze et al. (eds.), Flood Risk Management:
Hazards, Vulnerability and Mitigation Measures, 77–89.
© 2006 *Springer.*

distribution of daily precipitation depths on the upper Hron River basin in central Slovakia were estimated by three interpolation methods. From the set of daily precipitation maps time series of basin average precipitation was computed. Annual maximum 1-to-5 day basin average precipitation depths were extracted from it. These were statistically analyzed and 1-to-5 day basin average maximum annual precipitation depths for several return periods were derived and compared. The influence of the particular interpolation method on the design values was discussed.

2. SPATIAL INTERPOLATION OF PRECIPITATION FIELDS AND RAINFALL FREQUENCY ANALYSIS

Interpolation and geo-statistical techniques can help us to obtain spatial estimates of precipitation fields from point precipitation values, which are usually available from a limited number of meteorological stations. Various approaches were therefore developed to ensure acceptable interpolation results by incorporation of the impact of topography, for example: MTCLIM (Tabios and Salas 1985), PRISM (Daly and Neilson 1992), ANUSPLIN (Hutchinson and Bischof 1983).

Many papers dealt with evaluation various techniques for interpolation of precipitation, see e.g. Tabios and Salas (1985). Custer et al. (1996) compared ANUSPLIN-generated map (Hutchinson 1995) with expert-drawn precipitation map for Montana and concluded that ANUSPLIN had richer spatial pattern, due to the selection of relatively high DEM resolution. The Spatial Interpolation Comparison 97 (Dubois 1998) compared the results of interpolation of daily precipitation data for Switzerland obtained by more than 20 different methods, including inverse distance method, kriging, radial basis functions, neural networks, fuzzy logic interpolators and others (Dubois 1998). The estimates with the lowest RMSE were obtained by multiquadric functions with anisotropy and the study has demonstrated that performing a geostatistical analysis of data is helpful for selecting an appropriate interpolation method and its parameters. Incorporation of topography in this study did not lead to improved results for the tested methods. Parajka (1999) compared kriging and co-kriging methods with respect to expert-drawn precipitation map. The cross-validation analysis has shown that kriging estimates provided reasonable results in regions with sufficiently dense observations, but in mountainous regions with sparse data it did not reflect the impact of topographical patterns. Co-kriging method, which included the effect of topography, provided more accurate and realistic estimates.

The development of methods for design precipitation estimation in Europe and the world in general has recently been based on the application of some novel statistical techniques. It is worth mentioning the application of new types of distribution functions and parameter estimation methods which are different from those widely used before. We can mention some of many new guidelines for design precipitation estimation as KOSTRA for Germany (Malitz 1999; Bartels et al. 1997), FEH (1999) in Great

Britain, HIRDIS for New Zealand (Thomson 2002), in Switzerland Geiger et al. (1986). Also in Slovakia recent studies e.g. Faško et al. (2000), Jurčová and Kohnová (2002), Stehlová et al. (2001), already applied some of new methodologies concerning the statistical analysis of maximum precipitation depths.

As the previous examples illustrate, it is necessary to deal with issues involving new methodologies used for precipitation analysis. It is inevitable to acquire experience concerning the applicability of a number of contemporary methods under various physical/geographic conditions of Slovakia. These cited examples also show that relevance is paid to the analysis of precipitation with longer duration than 24 hours and that in such analysis there is a wide acceptance of interpolation, geostatistical and statistical methods. In this paper the influence of the choice of three such methods on the uncertainty of design precipitation will be tested.

3. DATA ANALYSIS

The upper Hron River basin with an area of 1766 km^2, which is located in a mountainous region central Slovakia, was chosen as a pilot basin for this case study. The minimum elevation of the basin is 340 m a.s.l., the maximum elevation is 2008 m a.s.l. and the mean elevation is 850 m a.s.l. 70 % of the basin area is covered by forest, 10 % by grassland, 17 % by agricultural land and 3 % by urban areas.

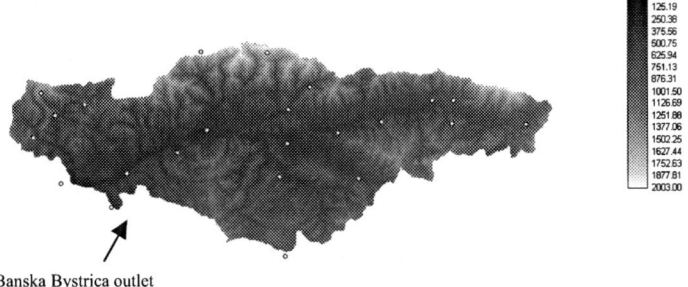

Banska Bystrica outlet

Figure 1. Digital elevation model of the upper Hron River basin with locations of rain-gauge station

Daily precipitation data from the period 1961-2000 from 23 rain-gauge stations were used (Table 1).

Table 1. Table of analysed precipitation stations

Station	Observation period	Elevation (m a.s.l.)
Chopok	1961-2000	2008
Telgárt	1951-2000	901
Šumiac	1961-2000	870
Pohorelá	1961-2000	764
Heľpa	1961-2000	662
Polomka	1961-2000	586
Beňuš	1951-2000	550
Pohronská Polhora	1961-2000	637
Brezno	1961-2000	487
Čierny Balog – Krám	1961-2000	530
Jarabá	1961-2000	839
Mýto pod Ďumbierom	1951-2000	610
Jasenie, na Bankovej	1961-2000	750
Jasenie pred Suchou	1961-2000	483
Brusno	1961-2000	406
Slovenská Ľupča	1951-2000	370
Môlča	1961-2000	459
Motyčky	1951-2000	688
Krížna	1961-2000	1570
Staré Hory	1961-2000	468
Banská Bystrica – Uľanka	1961-2000	398
Dolný Harmanec	1961-2000	481
Lom nad Rimavicou	1961-2000	1015

Variety of methods can be apply to take the point data from the gauges and estimating total precipitation over area; most are based on weighted average of gauge data or contouring gauge data. Given the mountainous character of the basin, the observed strong altitudinal zonality of the precipitation, and the developed orographic effects, it was concluded to test three methods for the estimation of the basin averaged precipitation using interpolation methods.

The general form of the equation for interpolation in the z(x,y) plane at a point z_i (Meijerink 1994) is:

$$z_0 = \sum_{i=1}^{n} w_i z_i,$$

(1)

where:

z_0 is the estimated value of the process at any point x_0 and y_0,

w_i is the weight of the sampling point i,

$z_i(x_i y_i)$ is the observed value of the attribute at point $x_i y_i$,

n is the number of sampling points considered.

The various interpolation techniques in principle differ in selection of the weights w (Tabios and Salas 1985).

Especially for daily data, complicated spatial interpolation (or the use of geo-statistical techniques) may not always be appropriate, because local storms in the summer months have a strong influence on daily precipitation totals, moreover storms can be smeared out between rainfall stations. Therefore the Thiessen polygons method was considered as one of the appropriate methods. It is assuming that the rainfall at any given point is equal to the rainfall at the nearest gauge; the area of influence of a gauge was defined by the Thiessen polygons. Thus the estimated value z_0

(x_0, y_0) was given by the observed value from the nearest measurement point with the smallest distance d_{0i} from it:

$$d_{0i} = \min(d_{01}, \ldots, d_{0n}) \tag{2}$$

and

$$d_{0i} = \sqrt{(x_0 - x_i)^2 + (y_0 - y_i)^2}, i = 1, \ldots, n \tag{3}$$

and the weights in equation (1) are selected as $w_j = 0$ for all $j \neq i$ and $w_i = 1$ for the point with the smallest distances d_{0j}.

Based on the fact, that one to five days extreme precipitation were foreseen to be analysed in the further steps, as the second method the inverse distance interpolation was selected for building interpolated rainfall surface over the basin. In the method a neighborhood about the interpolated point x_0, y_0 is identified and a weighted average is taken of the observation values x_i, y_i within this neighborhood. The weights w_i are a decreasing function of distance between the interpolated pint x_0, y_0 and the points with rainfall measurements x_i, y_i :

$$w_i = \frac{f(d_{0i})}{\sum_{i=1}^{n} f(d_{0i})}, \tag{4}$$

where d_{0i} is the distance between (x_0, y_0) and (x_i, y_i). The function $f(d_{0i})$ is defined as

$$f(d_{0i}) = \frac{1}{d_{0i}^{\beta}}, \tag{5}$$

with β as a parameter, which was taken from the interval <1;3>.

In the third method selected, the altitudinal zonality of the relief and precipitation in mountainous areas was taken into account. The basin was simply divided into elevation zones belonging to precipitation gauges; each gauge having its own elevation zone with varying width. The daily precipitation values from the stations belonging to the respective zones were used as the estimates of the areal precipitation in the zones.

By such procedures a set of daily precipitation maps was constructed for each method and day for the whole period of observation 1961-2000. Examples of such grid-based maps of daily precipitation depths are presented in Fig. 2 – Fig. 4. As next averaged daily rainfall series over the basin were computed from the grid maps for each interpolation method and day.

82

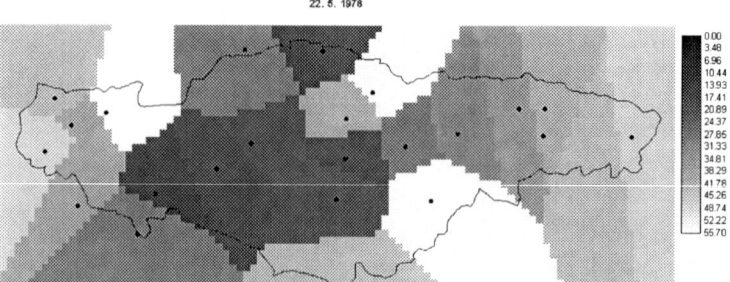

Figure 2. Grid map of daily rainfall [mm/day] determined by the Thiessen polygon method (May 22, 1978.

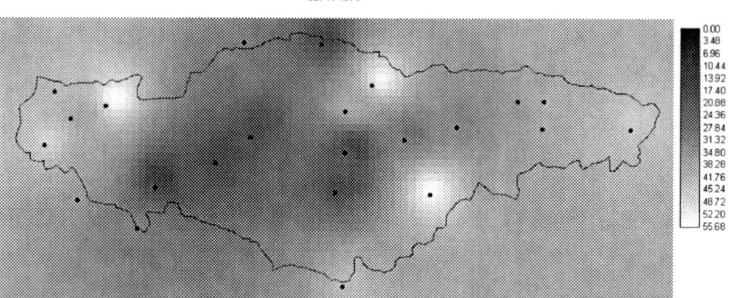

Figure 3. Grid map of daily rainfall [mm/day] determined by the inverse distance method (May 22, 1978)

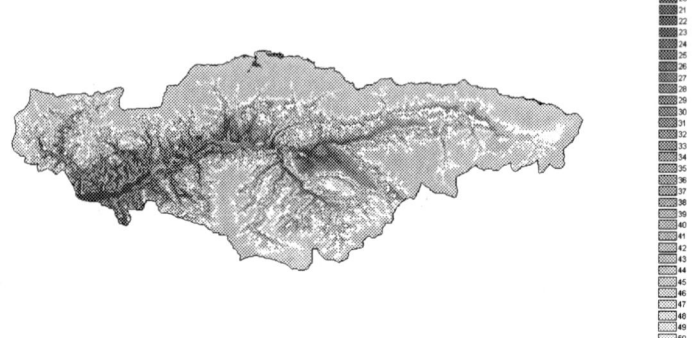

Figure 4. Grid map of daily rainfall [mm/day] determined by the method of altitudinal zones (May 22, 1978)

4. ESTIMATION OF DESIGN PRECIPITATION DEPTHS

Rainfall data from the warm season of the year were taken for the evaluation of the influence of the choice of the interpolation scheme on the uncertainty of estimation of basin average design precipitation. Annual maximum basin averages of 1, 2, 3, 4 and 5-day rainfall depths were extracted from these series. Only events with continuous rainfall during the indicated number of days were considered.

From the series of annual maximum 1- to 5- day rainfall depths N-year values were estimated by statistical analysis. Several distribution functions and parameter estimation methods were involved and compared: the Gumbel (EV1), the General Extreme Value (GEV), Pearson III (P3), logPearson III (LP3), General Logistic (GLO), Rossi (ME), Pareto (PA), Weibull (WB) and Lognormal (LN).

The DVWK (1999) statistical test was used to discriminate between the acceptability and rejection of the fit of a given distribution function to the data:

$$D + n\varpi^2 + (1 - r_p)$$

(6)

where:
- the quantity D is the maximum difference between the empirical and theoretical distribution in the probability space (Kolmogorov-Smirnov test value),
- the quantity ($n\varpi^2$) is a measure of the sum of the squares of the differences between the empirical and theoretical distribution function for all the data points in the probability space,
- the quantity r_p is the correlation coefficient between the ordered empirical sample data and the respective quantiles of the fitted distribution function.

The acceptable distributions should exhibit high r_p values, acceptable $n\varpi^2$ values and low D values. A detailed description of the procedure is given in DVWK (1999) and Kluge (1996). The best acceptable distributions for each rainfall duration and interpolation method are listed in Table 2.

Table 2. Best distribution functions selected according to DVWK procedure for all interpolation methods (the Gumbel (EV1), the General Extreme Value (GEV), Pearson III (P3), logPearson III (LP3), Rossi (ME), Weibull (WB) and Lognormal 3- parameter (LN3) distributions)

	Interpolation method		
Precipitation	Thiessen polygons	Inverse distance	Altitudinal zones
1-day	P3, WB3	P3, WB3	ME, LN3
2-day	GEV, LN3, ME	P3, WB3	GEV, LN3, LP3
3-day	EV1, ME, GEV	EV1, ME	GEV, LN3, LP3
4-day	GEV	GEV	EV1, GEV, ME
5-day	EV1, GEV, LN3	EV1, GEV	GEV, ME

It can be seen from Table 2 for the all interpolation methods (and as it was expected), that the number of statistically acceptable distribution functions was rather high in all cases. A comparison of the design precipitation values estimated from the functions showed that for the return periods considered in here (5, 10, 20, 50 and 100 years) they did not exhibit significant differences as illustrated for the inverse distance method in Table 3.

Table 3. Comparison of design basin average precipitation values estimated by statistically accepted different frequency distributions and parameter estimation methods for the inverse distance method. Parameter estimation methods: MOM – method of moments, MLM – maximum likelihood, PWM – probability weighted moments

	Return period of basin average rainfall				
	5	10	20	50	100
	Design basin average precipitation (mm)				
1 day					
P3(PWM)	37.4	42.9	48.1	54.7	59.5
WB(MM)	37.7	42.8	47.2	52.2	59.7
WB3(PWM)	**37.6**	**43.1**	**48.2**	**58.8**	**63**
2 days					
P3(MM)	49.4	53.9	57.8	62.4	65.6
WB(MLM)	**49.1**	**54.3**	**58.4**	**63.2**	**66.5**
WB3(PWM)	49.2	54.1	58.2	62.7	65.7
3 days					
E1(MLM)	58.5	65.8	75	84.9	88.8
E1(PWM)	58.9	66.4	72.8	83.2	90.1
ME(MLM)	**58**	**66.1**	**74.2**	**85.6**	**94.9**
4 days					
GEV(MM)	68.3	77.9	87.3	99.6	109
GEV(MLM)	68.2	77.3	86.1	97.4	106
GEV(PWM)	**67.9**	**77.8**	**87.9**	**102**	**113**
5 days					
GEV(MLM)	75.2	85.3	95.6	109	115
GEV(PWM)	74.9	85	94.7	106	114
E1(PWM)	**74.5**	**85**	**95.1**	**108**	**118**

Results from the statistical analysis were subsequently used to test the influence of the choice of the selected spatial interpolation schemes for this study on basin averaged design rainfall. It can be seen from Table 2, that for the basin average rainfall data derived from the inverse distance and the Thiessen polygon methods, types of acceptable distribution functions exhibits a better agreement, than between these both methods and the altitudinal zone method. This suggested that the data series from the former two methods show some degree of statistical similarity, whereas they seem to differ from the data from the third method. This tendency was confirmed by the subsequent analysis of the frequency curves. For these the best fitting distribution functions as suggested by the DVWK (1999) methods were used.

Frequency curves of annual maximum basin average 1- to 5- day precipitation depth derived from the three spatial interpolation schemes were compared in Figures 5 to 9. Most probably due to the fact, that convective precipitation can play a significant role in annual maximum daily precipitation values, for shorter durations the elevation zone method gave systematically higher values than the Thiessen polygon method and the inverse distance method. The differences are about 10 mm for the 100-year return period. For longer durations all methods gave comparable results. For all durations the latter two gave comparable results for all the data in this study. The inverse distance method has a better potential to reflect and project the actual spatial distribution of daily measured data into the spatially interpolated rainfall surface than the Thiessen polygon method, since the measurements reflect altitudinal zonality as well. This method is to be recommended as an appropriate interpolation scheme despite of the fact that the elevation zones method has a better potential to reflect the altitudinal zonality of precipitation especially for longer durations (but it tends to overestimate basin-averaged precipitation especially for daily values).

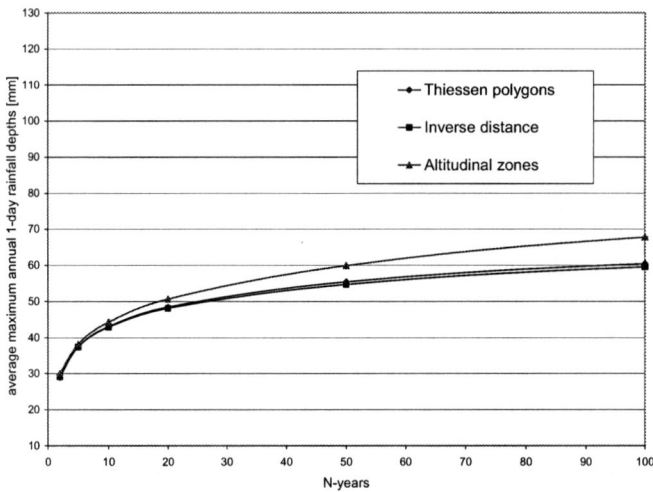

Figure 5. N-year values of basin average maximum annual 1-day rainfall depths

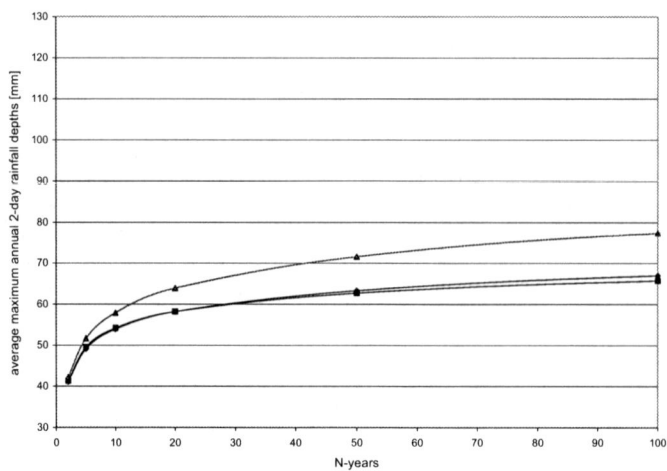

Figure 6. N-year values of basin average maximum annual 2-day rainfall depths

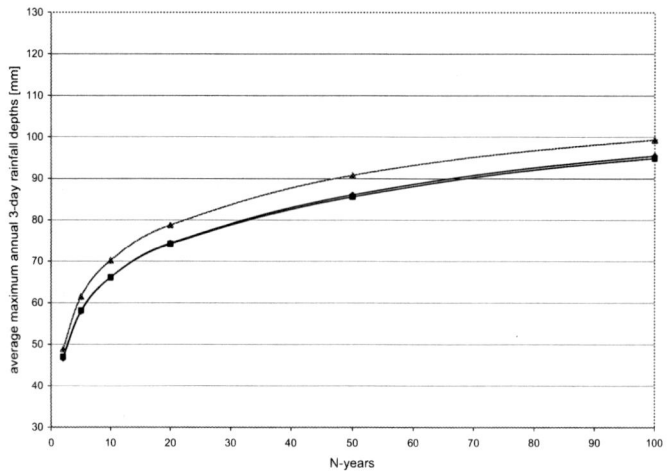

Figure 7. N-year values of basin average maximum annual 3-day rainfall depths

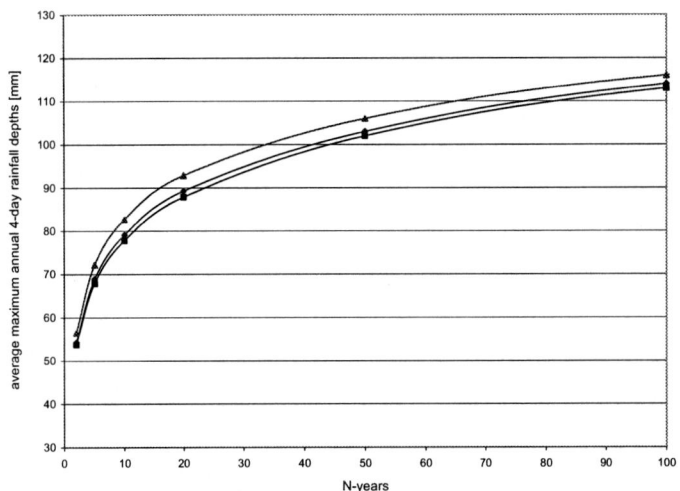

Figure 8. N-year values of basin average maximum annual 4-day rainfall depths

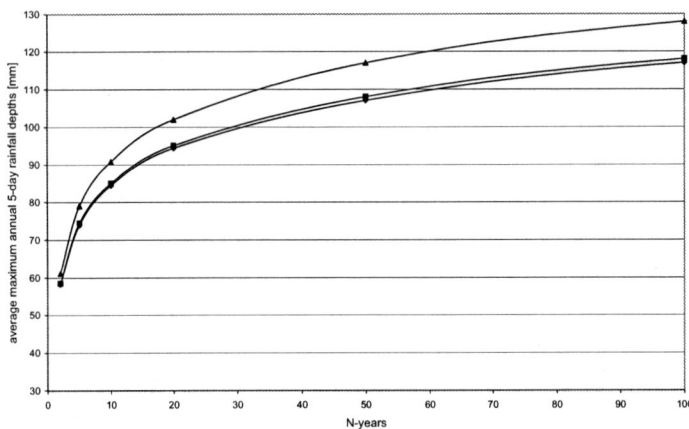

Figure 9. N-year values of basin average maximum annual 5-day rainfall depths

5. CONCLUSIONS

The aim of the study was to derive frequency curves of annual maximum basin average of 1-to 5- day precipitation depths for the upper Hron River basin. Three spatial interpolation schemes were selected for the analysis. The Thiessen polygon method and the inverse distance method gave systematically lower values for shorter durations of precipitation then the altitudinal zones method; most probably due to the fact, that convective precipitation can play a significant role in annual maximum daily precipitation values. For longer durations all methods gave comparable results. The inverse distance method has a better potential to reflect and project the actual spatial distribution of daily measured data into the spatially interpolated rainfall surface than the Thiessen polygon method, since the measurements reflect altitudinal zonality as well. Co-kriging, which can included the effect of topography, could probably also provide accurate and realistic estimates, but its application is tedious when compared with the inverse distance method. This method can be therefore recommended as an appropriate interpolation scheme for engineering applications.

ACKNOWLEDGEMENT

This work was supported by the Science and Technology Assistance Agency under Contract No. APVT-51-006502 and the Slovak Grant Agency under VEGA Project Nos. 1/1145/04 and 2/3085/23. The support is gratefully acknowledged.

References

Bartels H., Malitz G., Asmus S., Albrecht F.M., Dietzer B., Günther T. and Ertel H. (1997) *Starkniederschlagshöhen für Deutschland (KOSTRA)*. Selbstverlag des DWD, Offenbach am Main.

Custer G.S., Farnes P., Wilson J.P. and Snyder R.D. (1996) A comparison of hand- and spline-drawn precipitation maps for mountainous Montana. *Water Resources Bulletin* 32 (2), 393-405.

Daly C. and Neilson R.P. (1992) A digital topographic approach to modeling the distribution of precipitation in mountainous terrain. *Interdisciplinary Approaches in Hydrology and Hydrogeology*. American Institute of Hydrology, 447-454.

Dubois G. (1998) Spatial Interpolation Comparison 97: Foreword and Introduction. *Journal of Geographic Information and Decision Analysis* 2, 1-10.

DVWK Regeln 101/1999 (1999) *Wahl des Bemessungshochwassers*. Verlag Paul Parey, Hamburg.

Faško P., Lapin M., Šťastný P. and Vivoda P. (2000) Maximum daily sums of precipitation in Slovakia in the second half of the 20th century. Images of Weather and Climate. *Prace Geograficzne, fasc.* Cracow 108: 131-138.

FEH (1999) *Flood Estimation Handbook*. Part 3. Statistical procedures for flood frequency estimation, IH Wallingford, 325 pp.

Geiger H., Stehli A. and Castellazzi U. (1986) Regionalisierung der Starkniederschläge und Ermittlung typischer Niederschlagsganglinien, *Beitrage zur Geologie der Schweiz – Hydrologie* 33, 320 pp.

Hutchinson M.F. (1995) Interpolating mean rainfall using thin plate smoothing splines. *International Journal of GIS* 9 (5), 385 – 403.

Hutchinson M.F. and Bischof R.J. (1983) A new method for estimating the spatial distribution of mean seasonal and annual rainfall applied to the Hunter Valley, New South Wales. *Australian Meteorological magazine* 31, 179 – 184.

Jurčová S. and Kohnová S. (2002) Analysis of 5-day maximum precipitation rainfall in the upper Hron region. Diploma thesis, SvF STU Bratislava, 65 pp. (in Slovak).

Kluge Ch. (1996) Statistische Analyse von Hochwasserdurchfluessen. *Dresdner Berichte 7*, TU Dresden.

Malitz G. (1999) Starkniederschlag in Deutschland – Messergebnisse, statistische Auswertungen, Schätzungen. *Klimastatusbericht des DWD*, 35 – 41.

Meijerink A.M. et al. (1994) Introduction to the use of geographic information systems for practical hydrology. *ITC publication* 23, The Netherlands.

Parajka J. (1999) Mapping long-term mean annual precipitation in Slovakia using geostatistical procedures. Proceedings of the Int. Conf."Problems in Fluid Mechanics and Hydrology", Institute of Hydrodynamics AS CR, 424 – 430.

Stehlová K., Kohnová S. and Szolgay J. (2001) Analysis of 2-day precipitation depths in upper Hron region. *Acta Hydrologica Slovaca*. 2, No. 1, 167 -174 (in Slovak).

Tabios G.Q. and Salas J.D. (1985) A comparative analysis of techniques for spatial interpolation of precipitation. *Water Resources Bulletin* 21, 365 – 380.

Thomson C. S. (2002) The high intensity rainfall design system HIRDS. International Conference on Flood Estimation, Bern, Switzerland, 273 – 283.

Chapter 8

REGIONAL CLIMATE CHANGE
To be included in Future Flood Risk Analysis?

CHRISTIAN BERNHOFER[1], JOHANNES FRANKE[1],
VALERIE GOLDBERG[1], JÖRG SEEGERT[1],
WILFRIED KÜCHLER[2]
[1]*Technische Universität Dresden, Institute of Hydrology and Meteorology,
Member of the Dresden Flood Research Center (D-FRC), Dresden,
Germany,*
[2]*Sächsisches Landesamt für Umwelt und Geologie, Germany*

Keywords: Regional climate, climate analysis, climate change, regional downscaling,
rainfall, extreme precipitation, climate forecast, floods, flash floods.

1. INTRODUCTION

Floods and other seemingly weather related natural disasters have been increasing not only by reported damage but also by frequency and intensity (www.munichre.com). As global climate models predict a clear increase in temperature and weather extremes (IPCC 2001) due to increase in the atmospheric concentration of green house gases (GHG) it is obvious to relate the reported increase in flood risk to climate change. However, extreme precipitation is neither correctly modelled in existing global models nor correctly covered in existing statistical analysis of the past due to its inherent variability in space and time. This is specially true for the data base of future risk analysis as along with the spatial extend of typical flood prone catchments the need for regional and local estimates of extreme precipitation is increasingly important. New methods as well as new combinations of existing methods are required to address climate change related flood risk on the regional level.

2. BACKGROUND

Natural GHG concentrations are essential for surface climate. Their effect can be expressed as the deviation of mean temperature at the earth's surface from the effective radiation temperature of the earth-athmosphere

J. Schanze et al. (eds.), Flood Risk Management:
Hazards, Vulnerability and Mitigation Measures, 91–100.
© 2006 *Springer.*

system (equivalent to a mean radiative flux of about 342 Wm^{-2}). This effect amounts to an overall increase of about 33 K that is seen as prerequisite of the past ice age climate over the last 10 thousand years. Since 1800 due to industrialization the anthropogenic release of additional GHG amounts like CO_2, CH_4 or N_2O led to a relevant change in the atmospheric composition. The radiative forcing of additional CO_2 only amounts to 1.5 Wm^{-2}, of CH_4 to 0.5 Wm^{-2}, and of N_2O to 0.15 Wm^{-2}. This is well below 1% of the effective exchange of radiant fluxes of the earth-atmosphere system and therefore difficult to measure or model. The inverse effects by aerosols, feed backs by clouds and the biosphere increase further the challenges to correctly predict the climatic consequences of these additional GHG amounts.

Additional uncertainty is added by the estimates of future releases of GHG. Those depend drastically on economic development, energy needs and means of energy production, geographical distribution of production and wealth, and other factors completely beyond the scope of this paper. Therefore climate models rely on scenarios, which in turn reflect certain assumptions on economic changes. These scenarios are the main source of the uncertainty of the IPCC forecast (IPCC 2001): 1.4 K to 5.8 K temperature increase by 2100, an uncertainty of 4.4 K – while climate model uncertainty alone amount to only about 2 K (Figure 1).

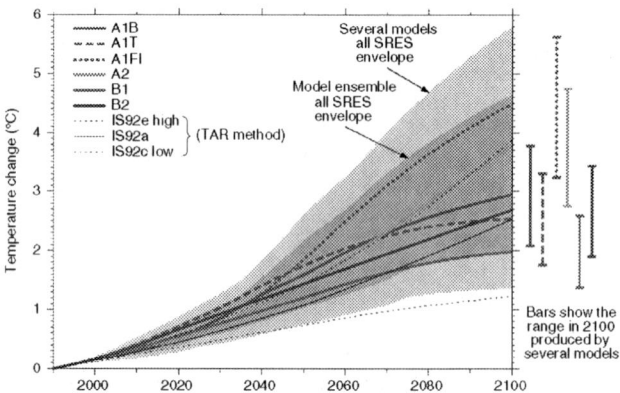

Figure 1. Predicted global changes in temperature for various scenarios predicted by different GCM (Global Circulation Models), from IPCC (2001)

Observed changes in global temperature are about 0.6 +/-0.2 K by 2000 and agree well with estimates due to the increase in GHG concentration (for CO_2 from 282 ppm by 1800 to 365 ppm by 2000). Reconstruction of temperatures since 1000 BC reveal that this increase is reverse to an overall decrease after the medieval "optimum" in the 12th century (Figure 2).

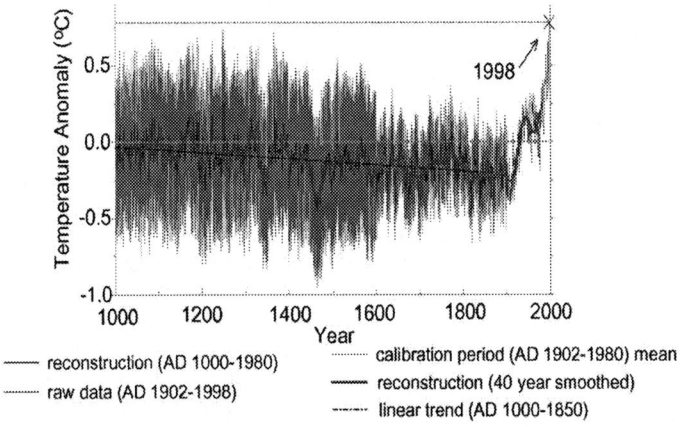

Figure 2. Mean annual temperatures of the northern hemisphere based on measurements and proxy data (from Mann et al. 1999)

Individual trends of climate records differ substantially from this global numbers, typically with larger changes in higher latitudes but large regional deviations. Predicted changes for 2100 show a similar dependence to latitude but due to their coarse resolution of several 100 km (e.g. 250 km for the German climate model ECHAM4-OPYC3; see www.dkrz.de) they cannot cover the regional deviations. This holds for temperature but is much more pronounced for precipitation and even more for extremes of precipitation. Those are typically not precisely enough covered in the short-term weather forecast, and they are currently beyond the scope of climate models. To overcome this shortcoming several methods were developed for spatial and temporal downscaling of GCM climate change scenarios and applied for regional impact assessment (Table 1).

Table 1. Three different methods of downscaling of GCM climate change scenarios

Method	Principle of initialisation and prediction	Advantage	Disadvantage
empirical	paleo-climatic analogies	fast and simple	uncertainty of paleo-climatic estimates, no information on the
semi-empirical	use of statistical relationships between large-scale circulation conditions and local	based on the most reliable GCM output (general circulation patterns)	complex statistics, sensitive to available time series of climate stations, linear effects
modelling	use of a dynamical meso-scale model nested into a GCM	physical basis of climate prognosis, non-linear coupling with regional land use and topography	error accumulation of the errors of the GCM and the nested model

3. MATERIAL AND METHODS

Here a semi-empirical method based on ECHAM4-OPYC3 output for the global scenario B2 (i.e., "regional-environmental"; see IPCC 2001) is used. The GCM predicts a change of +2 to +3 °C in Central Europe by 2100. The basis of the dynamic-statistical downscaling are climate records of all available stations for certain regions of Germany, where regional climate investigations have been intensified in projects like REKLIP for the Upper Rhine Valley (Parlow 1994), BayForKlim (Enders 1999) for Bavaria, and Blümel et al. (2001). Since several years the authors have gathered climate informations of the SE states of Germany (Saxony, Thuringia and Saxony-Anhalt) and processed into a ACCESS based data bank (Franke et al. 2004) with three main objectives: (1) data assimilation, verification, and if necessary homogenisation, of all available time series of climate standard quantities; (2) trend analyses of standard values and of derived complex climate quantities in the 1951–2000; and (3) as basis for the statistical-dynamic downscaling.

3.1. Data analysis

The data bank is described in detail in Franke et al. (2004) for Saxony and the tools described there had been applied to all data of SE Germany. Homogeneity was tested by methods of Schönwiese and Malcher (1985), Rapp and Schönwiese (1995) and Rapp (2000). Trend analysis was based on the data after verification, but it is almost impossible to distinguish between the verification of homogeneity and trend analysis (Herzog and Müller-Westermeier 1998), as in both cases data series are detected for time changes. To separate trends from the statistic variability of time series, the Mann-Kendall test was used to determine the significance level (SIG) of trends (Rapp 2000). After homogenisation of time series, the complete test cycle was repeated until homogeneity was reached. Linear trends were calculated after Rapp (2000) for climatological seasons, vegetation periods I, II, and complete years in the climatic normal periods (e.g. 1971–2000). The local trends of individual climate stations were interpolated by the method of 'natural neighbour' to area trends, using Thiessen polygons (Owen 1992). Comparison of area trends requires a regular distribution of climate stations in the region, and this condition is almost fulfilled for the area and time periods investigated (Figure 3).

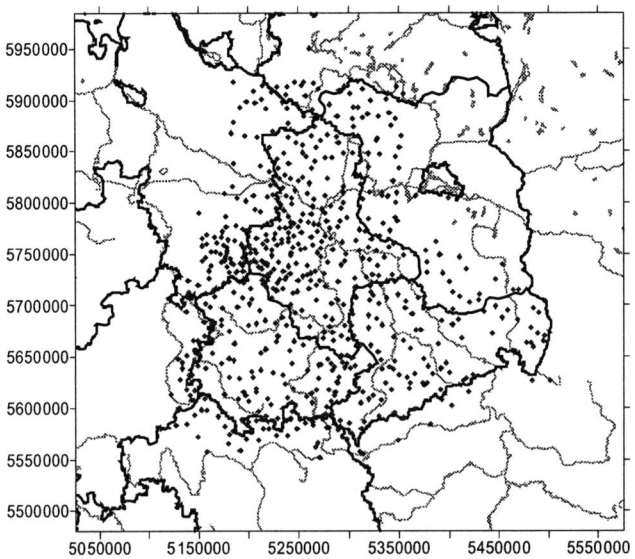

Figure 3. Locations with homogeneous records of daily rainfall in SE Germany included in the climate analysis (Gauß-Krüger coordinates)

3.2. Regional downscaling

A region like SE Germany is only represented by a few grid points in GCMs, which describe only mean trends. Statistical downscaling using weather classifications (Enke and Spekat 1997, Conway and Jones 1998, Schnur and Lettenmaier 1998) or dynamic meteorological models (Egger 1995, Fuentes and Heimann 1995) is required to connect large-scale climate trend information with specific characteristics (topography, land use, circulation patterns) of the region of interest. The statistical link to the observed weather patterns is achieved by regression, EOF (Empirical Orthogonal Functions), or ANN (Artificial Neural Networks). For downscaling in time stochastic weather generators are utilised. Weather pattern classification is done objectively on the basis of atmospheric pressure fields, typically a categorisation with a depth of 10 classes proofs to be adequate (Enke and Spekat 1997).

To forecast extreme values weather can be described by a chronology of circulation patterns, where weather conditions are related to seasonally dependent positive or negative deviations from the areal mean. Change of circulation pattern statistics (frequency, duration, sequence, internal characteristics) will be predicted reliably by the climate models, forming the background for a stochastic modelling within the predicted frequency distribution. These extremes can be used directly or instead, the frequency distribution can be used for an adapted "design precipitation" under conditions of climate change.

4. RESULTS

4.1. Climate analysis

From climate analysis maps of absolute values, and trends of single and derived quantities were created from the homogenised data. Additionally, frequencies of precipitation sums exceeded, snow height, and trends of persistence of droughts and snow cover were analysed. The most pronounced trend of precipitation (Figure 4) in the last 50 years was a maximum rainfall decrease of about 50% (SIG >80%) during summer in the lowland in the north-east of the investigated area. The trend in the low mountain range of southern Saxony was weak (max. +10%). The precipitation in the autumn, particularly in the low mountain range, and in winter (Figure 4), particularly in the lowland, generally increased by 20% (SIG >80%). This is a more moderate increase than in the western part of Germany (Schönwiese 1995), and may be caused partly by the increase in rainfall in relation to snowfall, due to higher temperatures in winter time, leading to smaller wind errors for liquid precipitation measurements. The winter increase does not compensate the decrease in summer in northern Saxony.

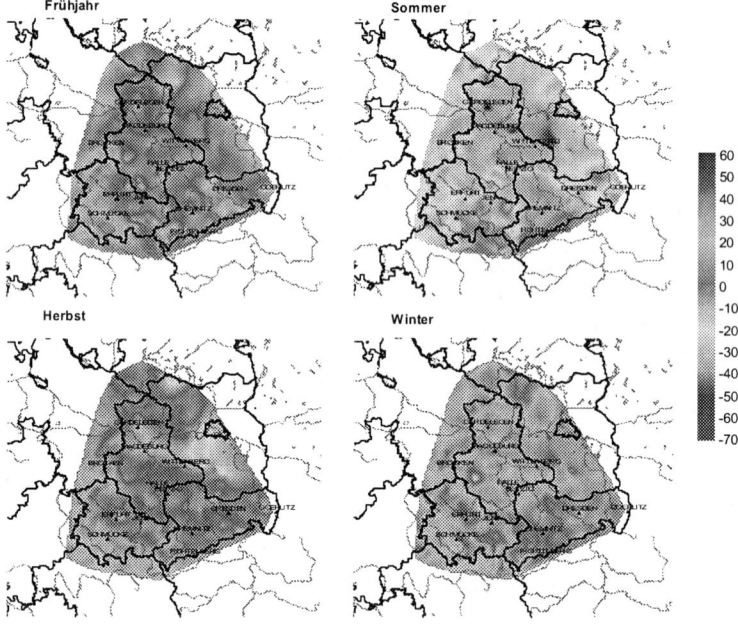

Figure 4. Observed change in seasonal rainfall [%] in SE Germany between 1951 – 2000, based on analysis of homogeneous records

During both vegetation periods (April–June and July–September) the persistence and frequency of droughts in the 2nd vegetation period (April to June) increased. Strong rainfall events (>20 mmd^{-1}) markedly influence the regional distribution of trends, as their frequency during the summer period increased 5-fold for 1971–2000, compared to 1961–1990. The trend of temperature (not shown) is positive in all seasons except autumn.

4.2. Regional downscaling

Enke et al. (2005) utilised the methods described above to construct precipitation maps for the decade 2041-2050. They show that the trend as observed in SE Germany will probably continue with dryer summers in S and E Germany (Figure 5) and wetter winters than currently observed.

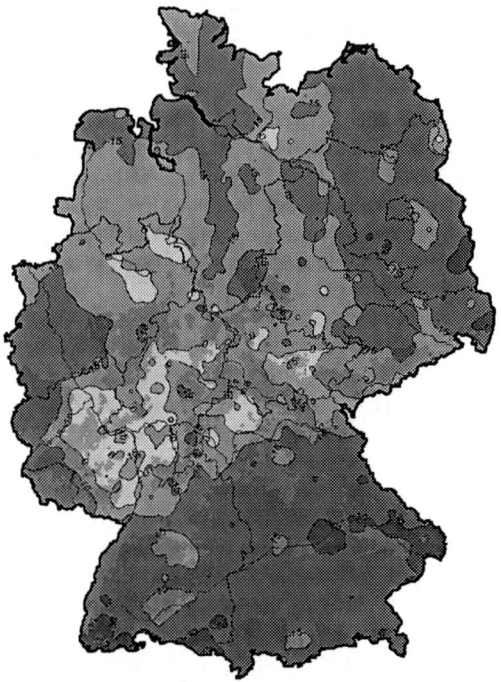

Figure 5. Predicted change in summer rainfall [%] in Germany for the decade 2041-2050, compared to 1981-2000 (from Enke et al. 2004)

They reveal a marked difference in the predicted precipitation between western and eastern parts of Germany: While in the west winter rainfall will compensate for summer losses, many regions in the east (already dryer today) will suffer from negative climatological water balances with serious potential effects for agriculture and forestry. At the same time the frequency and intensity of extreme events by convective rainfall is increasing, interrupted by long lasting droughts as observed in 2003. The consequence is an increase in local extremes that might cause flash floods with increasing probability especially in the mountainous regions in the south and east of Germany.

5. CONCLUSIONS AND OUTLOOK

Regional changes in climate are currently observed in the example region of SE Germany, reflecting a general change in Europe (Rapp 2000). Trends of temperature, precipitation and other climate variables between 1951 and 2000 document a significant regional climate change, with temperature as the most pronounced effect. Climate change due to an increase in GHG concentration is therefore no longer a possibility but a fact that has to be included in decision support. However, local and seasonal effects vary considerably, presumably due to changes in the patterns of atmospheric circulation as well as local effects of orography (Werner et al. 2001). The marked reduction in summer rainfall, and increased persistence and frequency of drought periods in the east and south of Germany, and the increase of strong summer rainfall events on the other, are clear signals of regional climate change. The simultaneous increase in drought periods and frequency of strong rainfall events, combined with a general decrease in rainfall amounts during the summer, pose new challenges on water management: flood protection on one hand, and supply of high-quality water on the other.

By combining the results of regional climate analysis and forecast, it will be possible to adapt guidelines for water resource management, agriculture, forestry, the energy sector, and tourism. The climate forecast for Germany, based on methods of statistical downscaling using climate model outputs and statistics of weather patterns (Enke et al. 2005) can also be extended to reflect changes in re-occurrence, duration and intensity of heavy rainfalls. This requires an analysis of the actual precipitation statistics with a high spatial and time resolution. In context to flood risk analysis the existing climate data base will be adapted for hourly or 10 min data for several small and medium catchments in Germany to test this approach. Finally, weather generators will be tuned to existing precipitation data and applied to future climate from statistical downscaling. This data can be utilised for regional water resources and flood risk purposes of the future. Special emphasis should be given to the inherent uncertainties of the methods, a task than can be solved either by ensemble forecasts or error propagation.

ACKNOWLEDGEMENT

Funding by the State Ministry of Environment and Agriculture of Saxony and Thuringia is acknowledged. The authors especially thank Dr. L.

Coufal of the Czech Hydrometeorological Institute (CHMI) in Prague for providing Czech data.

References

Bernhofer Ch. and Goldberg V. (2001) CLISAX – Statistische Untersuchungen regionaler Klimatrends in Sachsen. Final Report of the Saxon State Office for Environment and Geology, Dresden.

Bernhofer Ch., Goldberg V. and Franke J. (2002) CLISAX II – Assimilation von standardisierten und abgeleiteten Klimadaten für die Region Sachsen und Ausbau der Sächsischen Klimadatenbank. Final Report of the Saxon State Office for Environment and Geology, Dresden.

Blümel K., Klämt A., Malitz G., Matthäus H., Rachner M. and Richter D. (2001) Hydrometeorologische Untersuchungen zum Problem der Klimaveränderungen. *Ber. Dt. Wetterdienst* 219, Offenbach.

Chmielewski F.M. (2003) KLIVEG – Mögliche Auswirkungen klimatischer Veränderungen auf die Vegetationsentwicklung in Sachsen. Final Report of the Saxon State Office for Environment and Geology, Dresden.

Conway D. and Jones P.D. (1998) The use of weather types and air flow indices for GCM downscaling. *J. Hydrol.* 212/213, 348-361.

Egger J. (1995) Regional statistical-dynamical climate modelling tests. *Contr. Phys. Atmos.* 68: 281-289.

Enders G. (1999) Klimaänderungen in Bayern und ihre Auswirkungen. Final Report, Bayerischer Klimaforschungsverbund (BayForKlim), Ludwig Maximilian University, Munich.

Enke W. and Spekat A. (1997) Downscaling climate model outputs into local and regional weather elements by classification and regression. *Climate Research* 8, 195-207.

Enke W., Deutschländer T. and Schneider F. (2005) Results of 5 regional climate studies applying a weather pattern based downscaling method to ECHAM4 climate simulations. *Met. Z.* (in press).

Flemming G. (2001) Angewandte Klimatologie von Sachsen. Basis- und Zustandsklima im Überblick. *Tharandter Klimaprotokolle* 4, Dresden University of Technology.

Franke J., Goldberg V., Eichelmann U., Freydank E. and Bernhofer Ch. (2004) Statistical analysis of regional climate trends in Saxony. *Climate Research* 27 (2), 145-150.

Fuentes U. and Heimann D. (1996) Verification of statistical dynamical downscaling in the Alpine region. *Climate Research* 7, 151-168.

Herzog J. and Müller-Westermeier G. (1998) Homogenitätsprüfung und Homogenisierung klimatologischer Messreihen im Deutschen Wetterdienst. *Ber. Dt. Wetterdienst* 202, Offenbach.

HMU-MD (1973) Klima und Witterung im Erzgebirge. Hydrometeorologick´y Ústav CSSR, Meteorologischer Dienst der DDR (eds) *Abh. Meteorol. Dienst DDR 104*, Berlin.

IPCC (2001) Third Assessment Report of Working Group I of the IPCC. Summary for Policymakers. Shanghai.

Mann M.E., Bradley R.S. and Hughes M.K. (1999) Northern Hemisphere Temperatures During the Past Millennium: Inferences, Uncertainties, and Limitations. *Geophysical Research Letters* 26 (6): 759-762.

Owen S.J. (1992) An implementation of natural neighbour interpolation in three dimensions. MS thesis, Brigham Young University, Provo, UT.

Parlow E. (1994) Faktoren und Modelle für das Klima am Oberrhein – Ansätze und Ergebnisse des Regio-Klima-Projektes REKLIP. *Geogr. Rundsch.* 3, 160-167.

Rapp J. (2000) Konzeption, Problematik und Ergebnisse klimatologischer Trendanalysen für Europa und Deutschland. *Ber. Dt. Wetterdienst* 212, Offenbach.

Rapp J. and Schönwiese CD (1995) Klimatrend-Atlas Deutschland 1891–1990. *Frankfurt Geowiss. Arb.* 5, University of Frankfurt (Main).

Schnur R. and Lettenmaier D.P. (1998) A case study of statistical downscaling in Australia using weather classification by recursive partitioning. *J. Hydrol.* 212/213, 362-379.

Schönwiese C.D. (1995) *Klimaänderungen. Daten, Analysen, Prognosen.* Springer, Berlin.

Schönwiese C.D. and Malcher J. (1985) Nicht Stationarität oder Inhomogenität? Analyse klimatologischer Zeitreihen. *Wetter Leben* (Vienna) 37, 181-193.

Schrödter H. (1985) *Verdunstung – Anwendungsorientierte Messverfahren und Bestimmungsmethoden.* Springer, Berlin.

Schwanecke W. and Koch D. (1970) *Mittelgebirge und Hügelland der DDR. Klimastufen, zusammengestellt nach Ergebnissen der forstlichen Standorterkundung von 1956–1970.* Akad. Landwirt. Wiss. DDR, Potsdam.

Werner P.C. Gerstengarbe F.W. and Oesterle H. (2001) *Klimatypänderungen in Deutschland im 20. Jahrhundert.* Climate status report 2001, Deutscher Wetterdienst, Offenbach.

PART 3

FLOOD FORECASTING

Chapter 9

APPLICATION OF EO DATA IN FLOOD FORE-CASTING FOR THE CRISURI BASIN, ROMANIA

GHEORGHE STANCALIE[1], ANDREI DIAMANDI[1],
CIPRIAN CORBUS[2] AND SIMONA CATANA[1]
[1]*Romanian Meteorological Administration, Bucharest, Romania,*
[2]*National Institute for Hydrology and Water Mmanagement, Bucharest,
Romania*

Keywords: Earth observation, optical and radar satellite data, ASTER, MODIS, LANDSAT.

1. INTRODUCTION

The risk of flooding due to runoff is a major concern in many areas around the globe and especially in Romania. In the latest years river flooding and accompanying landslides, occurred quit frequently in Romania, some of which isolated, others-affecting wide areas of the country's territory.

One region, which suffers from flood damages on a regular basis, is the Crisuri basin in the Western part of Romania (Marsalek 2000). These losses included damages to houses, roads and railways, bridges, hydraulic structures, loss of domestic animals, and business losses.

The flood forecast and defence related information in Romania is presently based entirely upon the ground-observed data, which are mostly collected by non-automatic hydrometeorological stations. Such data are somewhat limited in terms of spatial distribution, temporal detail, and speed of collection and transmission, and these limitations already started to be overcome. (DESWAT project).

Orbital remote sensing of the Earth is presently capable of making fundamental contributions towards reducing the detrimental effects of extreme floods (Brakenridge et al. 1998). Effective flood warning requires frequent radar observations of the Earth's surface through cloud cover. In contrast, both optical and radar wavelengths will increasingly be used for disaster assessment and hazard reduction. These latter tasks are accomplished, in part, by accurate mapping of flooded lands, which is commonly done over periods of several or more days. The detection of

101

J. Schanze et al. (eds.), Flood Risk Management:
Hazards, Vulnerability and Mitigation Measures, 101–113.
© 2006 *Springer.*

new flood events and public warnings thereof is still experimental, but making rapid progress; radar sensors are preferred due to their cloud penetrating capability (Brakenridge et al. 2003). Relatively low spatial resolution, but wide-area and frequent coverage, are appropriate; the objective is to locate where within a region or watershed the flooding occurs, rather than to map the actual inundated areas. The rapid-response flood mapping and measurement provide information useful for disaster assessment, and has become a relatively common activity.

A wide variety of sensors have been used but there is much potential for development of advanced measurement capabilities that can better define flood severity and damage.

At the initiative of the Romanian Meteorological Administration, a project on "Monitoring of Extreme Flood Events in Romania and Hungary Using EO (Earth Observation) Data", proposed to the NATO Science for Peace (SfP) Programme started in 2002 (Brakenridge et al. 2001). The project, including representatives of Romania, Hungary and USA, considers the setting up of a satellite-based surveillance system connected to a dedicated GIS database that will offer a much more comprehensive evaluation of the extreme flood effects. The main goal of the project is to reduce flood damages in the study area by improved flood forecasting and flood defence.

Considering the necessity to improve the means and methods to forecast, assess and monitor flooding, the paper presents some capabilities offered by remotely sensed data, applied to the Crisuri basin in Romania.

2. STUDY AREA

The study area represents the Crisul Alb and Crisul Negru basin with an area of 14,900 km^2 on the Romanian territory. This basin comprises mountainous areas (38%), hilly areas (20%) and plains (42%), about 30% of the catchment being forested. Annual precipitation ranges from 600-800 mm/year in the plain and plateau areas to over 1200 mm/year in the mountainous areas. The orography of the Apuseni Mountains amplifies the precipitation on the western side of the mountain range. In the Crisuri Rivers Basin appear frequently large precipitation amounts in short time intervals and the frequency of such events seems to be increasing in recent years.

In terms of hydrography, there is a marked difference between high rates of mountain runoff and low rates of runoff in plains. Thus, runoff flood waves formed quickly in the mountain region part of the basin move rapidly to the plain situated near the Hungarian border, which is characterised by relatively slow flows and a potential for inundation. The hydrography of the study area is well monitoring. There are 62 hydrometric stations on the Crisul Alb and Crisul Negru and their tributaries; 7 of these stations have flow records longer than 80 years. The list of significant floods includes the events of June 1974, July-August 1980, March 1981, December 1995-January 1996, March 2000, April 2000 and April 2001.

To mitigate flood impacts in the study area, structural and non-structural measures have been undertaken in the past. Dikes along the Crisul Alb River and Crisul Negru River defend the Romanian area. These dikes were built in the 19th century for a 20-year design return period and further improved in later years. Currently, the dikes on the right bank of the Crisul Negru River and the Teuz River (43 km) are designed for a 50-year return period, and on the Crisul Alb, 67 km of dikes on the right bank and 59 km on the left bank are designed for a 100-year return period. In spite of these improvements, in April 2000, the right bank dike of the Crisul Negru broke near the village Tipari (a 130 m breach) and caused significant flooding, and damages in, the adjacent territory. Other structural flood protection measures include permanent retention storage facilities (total volume of 34×10^6 m^3) and temporary storage facilities (a total storage volume of almost 80×10^6 m^3).

3. DEVELOPMENT OF INTEGRATED METHODS, ENCOMPASSING EO DATA AND GIS FACILITIES, FOR FLOOD FORECASTING AND WARNING

EO images have wide applications in flood analysis, in such tasks as producing catchments maps, detecting water surface and soil moisture, detecting inundated areas, and assisting with remote flow measurement. Thus, image processing is important for developing such products and using them in flood analysis and management.

Medium and high spectral and spatial satellite resolution data, from the optical and microwave spectral domains (provided by the existing platforms like NOAA-AVHRR, DMSP, LANDSAT, SPOT, IRS, ERS, RADARSAT and by the newly available ones such as TERRA-MODIS, TERRA-ASTER and QUIKSCAT) could substantially contribute to determining the flood-prone areas and to the improvement of the existing flood forecasting and warning system in the Crisuri basin.

The geo-referenced information, obtained from optical and radar images could be used in determination of certain parameters required in flood forecasting and monitoring, such as the hydrographic network characteristics, water accumulation, snow cover, size of the flood-prone area, land cover/land use features, etc.

3.1. Identifying, delineating and mapping water and wetland areas

Different methods and algorithms for EO data processing and interpretation, needed for identifying, delineating and mapping water and excess soil moisture areas, were analysed and tested.

A Satellite Image Database (SID) provided by different platforms and sensors has been set up. The purpose of the SID is to gather information about the raw satellite scenes available as well as of the derived products and make it available in a simple format. This information is useful to test the processing and analysis algorithms for the water detection, mapping and analysis of flooding. The SID was build in Microsoft Works and will be available on-line on the file server, being updated as new satellite images are acquired. Each record of the database describe the characteristics of each satellite image: platform, sensor, date and time of data acquisition, duration of pass, spectral band, coordinates of the area covered, projection, calibration, size, bits/pixel, image file format, physical location (machine, directory), origin of data, type (raw/processed), type of processing applied, algorithm used, quick-look available, cloudiness.

3.1.1. Optical satellite data

For these types of satellite data, emphasis was placed on the new MODIS and ASTER images, provided by TERRA and AQUA platforms.

MODIS is a passive imaging spectral-radiometer carrying 490 detectors, arranged in 36 spectral bands that cover the visible and infrared spectrum. For our purpose a Level 1B, visible bands 1 and 2 (250 m resolution), calibrated radiances data (format MOD02QKM), have been used.

ASTER acquires 14 spectral bands and can be used to obtain detailed maps of surface. Each scene covers 60 x 60 km. The ASTER spectral bands are organized into three groups: 3 bands with 15 m resolution in VNIR, 6 bands with 30m resolution in SWIR and 5 bands with 90 m resolution in TIR. The data format use for MODIS and ASTER data is HDF-EOS, which is not always implicit among the data format of all image processing software. This is why it is necessary to read the image header or to use specific software that import the image and export it in a known format.

Technique for detection the surface water using ASTER and MODIS data. The ASTER data are used to delineate the water, using a common approach, like the un-supervised classification and then use as ground truth in figuring out the best way to delineate water in the MODIS data (Brakenridge et al. 2003). The idea is to get not just a binary delineation, water - no water, but also to determine what part of the MODIS pixel is water, in the edge pixels. The ASTER image (Figure 1) has to be first co-registered to WGS84 datum and UTM projection zone 34 and rotated with 10.433 degrees in order to co-locate and analyzed with MODIS image.

For the water delineation in the Aster image, the reflectance features of water in the green band and the absorption features in the near infrared band have been used. The water area, delineated for the selected subset (Tileagd and Lugasu Lakes) is shown in the blue color in the Figure 2 with a background in white color.

Figure 1. Subset of the TERRA/ASTER color composite image from 31/10/2001

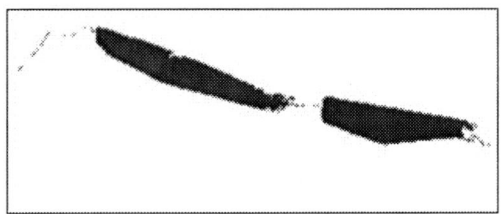

Figure 2. Close–up from the Tileagd and Lugasu lakes area (blue color) after the application of the detection water method on the ASTER image

In order to extract the water objects, the MODIS data products MOD02QKM (two bands, NIR and R, 250 m resolution) and MOD02HKM (which contains also a blue band, but at 500 m resolution) have been used. A MODIS image, which contains water, has been chosen, and an ASTER image as ground truth for the same area. The main steps for data processing are:

- Removal of the bow tie effect from MODIS image.
- Importing the two MODIS data (HKM and QKM) in ERDAS Imagine environment;
- Resampling of HKM image to 250 m resolution;
- The resolution merging of HKM and QKM data, in order to have an image with 3 bands at 250 m resolution;
- MODIS image geo-referencing (Figure 3) using the ASTER image (Figure 4) and choosing of a 300 m resolution (because of the eccentricity of the image);
- Performing the supervised classification in 4 classes (choosing parallelepiped limits as non-parametric rule and maximum likelihood as parametric rule for overlap);
- Creating a binary mask by merging the three non-water classes;
- Superposing the ASTER water mask over the MODIS image;
- Degrading the ASTER mask (15 m pixel) to 250 m pixel (as MODIS) in order to have percentage of water pixels instead of water – non-water pixels;
- Superposing the degraded ASTER mask over the MODIS water mask and obtaining a MODIS mask with the percentage of water pixels.

106

Figure 3. MODIS Image 31/10/2002   *Figure 4. ASTER Image 31/10/2002*

The algorithm serving to create a water mask using a threshold technique on multispectral MODIS images was adopted from the Dartmouth Flood Observatory (Hanover, USA) and further modified in early 2004 (Brakenridge et al. 2004). The 1.6 μm-channel MODIS data are preferred, because of the lowest water reflectance. However, this channel allows separating water from lowland areas, but not from snow in mountainous areas.

To mask out snowy areas, a threshold on the 0.87 μm channel has to be used.

The NDVI method. This method is based on the finding that large, clear water surfaces without vegetation have the Normalised Difference Vegetation Index (NDVI) < -0.2 and can be easily separated from shadows. However, using this threshold would mask out not only the shadows but also many pixels containing water, such as the pixels with high vegetation fractions, turbid waters, and mixed pixels along the coasts. To avoid the loss of such pixels, a NDVI threshold > -0.2 had to be used. After some experiments with different threshold values, a set of best-suited values was adopted.

There are some things that NDVI method can also get misclassified as water like:

- Cloud shadows (classification and sometimes NDVI);
- Clouds themselves (NDVI);
- Topography shadows (classification, sometimes NDVI);
- Urban areas or bare soil (classification method when trying to include classes of sediment or hazy water);
- The black border around geo-corrected image, forest fire scars (classification).

3.1.2. Radar satellite data

The functioning principle of the radar is already known, allowing the acquisition of satellite images (ERS, RADARSAT) irrespectively of weather conditions and the moment of day and night when the reception was made. In the microwave region the water presence could be appreciated by estimating the surface roughness, where water layers smooth surfaces dielectric constant is then heavily correlated to soil water content.

The data provided from the European ERS-1 and ERS-2 satellites and the Canadian RADARSAT one carry a Synthetic Aperture Radar (SAR) instrument have been used.

In case of SAR images the multi-temporal techniques was considered to identify and highlight the flooded areas. This technique uses black and white radar images of the same area taken on different dates and assigns them to the red, green and blue color channels in a false color image. The resulting multi-temporal image is able to reveals change in the ground surface by the presence of color in the image; the hue of a color indicating the date of change and the intensity of the color the degree of change. The proposed technique requires the use of a reference image from the archive, showing the « normal » situation. The multi-temporal image analysis, combined with the land cover/land use information allow the identification of the area covered by water (included the permanent water bodies) and then of the flooded areas.

The Figure 5 presents flowchart for the generation of the flood extent maps using satellite radar (SAR) images.

Using the methodology for the identification and mapping of the flooded areas is possible to monitor and investigate the flood evolution during different phases. The Figure 6 shows an example of the utilization of optical and radar data for the flood evolution monitoring, in the Crisul Alb basin using the RADARSAT image of 7.04.2000, during the flood event, comparing with the reference IRS image of 4.08.2000.

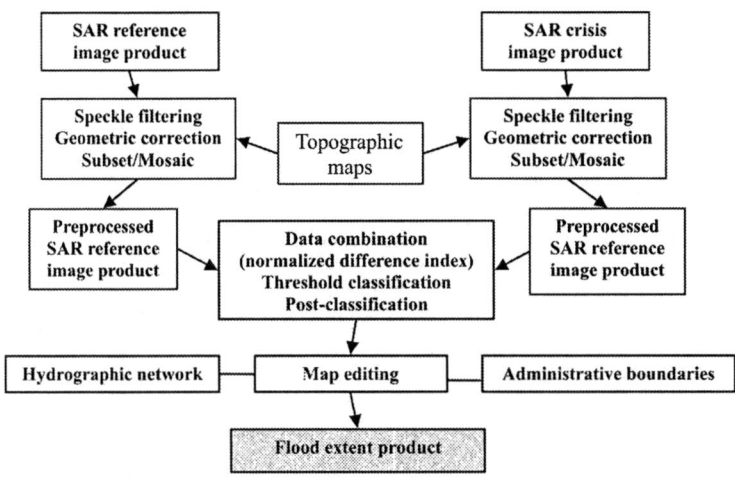

Figure 5. Flowchart for the estimation of the flood extent using satellite radar images

108

Figure 6. Flood evolution monitoring, in the Crisul Alb basin using RADARSAT image of 7/04/2000 (left), comparing with the reference IRS image of 4/08/2000

3.2. Potential of Satellite Radar Scatterometry for wetland detection

An innovative method, still experimental, based on radar scatterometry technique can detect the wetness anomalies and provide wetland monitoring (Brakenridge et al. 2003). The SeaWinds sensor on QuikSCAT, is a space-borne Ku-band scatterometer with a large swath but low spatial resolution (30 km cell), that provide every 2.5 days global coverage with the capability to see through clouds and darkness. The QuikSCAT data analysis is based on the polarization reversal of radar backscatter, measured by the scatterometer, over wetlands or flooded areas.

Displays are produced at Dartmouth Flood Observatory, Hanover, USA, for all continental land areas, are updated daily, and are currently available at the observatory web site (http://www.dartmouth.edu/~floods). Also being disseminated at this address and with the same frequency are 7-day animations of these displays, and other regional maps showing the location of strongly negative but un-differenced polarity ratios. These regional displays are useful in localizing areas of excess moisture receipts, and could be useful for hydrological forecasting modeling as input data.

3.3. Land cover/land use satellite-derived information

The land cover/land use classes derived from updating high spatial resolution satellite images are useful for the hydrological forecasting model input and dissemination of the obtained results.

The methodology for the achievement of the land cover/land use from high-resolution images is based on the observation of the following requirements:

- The structure of this type of information must be at the same time cartographic and statistic;
- It must be suited to be produced at various scales, so as to supply answers adapted to the different decision making levels;
- Up-dating of this piece of information must be performed fast and easily.
- The used methodology implies following the main stages below:
- Preliminary activities for data organizing and selection;
- Computer-assisted photo-interpretation and quality control of the obtained results;
- Digitization of the obtained maps (optional);
- Database validation at the level of the studied geographic area;
- Obtaining the final documents, in cartographic, statistic and tabular form.

Preliminary activities comprise collection and inventorying of the available cartographic documents and statistic data connected to the land cover: topographic, land survey, forestry, and other thematic maps at various scales.

To obtain the land cover map/land use map, satellite images with a fine geometrical resolution and rich multispectral information have to be used. For example, in case of IRS and SPOT data the preparation stage consisted in merging data obtained from the panchromatic channel, which supply the geometric fineness (spatial resolution of 5 m for the IRS, 10 m for the SPOT), with the multispectral data (LISS for IRS, XS for SPOT), which contain the multispectral richness.

A series of specific image processing operations were performed with the ERDAS Imagine and ENVI software. Those operations included: geo-referencing of the data, detection of cloud and water, image improvement (through using the histogram, contrast enhancing, slicking, selective contrast, combinations between spectral bands, re-sampling operation), statistic analyses (for the characterization of classes, the selection of the instructing samples, conceiving classifications). The computer-assisted photo-interpretation finalizes in the delimitation of homogeneous areas from images, in their identification and framing within a class of interest.

In Figure 7 a flowchart for the generation of the land cover/land use maps using high-resolution satellite data is presented. Discriminating and identifying the different land occupation classes rely on the classical procedures of image processing and leads to a detailed management of the land cover/land use, followed by a generalizing process, which includes:

(i) identification of each type of land occupation, function of the exogenous data, of the "true-land" data establishing a catalogue;
(ii) delimitation of areas suspected to represent a certain unity of the land;
(iii) expanding this delimitation over the ensemble of the image areas, which display resembling features.

110

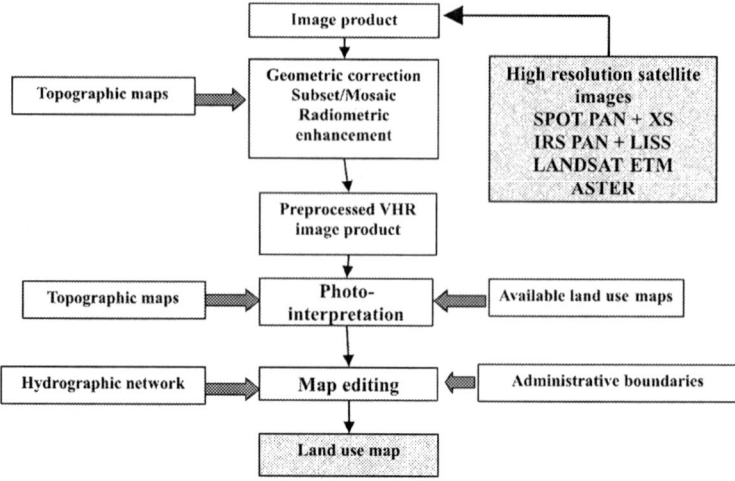

Figure 7. Flowchart for the generation of the land cover/land use maps

Validation of results from photo-interpretation, mapping (by checking through on land sampling at local and regional level) and building up the database aims at knowing the reliability level and the precision obtained for the delimitation of the units and their association to the classes in the catalogue.

The land cover maps are useful to classify the terrain function of the main types of land cover, thus allowing their characterization function of the land roughness and land impermeability degree (Stancalie et al. 2004). The notion of terrain impermeability is connected to the necessity to determine the absorption capacity and resistance to in-soil water infiltration. It must be mentioned that these parameters represent "potential capacities", having in view that the dry soil hypothesis is considered. For example, in view to establish the impermeability degree, specific coefficients, within a scale ranging from 0 to 1, have to be assigned for the identified land cover/use classes. These coefficients allow determining the mean impermeability coefficient by sub-basin, weighted with their mean slopes. To establish the water absorption capacity over a sub-basin, the land cover/use classes are associated with coefficients representing the precipitation amount, which a class is able to absorb. These coefficients are related with a runoff delay factor: the higher the coefficient, the higher the soil capacity to absorb water; also, the higher the water absorption capacity at the level of a spatial unit, the higher the precipitation necessary in a time unit for runoff occurrence.

The satellite based cartography of the land cover/land use is also important because it makes possible periodical updating and comparisons, and thus contribute to characterize the human presence and to provide elements on the vulnerability aspects, as well as the evaluation of the impact of the flooding (Stancalie 2004).

3.3   Snow cover mapping

The satellite data offer valuable information to identify and mapping areas with snow through differentiation from other bodies in the image having close spectral reflectance, especially the clouds and to delimitate and monitor the snow-covered areas evolution.

The achievement of the mentioned objectives, involve the processing of two kind of satellite images:
(i)   with high temporal resolution, but coarse spatial information (like NOAA-AVHRR,TERRA-MODIS or SPOT-VEGETATION);
(ii) with high spatial resolution but with more rare visiting periods (like LANDSAT ETM, SPOT, IRS, etc.).

The methods of snow cover delineation and mapping for the first category of satellite data uses mainly the visible and near IR channels. The basic procedures consist on contrasts enhancement, binary/multi-threshold segmentation of images. Reflectance of fresh snow is very high in the visible wavelengths, but decreases in the near-infrared wavelengths, especially as grain size increases. Fresh snow can have a reflectance up to about 80%, but its reflectance may decrease to below 40% after snow crystals metamorphose.Snow and cloud discrimination techniques are based on differences between cloud and snow/ice reflectance and emittance properties. Clouds typically have high reflectance in visible and near-infrared wavelengths, while reflectance of snow decreases in short-wave infrared wavelengths. In case of the second satellite data visible bands are preferred. For the LANDSAT 7 ETM+ data (15 m spatial resolution) the spectral bands used for the snow discrimination are: band 5 (red), band 4 (green) and band 2 (green-yellow). Rocks are reflective, while snow and ice are absorptive in short-wave infrared band 5. Vegetation and snow are reflective in near-infrared band 4. Snow is also reflective in visible band 2, but most rocks and vegetation are absorptive. Using the band-filter combination 5, 4, 2, for color composite, snow appears very clear in cyan, bare rock or soil in magenta and vegetated areas in green (Figure 8).

The results obtained with unsupervised classification procedure of the LANDSAT 7 ETM+ image allow determining the snow cover surface at the basin level or for different elevation stages, for various morphological conditions (slope, aspect) and land cover categories.

The mapping of snow cover becomes limited in areas where snow cover is obscured by forest canopies. Even in a continuously snow-covered area, much of the forested landscape will not be snow-covered. Furthermore, snow that falls onto the ground through the canopy may not be visible by the airborne or spatial instruments.

112

Figure 8. LANDSAT 7 ETM+ image (10.05.2002) Band-filter combination 5-R, 4-G, 2-B which offer good snow discrimination

4. CONCLUSIONS

Considering the necessity to improve the means and methods to forecast, assess and monitor flooding, the paper presents the capabilities offered by remotely sensed data and GIS techniques. Although satellite sensors cannot measure the hydrological parameters directly, optical and microwave satellite data supplied by the new European and American orbital platforms like the EOS-AM "Terra" and EOS–PM "Aqua", DMSP, Quikscat, SPOT, ERS, RADARSAT, Landsat7 can supply information and adequate parameters to contribute to the improvements of hydrological modeling and warning.

The basic concepts of the creation of products useful for the forecast assess and monitor the flood events can be summarized in the following stages:

- preparation of the EO data: radiometric, geometric correction and other basic image processing;
- combination of the interpretation techniques: automated and photo-interpretation tools, as well as radiometric techniques;
- multi-sensor approach: make the analysis process easier and provide the opportunity to improve the quantity and quality of the information obtained;
- multi-temporal approach: gives the possibility to monitor low frequency (land modification) and high frequency phenomena (evolution of the floods boundaries);
- integrated approach: the high level products are based on the EO derived information combined with other ancillary data, hydrologic/hydraulic models outputs using the GIS facilities.

References

Brakenridge R. G., Anderson E., Nghiem S. V., Caquard S. and Shabaneh T. B. (2003) Flood Warnings, Flood Disaster Assessments, and Flood Hazard Reduction: The Roles of Orbital Remote Sensing. Proc. of the 30th Intern. Symp. on Remote Sens. Environ., Honolulu, Hawai'i, 2003.

Brakenridge R.G., Stancalie G., Ungureanu V., Diamandi A., Streng O., Barbos A., Lucaciu M., Kerenyi J. and Szekeres J. (2001) Monitoring of extreme flood events in Romania and Hungary using EO data, NATO SfP project plan. September. Bucharest, Romania.

Brakenridge R.G, Stancalie G., Ungureanu V., Diamandi A., Streng O., Barbos A., Lucaciu M., Kerenyi J. and Szekeres J. (2003) Monitoring of extreme flood events in Romania and Hungary using EO data. Progress report, May. Hanover NH, USA.

Brakenridge R.G, Stancalie G., Ungureanu V., Diamandi A., Streng O., Barbos A., Lucaciu M., Kerenyi J. and Szekeres J. (2004) Monitoring of extreme flood events in Romania and Hungary using EO data. Progress report, May. Hanover NH, USA.

Brakenridge R. G., Tracy B.T. and Knox J. C. (1998) Orbital SAR remote sensing of a river flood wave. *Int. J. Remote Sensing* 19 (7), 1439 – 1445.

Marsalek J. (2000) Overview of flood issues in contemporary water management. Marsalek J., Watt W.E., Zeman E. and Sieker F. (eds.) Flood Issues in Contemporary Flood Management. *NATO Science Series 71*, 1-14, Kluwer Academic Publishers, Dordrecht/Boston/London.

Stancalie G. (2004) Contribution of Earth Observation Data to Flood Risk Analysis. XXII-nd Conference of the Danubian Countries on the Hydrological Forecasting and Hydrological Bases of Water Management. Brno, Aug. 30- Sep. 3, 2004. Proc. of the XXII-nd Conference of the Danubian Countries, Brno, Czech Republic (on CD).

Stancalie G., Alecu C., Catana S. and Simota M. (2000) Estimation of flooding risk indices using the Geographic Information System and remotely sensed data. Proc. of the XX-th Conference of the Danubian Countries on hydrological forecasting and hydrological bases of water management, Bratislava, Slovakia.

Stancalie G., Alecu C., Craciunescu V., Diamandi A., Oancea S. and Brakenridge R.G. (2004) Contribution of Earth Observation data to flood risk mapping in the framework of the NATO SfP "TIGRU" project. International Conference on Water Observation and Information System for Decision Support, Ohrid, FY Republic of Macedonia, May 25-29, 2004. Proc. of BALWOIS Conference, Ohrid, FY Republic of Macedonia. (on CD).

Chapter 10

FLOOD FORECASTING IN THE CRISUL ALB AND CRISUL NEGRU BASINS USING GIS DATA BASE

RODICA P. MIC[1], VALENTINA UNGUREANU[2],
GHEORGHE STANCALIE[3], CIPRIAN CORBUS[1]
AND VASILE CRACIUNESCU[3]
[1]*National Institute of Hydrology and Water Management, Bucharest,
Romania,* [2]*Romanian Waters Administration, Bucharest, Romania,*
[3]*Romanian Meteorological Administration, Bucharest, Romania*

Keywords: Flood hazards, vulnerability, risk, NATO, GIS, info-layers, land use mapping, flood forecasting, hydrological models, forecasting products.

1. INTRODUCTION

For the transboundary area between Romania and Hungary many destructive floods have occurred provoking severe inundation events and even ruptures of embankments. A close cooperation between these countries in searching the best and adequate methods and technology to impede these undesirable events is therefore a prerequisite for needed actions.

The information on EO data is used to test the processing and analysis algorithms aiming to establish an operational methodology for detecting and mapping the hazard, vulnerability and flood risk giving a comprehensive knowledge on the flood-prone areas concerning its GIS - referenced extension as well as the jeopardized socio-economic objectives.

This work is achieved in the framework of the NATO SfP 978016 project, "Monitoring of extreme flood events in Romania and Hungary using EO data", and it is developed for the Romanian – Hungarian transboundary area situated in the Crisul Alb and Crisul Negru basins.

J. Schanze et al. (eds.), Flood Risk Management:
Hazards, Vulnerability and Mitigation Measures, 115–126.
© 2006 *Springer.*

2. DESCRIPTION OF THE VIDRA HYDROLOGICAL MODEL

The *VIDRA* model has been developed in the National Institute of Hydrology and Water Management (Bucharest, Romania) and is available as a HOMS component - Hydrological Models for Water - Resources System Design and Operation. It is a conceptual model with physically based semi-distributed parameters for the simulation of the flood hydrograph and optimal reservoir operations. The model computes the discharge hydrographs on sub-basins, their integration and routing on the main river and on the tributaries and estimates the flood mitigation through reservoirs. It is structured in a succession of the following components (Şerban 1984):

a) Computing, for each sub-basin, of the *average rainfall* by weighting the values of rainfall recorded at the rain gauging stations and the assessment of the water release from the snow cover - *snow melt* (based on degree – day method);

b) Effective rainfall computing model (deterministic reservoir model (Figure 1) that allows the extraction from the total water flow which enters the basin (precipitation and water release from the snow cover) of the losses by evapotranspiration and infiltration, finally resulting the effective rainfall which contributes to the runoff formation

c) Integration of the effective rainfall on the slopes and in the primary river network finally resulting in the discharge hydrograph in each sub-basin; as a transfer function of the hydrographical system, the instantaneous *unit hydrograph*;

d) *Routing model* (Muskingum transfer function) that carries out the integration of the flow formed in sub-basins and its propagation in the river bed;

e) *Attenuation model* (Puls method) of the floods through reservoirs or lateral polders on the river beds

VIDRA model is completed with *CORA procedure* of the hydrological forecasts updating that, according to the errors between the calculated and measured discharges, corrects the simulated discharges over the entire forecast interval. This procedure detects the error type (amplitude, phase or form) and consequently carries out a rough updating of the forecast, based on the entry variables of the model or a fine updating based on the measured discharge.

The parameters of the *VIDRA* model (Şerban 1984) are corresponding to the model components:

a) *For the snow-melt component:*

- T_e - the equilibrium temperature whereof any heat exchange between the snow cover and the environment (near 0°C);

- M - the melting factor or the degree - day factor (*mm / °C / day*), which is estimated from the scientific references (Şerban 1984) in function of month and forest cover coefficient.

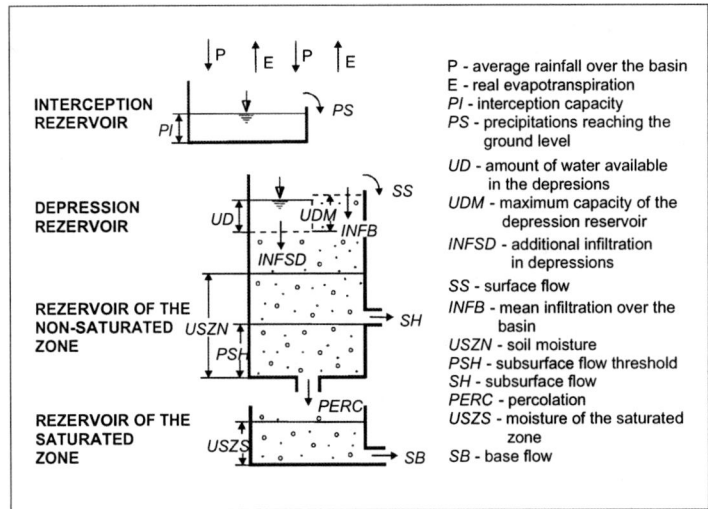

Figure 1. Conceptual scheme of the effective rainfall model

b) *To average rainfall on each sub-basin,*
   - the weights of the rainfall gauging stations.
c) *To calculate the effective rainfall on each sub-basin:*
   - PI - the interception capacity (*mm*);
   - FOM - the maximum infiltration capacity of soil (*mm/hour*);
   - FC - the minimum infiltration capacity of soil (*mm/hour*);
   - CH - the hypodermic runoff parameter;
   - UDM - the nominal capacity of the depression (*mm*);
   - PS - the hypodermic runoff threshold (*mm*);
   - $USZN$ - the nominal capacity of the reservoir corresponding to the non-saturated zone (*mm*);
   - CB – the base runoff parameter;
   - PSB – the base runoff threshold (*mm*).
d) *To calculate the discharge hydrograph for each sub-basin:*
   - K_r - the recession coefficient of the hydrograph;
   - T_L- the delay time between the rainfall gravity centre and surface flow hydrograph gravity centre.
     For each sub-basin three pair of unit hydrograph are considered; corresponding to high, medium and low intensity of rain (*mm/hour*).
e) *Routing hydrograph along the riverbed:*
   - K - a parameter associated to the discharge routing hydrograph. It varies depending on the discharge being routed along the river reach;
   - X – the attenuation coefficient.
f) *Floodwave attenuation through reservoirs* and their conjunctive operation,

- characteristics of the reservoirs (characteristic levels, reservoir capacity curve etc.) and of the outlets (characteristic levels, dimensions, discharge coefficient etc.).

From the total number of parameters only about 10 parameters are usually changed in the calibration action.

The *VIDRA* model allows the stimulation of the runoff in small basins (with area beginning from about 100 km^2) as well as in the medium and large size basins. The large basins are divided into homogenous units (sub-basins). The model computes the discharge hydrographs on sub-basins, their integration and routing on the main river and on the tributaries and estimates the flood mitigation through reservoirs.

The model runs usually with a hourly time step, because the model parameters corresponding to the effective rainfall component are established hourly, but the model can optionally runs with 3, 6, 12 or 24 hours time step (both for input and output data), computing program having a subroutine to share the input data hourly.

The input data of the *VIDRA* model are: precipitation and air temperatures as meteorological data, soil moisture estimation, the water equivalent of the snow cover at the beginning of the forecast interval and start discharges for each sub-basins and the discharges recorded at the control gauging stations, the last data being needed to calibrate, verify and updating the outputs in operational forecasting. Also for he attenuation through reservoir component the initial water level in reservoir is needed.

The output data of the *VIDRA* model refer to:

- Data of each sub-basin: components of the hydrological balance (the average precipitation, the depth of runoff and the runoff coefficient as a ratio between the depths of runoff and the precipitation) and the discharge hydrograph simulated and recorded (for the sub-basins where there is gauging station);

- Data of each river reach: components of the hydrological balance (the flood volume, the depth of runoff and the runoff coefficient for sub-basins and at the outlet of the reach) and the discharge hydrographs simulated and recorded (for the sub-basins and outlet (where there is gauging station);

- Data obtained as a result of the flood attenuation through reservoir (the outflow hydrographs from the reservoirs, the variation of the water level in the reservoir, the discharge gate operation at the high waters).

- The results obtained after running *VIDRA* model can be stored in ASCII files that can be viewed on the screen. The discharge hydrographs (recorded and simulated) can be visualized as values or as graphs (Figure 2).

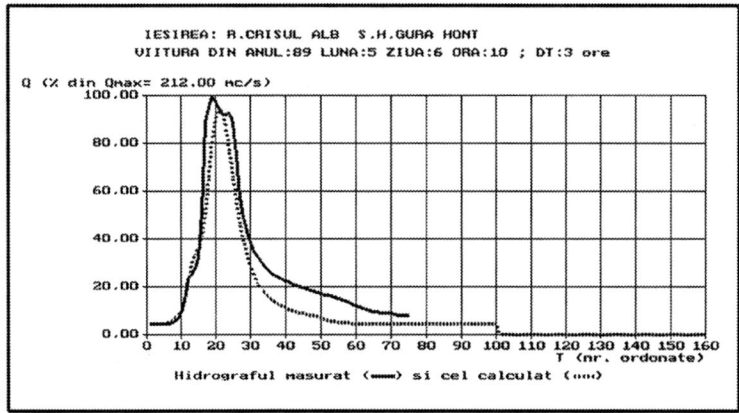

Figure 2. Recorded (---) and calculated (- - -) hydrographs - Gurahont station on Crisul Alb River

In order to calibrate the *VIDRA* model parameters for a river basin the major rainfall – runoff events are selected at least 5 events. Forward the initial values of the parameters corresponding to some of the model components (effective rainfall computing model, unit hydrograph model and routing model), are evaluated from generalized relationships of that parameters function of the morphometric characteristics of the river basin or of the river reaches (Diaconu and Şerban 1994).

2. DESIGN AND BUILDING A GIS FOR THE CRISUL ALB AND CRISUL NEGRU BASINS OF ROMANIA

The GIS database for the Crisul Alb and Crisul Negru basins was conceived to correspond to the *VIDRA* model requirements (some river basin parameters) and to the flood forecasting dissemination.

2.1. Identification of the corresponding information layers, which are Integrated into the GIS datatbase

The geo-referenced database comprises the following thematic layouts, organized as info-layers, obtained from topographic maps and from satellite data: basin and sub-basins boundaries; river hydrographical network; railways and main roads; villages and towns; digital elevation model (DEM); main land cover/land use types derived from satellite images; location of the meteorological, hydrological and rain gauge stations; soil texture.

120

2.2. Specification of the products derived from GIS info-layers used for the VIDRA model

The main tasks for the GIS database are: acquisition, storage, analysis and interpretation of data handling and preparation for a data rapid access, the updating of the information (temporal modification), generation of value-added information.

The products derived from the GIS database are useful for the *VIDRA* hydrological forecasting model (in terms of input parameters) and dissemination and visualization (in terms of the spatial representation of the flood forecasting results). The main products issues from the GIS vector and raster info-layers are: space maps; updated land cover/land use maps; DEM and derived topographic parameters (mean basin and sub-basin slopes, altitude); GIS layers with hydro-meteorological stations network and hydrographical network, including the existing dams, dikes, and canals.

The main physical-geographic characteristics which could be extracted from the GIS database for the rivers and sub-basins of the Crisul Alb and Crisul Negru river basins are: river's length; reception basin and sub-basins surfaces; mean altitude of the sub-basins; mean slope of the sub-basins; forested area or the forestry coefficient;

The Figure 3 presents the general land cover/land use mapping procedure used for the Crisul Alb and Crisul Negru basins.

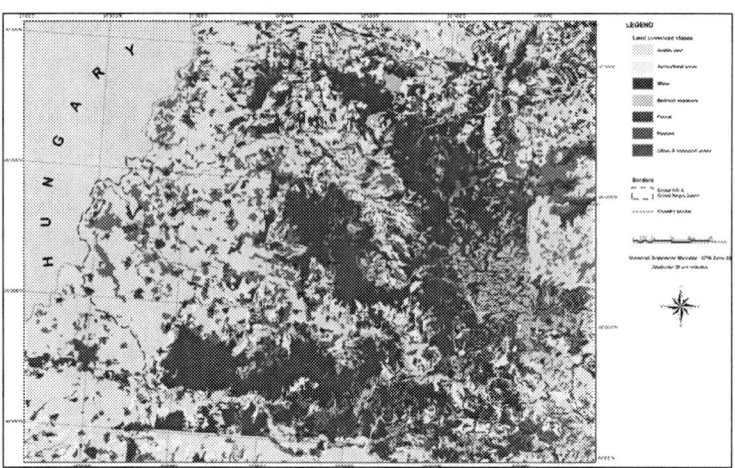

Figure 3. The general land cover/land use mapping procedure

3. CALIBRATION OF THE VIDRA MODEL FOR THE CRISUL ALB AND CRISUL NEGRU RIVER BASINS USING THE HISTORICAL DATABASE FOR THE MAJOR RAINFALL – RUNOFF EVENTS

The general strategy of flood wave simulation in large catchments requires double modelling, i.e. both topological modelling of the catchment and rainfall-runoff process modelling. Consequently the main calibration steps developed for the VIDRA model are the following:
- Dividing the basin into homogenous units (sub-basins) and the river network into homogenous reaches (sectors);
- Estimating the initial model parameters using the parameter generalization relationships;
- Calibration the overall model parameters using the historical database for the major rainfall – runoff events.

3.1. Modelling of the rivers basins – basin sketches

In view to divide the catchment in sub-basins and the river network in reaches, for the Crisul Alb and Crisul Negru River Basins.

3.2. Estimation of the initial model parameters using the generalization relationships

In order to determine the initial values of the parameters corresponding to some of the model components, generalized relationships of that parameters function of the morphometric characteristics of the river basins or of the river reaches are used. These relationships have been obtain using the hydrometeorological data recorded at the hydrometric stations, located in the Crisul Alb and Crisul Negru river basins for at least 5 flood events having maximum discharges larger than the warning discharge.

3.2.1. Rationalization relationships of the infiltration structure parameters

The infiltration parameters have been correlated with the soil texture (Table 1), basin area size, mean slope of the basin and forest cover coefficient. These basin parameters have been obtained using GIS database.

Parameter PI have been determined in terms of the forest cover coefficient of the basin, c_p, ($PI = 2 + 6c_p$) and parameter CB was determinate from the Table 2. For the others parameters the relationships of the regionalisation are presented in the Figures 4 - 7. Because for the parameter *PSB* it has not been obtained a regionalisation relationship, it will be finally adopted through the overall calibration of the flood simulation model.

Table 1. Classification of soils on the hydrologic type

Type	Texture	Infiltration rate
A	Coarse	Very great
B	Medium	Great
C	Medium fine	Average
D	Fine	Low
E	Very fine	Very low

Table 2. Coefficient of the base runoff CB function of the basin surface (F) and the land slope (I_b)

F (km^2)	I_b ($^0/_{00}$)				
	<100	100-200	200-300	300-400	>400
<500	0.05-0.08	0.08-0.11	0.11-0.14	0.14-0.17	0.17-0.20
500-1000	0.04-0.07	0.07-0.10	0.10-0.13	0.13-0.15	
>1000	0.03-0.06	0.06-0.08	0.08-0.10		

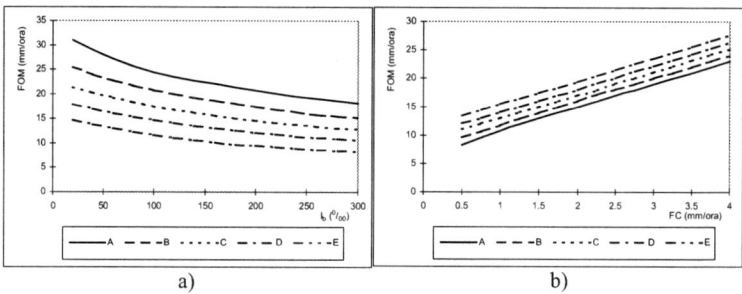

a)   b)

Figure 4. Variation of the maximum infiltration capacity (a) and of the minimum infiltration capacity (b) function of the soil texture and mean slope (Ib)

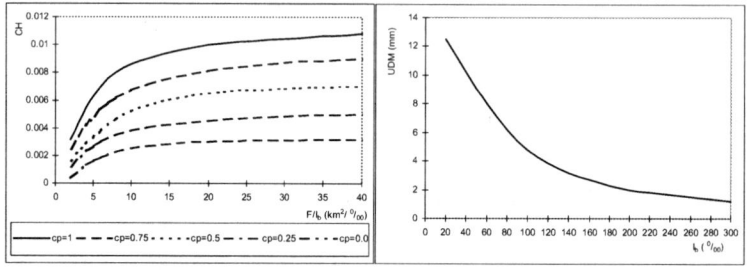

Figure 5. Variation of the hypodermic runoff parameter function of the F/I$_b$ and the forest cover coefficient cp

Figure 6. Variation of the nominal capacity of the depression function of the mean slope I$_b$

Using the regionalisation relationships the parameters of the infiltration structure for all the Crisul Alb and Crisul Negru sub-basins have been determined.

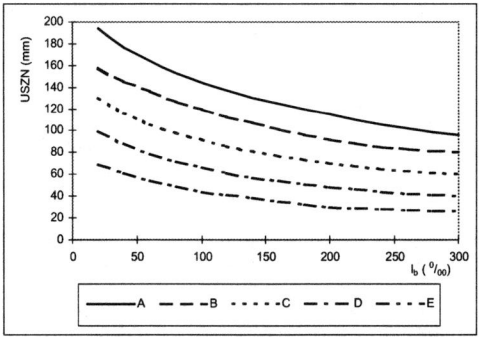

Figure 7. Variation of the nominal capacity of the reservoir corresponding to the non-saturated zone function of the soil texture and mean slope

3.2.2. Determination of the parameters of the unit hydrograph

The parameters of the unit hydrograph (IUH) for the IMH model are determined using the following relationships:

$$T = \frac{5}{3}\left(M_1 - \sqrt{1,2m_2 - 0,2M_1^2} \right) \text{ and } K = M_1 - \frac{T}{2} \tag{1}$$

$$M_1 = T_L \text{ and } m_2 = S_2 T_L^2 \tag{2}$$

$$K_r = e^{-\Delta t / K} \tag{3}$$

where: T is the time interval between the end of rainfall and the moment when the water entered in the main river network, K - the parameter of the decrease branch of the hydrograph; M_1 and m_2 - the moments of the unit hydrograph, by first order taken as against the origin of the unit hydrograph and by the second order taken as against the centre of he area of the unit hydrograph, computed in terms of the shape factor of the IUH (S_2) and the leg time between the mass weight of the rainfall hyetogramme and the mass weight of the surface flow hydrograph (T_L); K_r the recession coefficient.

The unit hydrograph moments (M_1 and m_2) have been determined using the recorded data at the selected gauging stations or by the regionalization relationships in terms of the morphometric characteristics of the basins (length and river slope) and rain intensity for the parameters T_L (Figure 8) and S_2 (Figure 9). The basin characteristics (length and river slope) have been obtained using GIS database.

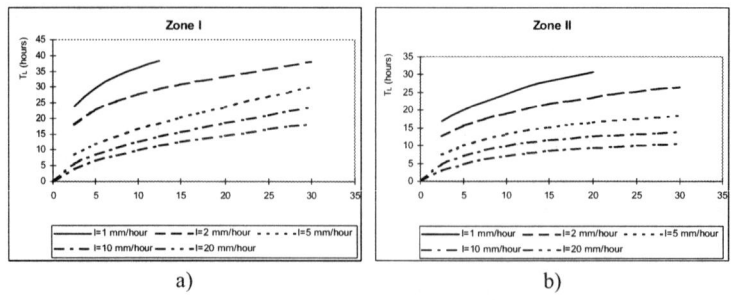

a)   b)

Figure 8. Regionalisation relationships for the parameter T_L

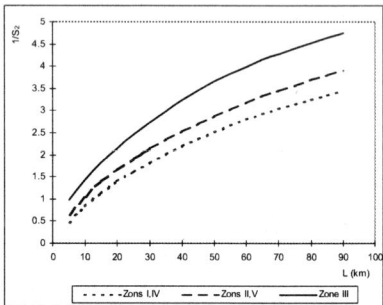

Figure 9. Regionalization relationships for the parameter $1/S_2$

The above presented regionalization relationships have been used for the determination of the unit hydrograph parameters in the ungauged sections from the three river basins.

3.2.3. Determination of the parameters of the flood routing through the river reaches

The parameters X and K of the model used for the flood routing have been determined for the characteristic river reaches having Δx length, using the hydrographs recorded at the gauging stations and the morphometric characteristics of the riverbed.

The initial estimation of the attenuation parameter X have can be made depending on the morphometric characteristics of the river reaches conform with the following table.

Characteristics of the river reach	X
Embankment river reaches or channels	0,4...0,5
River reaches having little developed flood plains	0,2...0,4
River reaches having well developed flood plains with the slope above 0,5%	0,0...0,2
River reaches having very large flooding areas with the slope under 0,5%	-3,0...0,0

Parameter K, which represents the routing time of flood along the considered river reach, has been computed for a certain river reach between two river stations using the hydrographs of the biggest floods recorded at the two successive stations.

Further on, the rationalization relationships have been carried out and presented in the Figures 10 and 11. The parameter K for all other reaches has been assessed by applying the formula:

$$K = L / v_K$$

where: L is the length of the reach in km; v_K - the routing velocity corresponding to the parameter K;

Parameter v_K have been determined in terms of the flood routing velocity (v_p) and the slope of the river reach (I_s).

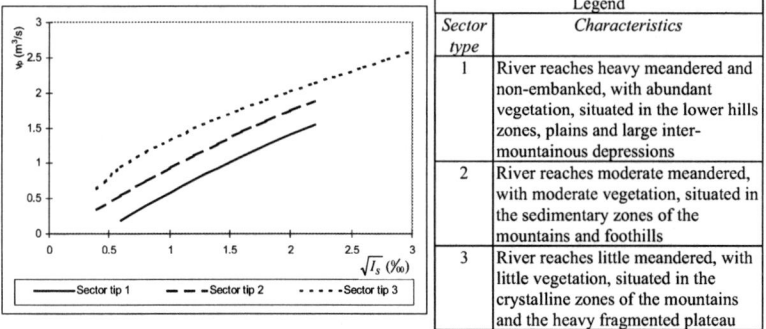

Figure 10. Variation of the routing velocity in function of the slope and the characteristics of the river reach

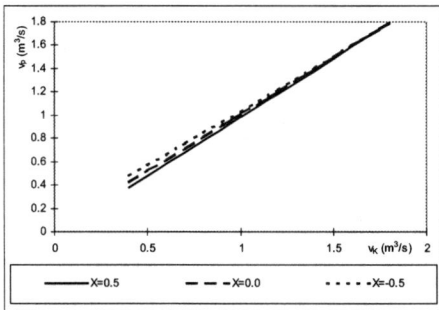

Figure 11. Variation of the velocity v_K in terms of the velocity v_p and the attenuation parameter X

4. CONCLUSIONS

Analysing the calibration results the *VIDRA* model can be considered a good model for flood forecasting. The validation made on the flood event from the spring 2000 and 2001 confirms the calibrated parameters.

The results of the simulations obtained using the *VIDRA* model for a non flood event show that the model is more appropriate for the floods events, for the non-flood events the errors are greater than those accepted for hydrological forecasts. Also, the simulations using different flood events with diverse characteristics of the generating factors show that the model is more adequate for the floods generated by precipitation evenly distributed over the basin.

The overall errors are both due to the model errors and to the availability and accuracy of the input data. The main error is found as being stemmed by lack in the spatial distribution of the precipitation over the basin.

References

Diaconu C. and Şerban P. (1994) *Hydrological Syntheses and regionalisations.* Technical Publishers, Bucharest.

Şerban P. (1984) Mathematical Models for the Flood Waves Forecast in Hydraulic Structured Basins, *Hydrological Studies* 51, Bucharest.

Şerban P. (1984) Conceptual Model for the Determination of the Unit Instantaneous Hydrograph, *Hidrotehnica* 2, Bucharest.

Chapter 11

FLOOD MONITORING USING ON-LINE SUPPORT SYSTEM FOR SPATIAL INFORMATION MANAGEMENT

VASILE CRACIUNESCU[1], GHEORGHE STANCALIE[1], ANDREI DIAMANDI[1] AND RODICA P. MIC[2]

[1]*Romanian Meteorological Administration, Bucharest, Romania,*
[2]*National Institute for Hydrology and Water Management, Bucharest, Romania*

Keywords: Spatial information management, on-line information system, GIS database, earth observation.

1. INTRODUCTION

Flood management evolves and changes as more knowledge and technology becomes available to the environmental community. One of the most powerful tools to emerge in the hydrological field is the Geographic Information System (GIS), which allows for the collection and analysis of environmental data. The decision process starts with observed data that supports the creation of information through modelling, the information evolves into knowledge through visualization and analysis, and finally the knowledge supports hydrological decisions.

In the last years the integrated remote sensing data started to play an important role in the creation or in updating the existing GIS databases. Earth Observation (EO) images have wide applications in flood analysis, in such tasks as producing catchments maps, detecting water surface and soil moisture, detecting inundated areas, and assisting with remote flow measurement. Thus, image processing is important for developing such products and using them in flood analysis and management (Brakenridge et al. 2001).

The distribution of the graphic and cartographic products (derived using the GIS facilities and based on remote sensing data, maps and field surveys) to the interested authorities, media and public opinion is an important issue in the framework of the NATO SfP "TIGRU" project "Monitoring of extreme flood events in Romania and Hungary using EO data". These products will contribute to preventive consideration of

127

J. Schanze et al. (eds.), Flood Risk Management:
Hazards, Vulnerability and Mitigation Measures, 127–137.
© 2006 *Springer.*

flooding in land development and special planning in the flood-prone areas, and for optimising the distribution of flood-related spatial information to end-users (Brakenridge et al. 2003).

The paper presents the design and the main function of the flood monitoring on-line support system for spatial information management, as well as the preliminary results of the implementation.

2.   STUDY AREA: CRISUL/KŐRŐS BASIN

The study area represents the Crisul Alb/Negru/Kőrős transboundary basin spanning across the Romanian–Hungarian border, with a total area of 26,600 km^2 (14,900 km^2 on the Romanian territory).

In Romania, the catchment (Figure 1) comprises mountainous areas (38%), hilly areas (20%) and plains (42%). About 30% of the catchment is forested. On the Hungarian side, the catchment relief represents plains. Annual precipitation ranges from 600-800 mm/year in the plain and plateau areas to over 1200 mm/year in the mountainous areas of Romania. This precipitation distribution can be explained by the fact that humid air masses brought by fronts from the Icelandic Low frequently enter this area. The orography of the area (Apuseni Mountains) amplifies the precipitation on the western side of the mountain range. Thus, the Crisuri Rivers Basin frequently experiences large precipitation amounts in short time intervals and the frequency of such events seem to be increasing in recent years (Brakenridge et al. 2001).

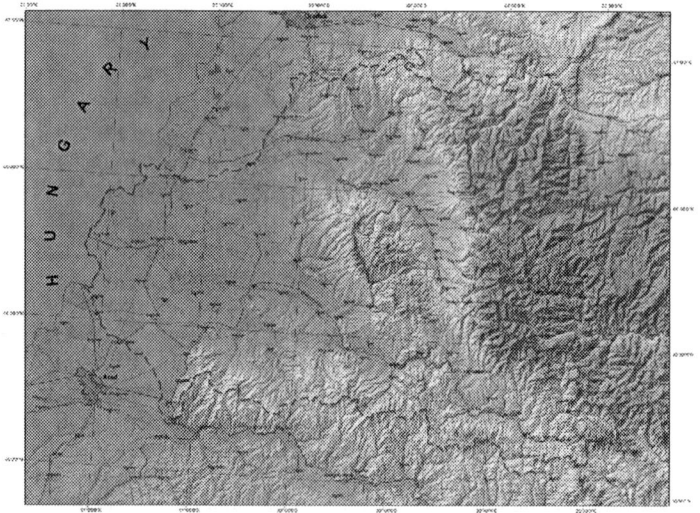

Figure 1. Romanian part of the Crisul Alb/Negru basin

3. ON-LINE SUPPORT SYSTEM FOR SPATIAL INFORMATION MANAGEMENT

A flood forecasting and warning system is already active in the study area. The existing system does not include a spatial component of the phenomena (pertaining to geographic distances or patterns) both in the pre and post crises phases. The purpose of the development of a dedicated sub-system based on remote sensing and GIS technology is to contribute to regional quantitative risk assessment for monitoring and hydrological validating risk simulations, in the Romanian – Hungarian transboundary test-area. Also an important result will be the preventive consideration of flood events when determining land development and in special planning of the flood-prone areas (Brakenridge et al. 2003).

The main function of this sub-system will be:

- Acquisition, storage, analysis and interpretation of data;
- Management and exchange of raster and vector graphic information, and also of related attribute data for the flood monitoring activities;
- Handling and preparation for a data rapid access;
- Updating of the information (temporal modification);
- Data restoring, including the elaboration of thematic documents;
- Generation of value-added information (complex indices for flood prevention, risk maps);
- Distribution of the derived products to the interested authorities, media, etc.

Figure 2 presents the proposed dedicated sub-system based on remote sensing and GIS with the data flows and the links with the data suppliers and end-users.

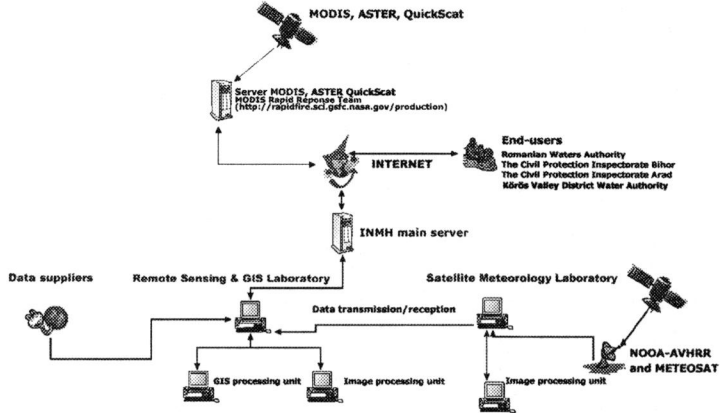

Figure 2. Dedicated sub-system based on remote sensing and GIS technology

3.1. Construction of the GIS database

The structure of the dedicated GIS database was planned in order to be used for the study, evaluation and management of information, which contribute to flooding occurrence and development, as well as for the assessment of damages inflicted by flooding effects. In this regard the database represented by the spatial geo-referential information ensemble (satellite images, thematic maps and series of the meteorological and hydrological parameters, other exogenous data) is structured as a set of file-distributed quantitative and qualitative data focused on the relational structure between the info-layers. The GIS database is connected with the hydrological database, which allows synthetic representations of the hydrological risk using separately, or combined parameters (Brakenridge et al. 2004, Stancalie et al. 2003).

It was decided to develop a GIS database for the whole study area of the Crisul Alb, Crisul Negru and Kőrős basins using different cartographic documents at the scale 1:100 000. The construction of this GIS is based, mainly on classical mapping documents, particularly represented by maps and topographic plans. Most of the thematic layers were extracted from this classical mapping support. Due to the fact that, in most of the cases, the information on the maps is old-fashioned, it is imposed to update it on the basis of the recent satellite images (e.g. the hydrographic network, land cover/land use) or by field measurements (e.g. dikes and canals network). The topographic maps at 1: 100,000 in Gauss-Kruger projection (zone 34) present the necessary information to serve as support for the construction of the GIS database for the whole study area.

The GIS database contains the following info-layers:
* Sub-basin and basin limits;
* Land topography (90 meters DEM);
* Hydrographic network, dikes and canals network;
* Communication ways network (roads, railways);
* Localities;
* Meteorological stations network, rain-gauging network, hydrometric stations network;
* Land cover/land use, updated from satellite images.

In the figure 3 are presented the GIS info-layers related with the hydrographical network, the road and railways network for the Crisul Alb and Crisul Negru Romanian basins.

The preparation of the info-layers that constitute the digital geographic information database or the geo-spatial information was realized by:
* Identification of the reference points;
* Scanning of the cartographic documents (on paper);
* Integration of the geo-spatial information in the thematic info-layers;
* Association of attributes for different geographic objects (watercourses, hydro-meteorological stations, villages, roads, etc.).

Figure 3. GIS info-layers for the Crisul Alb and Crisul Negru basins

For the acquisition of the digital geographic data it was necessary to define the specifications of the information layers related with:

- The scale of the cartographic documents or image data;
- The type of the geographic objects, which constitute the layers (represented by layers in vector, tin or raster format);
- The attributes which characterized them.

For the most considered flood vulnerable area, situated in the plain of the Crisul Alb/Negru/Kőrős basins, limited at its Eastern part by the Ineu – Talpos and at its northern part by the Crisul Repede basin (Figure 4), a more precise GIS database was constructed using 1:5.000 and 1:10.000 topographic plans.

One of the most important products obtained for this area is a precise digital elevation model. For this purpose the shape with elevation information's extracted from individual map sheets have been merged and corrected and then interpolated to obtain the digital elevation model (DEM).

Interpolation methods produce a regularly spaced, rectangular array of Z values from irregularly spaced XYZ data. The term "irregularly spaced" means that the points follow no particular pattern over the extent of the map, so there are many "holes" where data are missing. Interpolation fills in these holes by extrapolating or interpolating Z values at those locations where no data exists (Lee and Schachter 1980, Isaaks and Srivastava 1989).

The interpolation methods tested were based on the Kriging, Triangulated Irregular Network (TIN), Minimum Curvature and Natural Neighbour algorithms. The best result was obtained for the Kriging method. The digital elevation model was then used for deriving terrain slope, aspect and curvature maps.

132

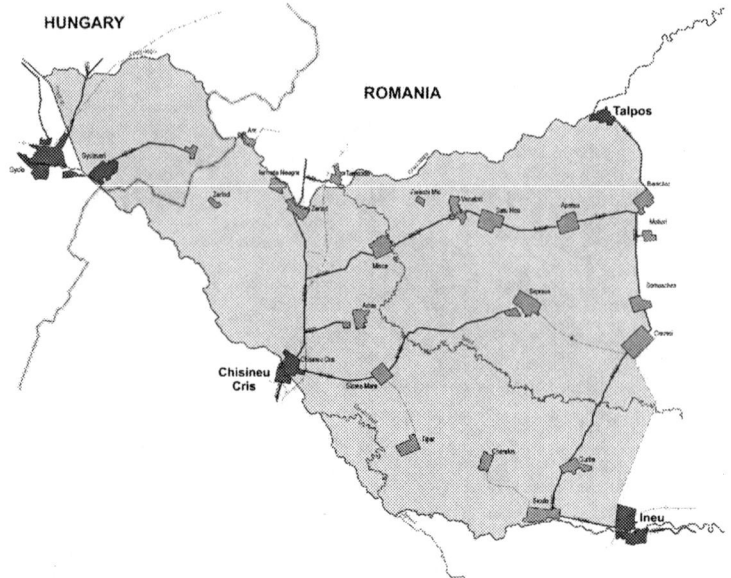

Figure 4. Vulnerable area in Crisul Alb/Negru/ Kőrős basin

3.2. The methodology for the land cover/land use updated maps and integration into the GIS database

The methodology for the achievement of the land cover/land use from high-resolution images developed within the project SAT-GIS Group is based on the observation of the following requirements:

- The structure of this type of information must be at the same time cartographic and statistic;
- It must be suited to be produced at various scales, so as to supply answers adapted to the different decision making levels;
- Updating of information must be performed fast and easily.

The used methodology implies following the main stages below:
- Preliminary activities for data organizing and selection;
- Computer-assisted photo-interpretation and quality control of the obtained results;
- Digitisation of the obtained maps (optional);
- Database validation at the level of the studied geographic area;
- Obtaining the final documents in cartographic, statistic and tabular form.

Preliminary activities comprise collection and inventorying of the available cartographic documents and statistic data connected to the land cover: topographic, land survey, forestry, and other thematic maps at various scales (Brakenridge et al. 2003).

To obtain the land cover map, satellite images used must have a fine geometrical resolution and rich multispectral information. The computer-assisted photo-interpretation finalizes in the delimitation of homogeneous areas from images, in their identification and framing within a class of interest. Discriminating and identifying the different land occupation classes rely on the procedures of image processing, such as: adaptation of the contrast-brightness ratio, equalling the histogram of the levels of grey, computing the vegetation indices (especially the normalized difference vegetation index, the brightness index and the normalized vegetation index), the analysis by the main component, radiometric transformation by vicinities (filters), radiometric statistic analyses, automated and supervised classifications. The result of computer-assisted photo-interpretation is a detailed management of the land cover/land use, followed by a generalizing process, which includes:

- Identification of each type of land occupation, function of the exogenous data, of the "true-land" data establishing a catalogue;
- Delimitation of areas suspected to represent a certain unity of the land;
- Expanding this delimitation over the ensemble of the image areas.

For updated the land cover/land use in the study area, TERRA-AQUA/ASTER data have been used. These satellite data proved to be suitable for detailed maps of land cover/land use, especially the visible and near infrared bands (1, 2, 3B) with 15 m resolution. The ASTER data were obtained from the Earth Observing System Data Gateway, by courtesy of Prof. G. R Brakenridge from DFO, USA (Stancalie et al. 2004). The figure 5 presents an example of the ASTER color composite images over the study area.

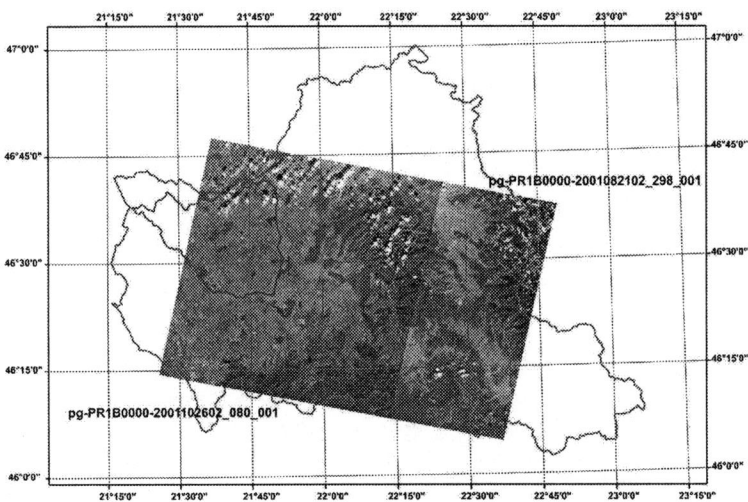

Figure 5. Example of ASTER data used to obtain land cover/use maps

The method for the land cover/land use mapping based on TERRA/ASTER data includes the following steps:

i) Geo-referencing of the ASTER data

The images are co-registered to WGS84 datum and UTM projection zone 34 and rotated with different angles in order to co-locate and analysed with topographic maps.

ii) Detection of cloud and water

Clouds obstruct the viewing of the earth's surface by satellites operating in the visual and infrared spectrum. On the basis of their particular spectral behaviour, clouds have to be detected and eliminated from the image before further processing. Thus, a reliable cloud mask is an essential and early step.

The water in streams, coastal areas especially near river mouths are more problematic. The water can be shallow and transparent to the bottom or contaminated with silt and vegetative matter. This type of water can have the same spectral characteristics as wet or shadowed land. Most land surface (including bare soil and vegetation) is brighter in ASTER band 3 than band 1 and the differences increases linearly with increasing spectral information in band 3. Spectral signatures of the water in the ASTER visible and near infrared bands are used to map surface water. Thus water, clouds and cloud shadows could be separated and the land patterns can be emphasized for the classification procedures.

iii) Unsupervised classification

The next step involves the data classification procedures. When performing an unsupervised classification it is necessary to find the right number of classes that are to be found. Too many, and the image will not differ noticeable from the original, too few and the selection will be too coarse. The unsupervised classifications were made for 30, 50 and 90 classes, with a specified number of iterations. The resulted classes were finally regrouped into 6 classes: forest, pasture, winter crops, summer crops, cities/villages and water.

iv) Supervised classification

The supervised classification is based on training areas using the "region of interest" tool in the ENVI 3.6 software. Supervised classification requires a priori knowledge of the number of classes, as well as knowledge concerning statistical aspects of the classes. The procedure starts with establishing training samples, which are areas that are assumed or verified to be of a particular type. The parallelepiped classification was used after the water; cloud and shadow masks were applied on the ASTER initial image. In the final land cover map 6 classes (forest, cities/villages, pasture, winter crops, summer crops, and water) have been selected.

The localities were digitized as an info-layer and overlapped on the classification. The digitization was made on the topographical maps and satellite images. The topographical maps 1:50 000 represent the situation of the years 1969-1972, they are analyzed and used as a support. Satellites images (recently acquired - 2000-2002) are used to obtain the actual extend of the localities.

v) Manual Correction

After classifications, were made manual corrections using ENVI software in order to remove the incorrect pixels.

vi) Validation of results

Validation included methods of checking through on land sampling at local and regional level, focused on the classification precision, the number and surface of the charted areas, the exactness of the geographic boundaries and the homogeneousness of the occupation structure.

The land cover/land use information are integrated in the thematic database organized in GIS environment allowing obtaining thematic maps at different scales and generalizing levels respectively.

3.3. Spatial data preparation for rapid access

The project objectives involve working with different types of spatial geo-data (scanned maps, satellite images, vector files, digital elevation models) in different file formats and coordinate systems, processed by the project partners on Window, Linux, Solaris computing platforms and software environments. To make all the work easy available to the participants and end-users, a detailed specification package was developed. These ensure the fact that every piece of information uses the same file format (ESRI shapefile for vector data; ESRI grids for digital elevation model; ERDAS .img for maps and satellite images) and the same geographic coordinate system (UTM Zone 34/WGS84).

At this point one of the most important tasks was to build a Satellite Image Database (SID), to gather information about the raw satellite scenes available as well as of the derived products and make it available in a simple format.

The SID was build in MySQL and is available on-line on a server, being updated as new satellite images are acquired. Each record of the database describe the characteristics of each satellite image: platform, sensor, date & time of data acquisition, duration of pass, spectral band, coordinates of the area covered, projection, calibration, size, bits/pixel, image file format, physical location (machine, directory), origin of data, type (raw/processed), type of processing applied, algorithm used, quick-look available, cloudiness. Queries are very easy to conduct using the web interface.

3.4. Spatial data dissemination

One of the most important functions of the dedicated sub-system involves distribution of the project results to the participants, end-users and public opinion. The easiest way to distribute the spatial and tabular attribute data is by setting up a FTP server was the information could be store and accessed.

From the end-user point of view, this approach has two major disadvantages:

- When the database grow the relevant information are more difficult to find;
- The data is stored in a common GIS file format and this implies special software and training for the user to be able to read and to analyse the information.

Another option is to distribute spatial and tabular attribute data over an Internet Web-based network. This is a powerful and effective communication method that overcomes the disadvantages of the first approach. By doing so, all interested agencies and end-users can have data access without being a technical expert.

Generally, viewing GIS data on the Web involves a three-tiered architecture:

- A spatial server that can efficiently communicate with a Web server and is capable of sending and receiving requests for different types of data from a Web browser environment;
- A mapping file format that can be embedded into a Web page;
- A Web-based application in which maps can be viewed and queried by an end-user/client via a Web browser.

Publishing the data on the Web using this approach would not change the existing data workflow – how the data is created, maintained, and used by desktop applications (Hendry 2004). This means that the mapserver dynamically generates maps from the files stored in a certainly folder every time a user send a request.

A prototype to distribute spatial data via Web interface for the Crisul Alb/Crisul Negru/Kőrős basin is currently under construction and the first version will be available in early 2005. The Web-based application is developed using standard technologies such as HTML, XML, JavaScript, PHP, SVG, COM and supports the Open GIS Consortium (OGC) and the Open Web Services specifications.

4. CONCLUSIONS

The development of a dedicated sub-system based on remote sensing and GIS technology will improve the flood management and will aid the implementation of flood mitigation programs in the Romanian-Hungarian transboundary test-area. This sub-system will allow the storage, management and exchange of raster and vector graphic information, and also of related attribute data for the flood monitoring activities. This dedicated information sub-system will contribute to regional quantitative risk assessment (using flood hazard and vulnerability characteristics) for monitoring and hydrological validating risk simulations.

An important result will be the preventive consideration of flood events when determining land development and in spatial planning of the flood-prone areas.

The GIS database will be implemented for selected areas in the transboundary Crisul Alb and Crisul Negru basins at Crisuri Rivers Authority, Oradea, and District Inspectorate of Civil Protection Bihor and Arad. The GIS database will be set up at the local operational hydrological service and at the District Inspectorate of Civil Protection operational units. The communication of the information between NIMH and the Crisuri Rivers Authority, as well as between Water Authority – District Inspectorate of Civil Protection – public authorities, will use the FTP or the e-mail for the simple mail transfer protocol to upload and download data and other information.

References

Brakenridge R.G, Stancalie G., Ungureanu V., Diamandi A., Streng O., Barbos A., Lucaciu M., Kerenyi J. and Szekeres J. (2001) Monitoring of extreme flood events in Romania and Hungary using EO data. Project plan, September, Bucharest, Romania.

Brakenridge R.G., Stancalie G., Ungureanu V., Diamandi A., Streng O., Barbos A., Lucaciu M., Kerenyi J. and Szekeres J. (2003) Monitoring of extreme flood events in Romania and Hungary using EO data. Progress report, May, Hanover NH, USA.

Brakenridge R.G, Stancalie G., Ungureanu V., Diamandi A., Streng O., Barbos A., Lucaciu M., Kerenyi J. and Szekeres J. (2004) Monitoring of extreme flood events in Romania and Hungary using EO data. Progress report, May, Hanover NH, USA.

Hendry F. (2004) Best Practices for Web Mapping Design. Proc. of the 2nd MapServer Users Meeting, Otawa, Canada, June 9-11, 2004, (on CD).

Isaaks E.H. and Srivastava R.M. (1989) *An Introduction to Applied Geostatistics.* Oxford University Press, New York, 561 pp.

Lee D.T. and Schachter B.J. (1980) Two Algorithms for Constructing a Delaunay Triangulation. *International Journal of Computer and Information Sciences* 9 (3), 219-242.

Stancalie G., Alecu C., Craciunescu V., Diamandi A., Oancea S. and Brakenridge R.G. (2004) Contribution of Earth Observation data to flood risk mapping in the framework of the NATO SfP "TIGRU" project. Proc. of BALWOIS International Conference on Water Observation and Information System for Decision Support, May 25-29, 2004, Ohrid, FY Republic of Macedonia, (on CD).

Stancalie G., Diamandi A., Ungureanu V. and Stanescu V.A. (2003) Sub-system based on remote sensing and GIS technology for flood and related effects management in the framework of the NATO SfP "TIGRU" project. First Annual Session of the NIHWM, Proc. of 1st Annual Session of the NIHWM, Sep. 22-25, 2003, Bucharest, Romania (on CD).

Chapter 12

FLOOD WARNING CONCEPT FOR TURKEY: TECHNICAL AND ORGANISATIONAL ASPECTS

THOMAS EINFALT
einfalt&hydrotec GbR, Luebeck, Germany

Keywords: Flood warning systems, rainfall forecasting, flood forecasting, radar, organisation of flood management, data management.

1. INTRODUCTION

Operational applications on flood warning require a well-tailored technical concept as well as the consideration of a number of organisational points. Both items are of equal importance for a successful implementation of a flood warning system. Therefore, the first part of this paper is devoted to technical aspects and the second part to organisational ones.

2. TECHNICAL ASPECTS

2.1. Background

In Turkey, the implementation of a flood forecasting and warning system is being conducted by the General Directorate of State Hydraulic Works (DSI) in the framework of the Turkey Emergency Flood and Earthquake Relief (TEFER) project. DSI has selected four pilot catchments for the establishment of the forecasting system: West Black Sea, Susurluk, Gediz and Buyuk Menderes, with catchment areas ranging from 18,000 to 30,000 km^2.

Three new radars have been installed, together with over 200 raingauges and more than 100 water level gauges, all for online monitoring purposes. The data transmission is taking place via satellite and has been operational since 2003.

The flood forecasting system takes real time monitoring data of the regional meteorology and the catchment status, and produces forecasts of the flood state of the catchment. The forecasting system is based on MIKE FLOODWATCH and SCOUT. MIKE FLOODWATCH is a GIS based

J. Schanze et al. (eds.), Flood Risk Management:
Hazards, Vulnerability and Mitigation Measures, 139–147.
© 2006 *Springer.*

decision support system for flood management, with MIKE 11 at its core. SCOUT integrates real time numerical weather prediction, radar and raingauge data to produce rainfall forecasts. The system combines the compilation of real time data with rainfall and flood forecasting and presentations of the information and results.

2.2. Rainfall forecasting

Rainfall forecasting is based on three different types of data used in real time: numerical weather prediction (NWP), radar data from three radar stations and raingauge data from more than 100 raingauge stations. The NWP data, the radar data and a part of the raingauge data are provided by the Turkish Weather Service DMI, the other data are stemming from DSI stations.

Before being used, all incoming data are quality controlled: raingauge data are checked for extreme values and hidden missing data (Maul-Kötter an Einfalt 1998). Radar data are checked and corrected for bright band, ground clutter, anomalous propagation, vertical profile and adjusted to raingauges. A part of this work is done on the radar workstation at DMI by SIGMET IRIS software, the other part is performed by SCOUT (Einfalt et al. 2000, Golz et al. 2003). Both data quality checks and potential corrections are of highest importance for the quality of resulting model outcome.

2.2.1 Radar rainfall forecasting

SCOUT is a feature tracking approach to determine echo motion and was first implemented in a suburban county near Paris to control the sewer network (Einfalt et al. 1990). SCOUT is based on the mass centroid method, deriving the displacement vector between consecutive radar scans from the distance of the mass centres of two corresponding radar echoes. The centres are assumed to be representative for individual convective cells or storms. Distinctive features of radar echoes in consecutive scans are identified and recognised, admitting a reasonable degree of change in each of the features. Thus, the history of the echoes is considered. Individual displacement vectors are applied to extrapolate each echo separately. A comparison between the previous forecasts and the actual measurements provides a means for a quality estimation of the current forecast.

As a result, SCOUT provides forecast images and catchment specific time series for the individual catchments for which it has been configured.

2.2.2 Rainfall forecasting with the help of raingauges

In a real time environment, radar data are not always available. Therefore measurements from a raingauge network can work as a fallback strategy for getting rainfall information. For the support of the forecast module in SCOUT, an approach has been developed, linking raingauge data from the raingauge network, mesoscale numerical modelling, and extreme value statistics for a rainfall forecast over 72 hours lead time.

The raingauge based forecast uses the locally measured information through the first hour of lead time by applying a spatio-temporal rainfall analysis. This analysis procedure takes advantage of the fact that rainfields arrive at different instants in time at the different raingauge stations. Thus, arrival time at the stations implies a direction and a speed of movement for the rainfield (fig. 1). A least squares algorithm computes the most suitable movement.

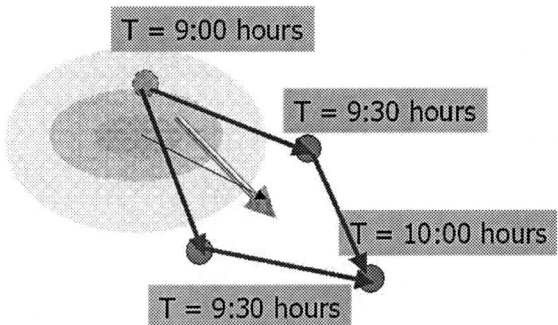

Figure 1. Determination of the displacement of rainfields based on raingauges

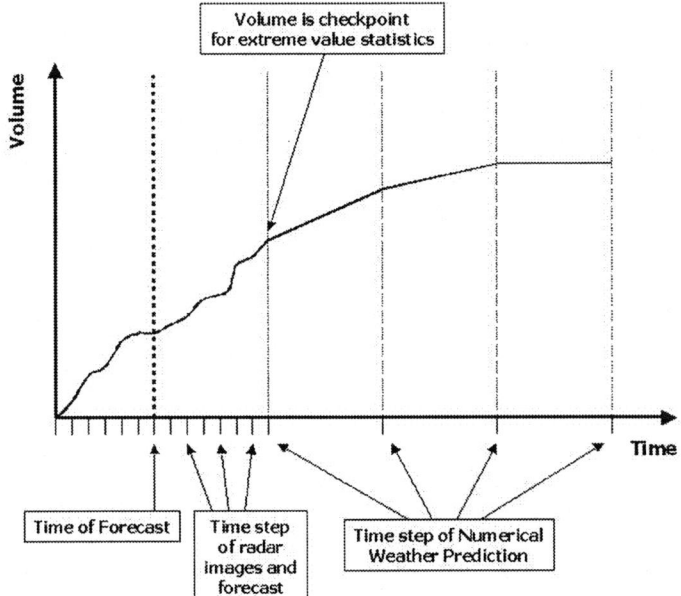

Figure 2. Integration of rainfall time series with NWP data

142

The different modules combine the maximum information available for the catchment (fig. 2). The radar provides a spatial view of the rainfall over the catchment and is used for nowcasting purposes up to one hour, the raingauge network describes the actual state in the catchment and can be used for nowcasting purposes up to one hour, if no radar forecast is available.

After approximately one hour, the reliability of radar or raingauge based forecasts tends to be more uncertain, and the numerical model results, provided by the ECMWF model used by DMI, are included for the following time period. In this way, rainfall measurements during the current NWP forecast period can be taken into account for the 72-hour forecast.

2.3. Flood forecasting

For flood forecasting, FLOODWATCH has for the first time been operationally coupled simultaneously to input from raingauge data, from SCOUT radar data measurements and forecasts of three different radars, and from numerical weather prediction results.

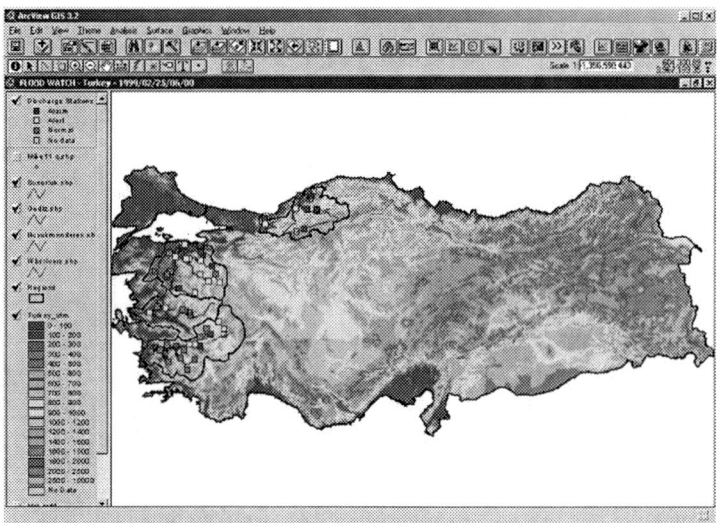

Figure 3. FLOOD WATCH Application for Turkey

The flood forecasting system is set up within MIKE FLOOD WATCH (Barbero et al. 2001, Jørgensen and Madsen 1997) and comprises the following "push button" operations (figure 3):
- Real time data assimilation and preparation
- Hydrologic-Hydrodynamic module, describing the rainfall-runoff process, and the propagation of flood waves through the river system
- Flood Forecasting module, incorporating an updating procedure to

ensure maximum and most effective use is made of the available real time information on the catchment

- Flood Mapping module, providing real time flood maps of selected flood prone areas, showing the area and depth of the actual and historical flooding
- Presentation of forecasts in a variety of formats (tables, plots, maps), and uploading to the Internet

2.3.1 Hydrologic module

The hydrologic module of the forecasting system takes forecast rainfall as its primary input, and simulates the land phase of the hydrologic cycle to forecast the runoff to the main rivers. The module is based on the NAM (rainfall-runoff) component of the MIKE 11 modelling system. NAM operates by continuously accounting the moisture content in four interrelated storages representing the physical elements of the catchment:

- Snow layer (distributed by altitude)
- Surface zone (vegetation, small channels and lakes)
- Root zone (the depth from which plants draw water)
- Ground water

The forecast rainfall over each subcatchment (average area around $1,000km^2$) is received in real time from SCOUT. With real time evaporation and temperature (for snowmelt), the module forecasts the total catchment runoff and effective precipitation for the hydrodynamic module. The module has been calibrated against available historical discharges from 1995 to 2000.

2.3.2 Hydrodynamic module

The Hydrodynamic Module has been set up using MIKE 11, with an automatic linkage to the NAM rainfall-runoff module to receive the runoff from the dry areas of the sub-catchments and the direct effective precipitation on lakes and flooded areas. The module uses an implicit finite difference scheme for the computation of steady and unsteady flows in open channels. The module describes critical and subcritical flows through a numerical scheme which adapts to local temporally and spatially varying flow conditions. The high order wave formulation applied is particularly suited to flood wave propagation through steep rivers in the upstream catchment. Advanced computational modules are applied to the description of flows through fixed and movable hydraulic structures.

2.3.3 Other modules

Other modules on the hydrological side of the forecast system are the flood forecast module, the flood mapping module and the flood warning module. All of them have been set up to run in close communication for the four catchments that are supervised. Warning informations are issued via web-based services as well as inside the Flood Control Centre in Ankara.

3. ORGANISATIONAL ASPECTS

The organisational aspects touch both, technical and human communication and their organisation. The above described system will only run successfully, if
a)   all required data are arriving in a timely manner,
b)   a proper transfer of knowledge has taken place,
c)   failures are recognised and actions are taken accordingly,
d)   the staff is well organised and motivated.

3.1.   Data arrival

In a complex flood forecasting system, the data transfer is a crucial and by no means trivial task. It has to be secured that data from different sources are arriving, in time and without missing parts. Even if an online system has to be able to cope with missing data, the amount of available data with a good quality determines the accuracy of the forecast.

The system in Turkey has two main channels for data transport : a dedicated line with 2 Mbit/s transfer speed for the transfer of radar and NWP data from DMI, and a satellite based communication system for the retrieval of all outstation data (raingauges, flow gauges, climatological measurements).

Other established means of communication are telephone (GSM, ISDN, DSL), the internet and radio. It should be noted that dedicated communication channels provide a higher and faster data availability and safety than dial-in communication.

3.2.   Transfer of knowledge

When introducing a model based flood forecasting system using radar data, this represents for most services a complete change in work flow and availability of information. On the one hand model results are available and possibilities of comparison with measurements do exist on a time series or GIS basis. On the other hand radar rainfall measurement is a new technology with different characteristics than the usual raingauge values.

In general, a change of philosophy is required recognising that neither the traditional measurement nor the radar measurement are capable of telling "the truth" – both only provide information from a certain ("peephole") angle.

Educational needs on radar measurements, radar data and radar derived information were one basic but very important aspect that had to be covered during the project. Besides the aspect of data preprocessing, it was absolutely essential that the staff learned what they can get with radar data and what they will **not** get.

To put it in simple terms, they will get data with a higher spatial density than the majority of them have been available at present, but they had to understand

- different kinds of errors (than they are used to) in the data;
- the additional effects of the data for the hydrological and hydrodynamic model; and
- another view of uncertainty and hydrological certainty than in the "raingauge age".

Still, time is required to "learn by doing" until each staff member will feel comfortable with all aspects of the system. This comfortable feeling is particularly important in order to have the system accepted by the staff.

3.3. Recognition of failures

Staff must be trained in such a way that failures of parts of the system or of a staff member are recognised and corrected by them.

The first point is closely linked to training and the knowledge on "usual" system behaviour. In particular data errors and missing data are prone to cause problems in a flood forecasting system. Therefore, these failure causes need to be correctly analysed when the system is not working as expected. When in doubt, and the team in charge is not finding a solution, an expert for the system must be reachable.

The second point refers to trust among the operators and of the hierarchy towards the operators. In an operational environment, it is not unusual that an action may be taken wrongly. What is important in such a situation is the identification of the fact (wrong action) and the immediate support of the staff member for now taking the correct action. Time is the crucial parameter in operation, and any activity slowing down effective management, such as lenghty discussions among staff, may lead to increased damage. A well written system documentation along with prescribed procedures for the potential emergency cases provide important supports for action in rare events.

3.4. Staff organisation

It has been very useful if staff attribution is done clearly and very early in the project, and if the persons are not changing frequently. This means that there is a well defined organisation scheme for flood management, where responsibilities are clear. Ideally, a structure dealing with flood forecasting is explicitly created, like it is the case in the Seine-Saint Denis urban real time control system in France.

146

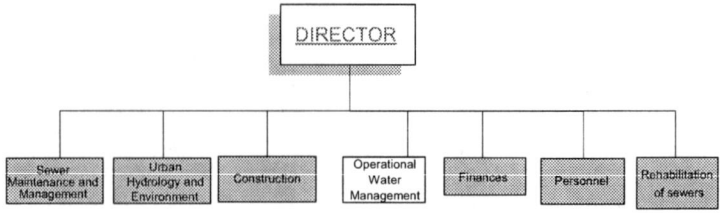

Figure 4. Organisation of the RTC system in Seine-Saint Denis (simplified)

As a minimum requirement in flood situations, the staff responsible for the forecast system must be free from other tasks in order to completely concentrate on management of the flood prone situation. It should be recalled that a scheme for having staff assigned outside the normal office hours is needed, including the assignment of persons on a temporal level (e.g. in form of work plans) and on a financial level (e.g. extra payment for night and weekend service).

4. CONCLUSIONS

In an operational context like flood forecasting, the convincing technical solution is as important as the organisation of the flood forecasting service. Semi-technical aspects such as availability of data and the acceptance of the system by the operators are crucial in this aspect – else the system will not be used.

The three most important items in this context are:

- Availability and good quality of input data and a well calibrated model,
- A good training of the operators on several levels (overview, technical training, hands-on training, supervised stand alone training, information of the hierarchy),
- A suitable service organisation structure, permitting the operators to be free from other tasks in flood periods and, ideally, to be a dedicated service for flood forecasting.

ACKNOWLEDGEMENTS

The authors wish to express their appreciation for the essential contributions from Mr Fayik Turan of DSI and Mr Niyazi Yaman of DMI, and their colleagues in Ankara and the regions, in providing extensive data and information for setting up the forecasting system to meet the particular conditions for Turkey.

References

Barbero S., Rabuffetti D., Wilson G. and Buffo M. (2001) Development of a Physically-Based Flood Forecasting System: MIKE Flood Watch in the Piemonte Region. DHI User Conference, Helsingør, Denmark, June 2001.

Einfalt T., Denoeux T. and Jacquet G. (1990) A Radar Rainfall Forecasting Method Designed for Hydrological Purposes. *Journal of Hydrology* 114, 229-244.

Einfalt T., Maul-Kötter B. and Spies S. (2000) A radar data quality control scheme used in hydrology. *Physics and Chemistry of the Earth, Part B* 25 (10 – 12), 1141-1146.

Golz C., Einfalt T., Gabella M. and Germann U. (2003) Quality Control Algorithms for Rainfall Measurements. Oral presentation at the 6th International Workshop on Precipitation in Urban Areas, Pontresina, Switzerland.

Jørgensen G. and Madsen J.H. (1997) Development of a Flood Forecasting System in Bangladesh. Oral presentation at the Operational Water Management Conference, Copenhagen.

Maul-Kötter B. and Einfalt T. (1998) Correction and preparation of continuously measured raingauge data: a standard method in North Rhine-Westphalia. *Water Science and Technology* 37 (11).

PART 4

VULNERABILITY AND FLOOD DAMAGES

Chapter 13

FLOOD DAMAGE, VULNERABILITY AND RISK PERCEPTION – CHALLENGES FOR FLOOD DAMAGE RESEARCH

FRANK MESSNER AND VOLKER MEYER
UFZ –Centre for Environmental Research Leipzig-Halle, Member of the Dresden Flood Research Center (D-FRC), Dresden, Germany

Keywords: Flood damage analysis, flood vulnerability, risk perception, cost-benefit analysis, integrated assessment.

1.   INTRODUCTION

In this contribution it is argued that the current challenge in flood damage research consists in developing a better understanding of the interrelations and social dynamics of flood risk perception, preparedness, vulnerability, flood damage and flood management, and to take this into account in a modern design of flood damage analysis and flood risk management. Accordingly, the sections of this contribution are organised as follows: In the next section the relationship between flood damage, vulnerability and risk perception is analysed and clarified. Section three deals with state-of-the-art approaches to flood damage analysis. The fourth section discusses the shortcomings of the current approaches with a special focus on the disregard for socio-economic factors and methods. Finally, the contribution concludes with an outlook, presenting current EU research efforts to improve state-of-the-art approaches to flood damage analysis.

2.   THE RELATIONSHIP OF FLOOD DAMAGE, VULNERABILITY AND RISK PERCEPTION

The relationship between flood damage, vulnerability and risk perception has been recognised in a small scientific community. However, neither its relevance regarding the methods of flood damage analysis, nor its significance for the level of public flood protection and flood risk management has been widely acknowledged. It is the purpose of this section to shed some light on the convoluted relationship of these notions.

J. Schanze et al. (eds.), Flood Risk Management:
Hazards, Vulnerability and Mitigation Measures, 149–167.
© 2006 *Springer.*

Since the central terms to be used in this discussion are highly controversial in the vulnerability debate, it is essential to start with some fundamental definitions in the beginning.

2.1. Flood damage

The actual amount of flood damage generated by a specific flood event is time and again a driving force that stimulates politicians to strengthen flood policy measures – usually soon after flood events. Flood damage refers to all varieties of harm caused by flooding. It encompasses a wide range of harmful effects on humans, their health and their belongings, on public infrastructure, cultural heritage, ecological systems, industrial production and the competitive strength of the affected economy. Some of these damages can be specified in monetary terms, others – the so called intangibles – are usually recorded by non-monetary measures like number of lives lost or square meters of ecosystems affected by pollution. Flood damage effects can be further categorised into direct and indirect effects. Direct flood damage covers all varieties of harm which relate to the immediate physical contact of flood water to humans, property and the environment. This includes, for example, damage to buildings, economic goods and dykes, loss of standing crops and livestock in agriculture, loss of human life, immediate health impacts, and contamination of ecological systems. Indirect or consequential effects comprise damage, which occurs as a further consequence of the flood and the disruptions of economic and social activities. This damage can affect areas quite a bit larger than those actually inundated. One prominent example is the loss of economic production due to destroyed facilities, lack of energy and telecommunication supplies, and the interruption of supply with intermediary goods. Other examples are the loss of time and profits due to traffic disruptions, disturbance of markets after floods (e.g. higher prices for food or decreased prices for real estate near floodplains), reduced productivity with the consequence of decreased competitiveness of selected economic sectors or regions and the disadvantages connected with reduced market and public services (Smith and Ward 1998, 34ff.; Green et al.1994, 39ff.).

2.2. Vulnerability

The actual amount of flood damage of a specific flood event depends on the vulnerability of the affected socio-economic and ecological systems, i.e., broadly defined, on their potential to be harmed by a hazardous event (Cutter 1996, Mitchell 1989). Generally speaking, an element at risk of being harmed is the more vulnerable, the more it is exposed to a hazard and the more it is susceptible to its forces and impacts.[i] Therefore, any flood vulnerability analysis requires information regarding these factors, which can be specified in terms of element-at-risk indicators, exposure indicators and susceptibility indicators (see Figure 1). In this regard, natural and social science indicators are highly significant.

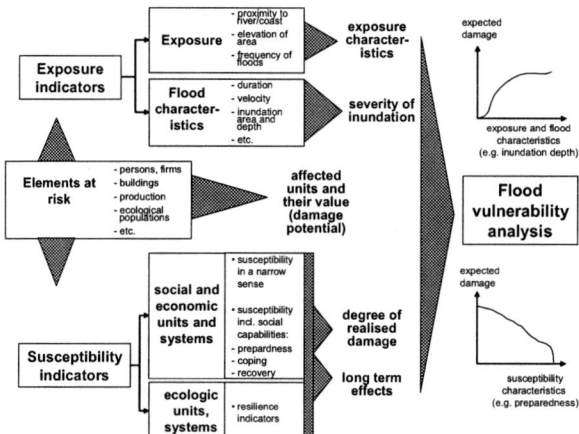

Figure 1. Indicators to be used in flood vulnerability analysis

2.2.1. Element-at-risk indicators

As shown in the centre of Figure 1, the subject matter of any flood vulnerability analysis is the group of elements which are at risk of being harmed by flood events. Element-at-risk indicators specify the amount of social, economic or ecological units or systems which are at risk of being affected regarding all kinds of hazards in a specific area, e.g. persons, households, firms, economic production, private and public buildings, public infrastructure, cultural assets, ecological species and landscapes located in a hazardous area or connected to it. Based on information regarding which and how many elements are at risk of being affected by flood events, the magnitude of damage can be estimated in monetary and non-monetary units, which reflects in total the maximum possible flood damage. This is also called damage potential. And, because every element at risk is more or less exposed to flood events and more or less susceptible to them, exposure and susceptibility indicators are always related to element-at-risk indicators and contribute significantly to the analysis of flood vulnerability.

2.2.2. Exposure indicators

As regards exposure indicators, two categories must be discerned. The first one is needed to typify the kind of exposure of different elements at risk. Indicators supply information about the location of the various elements at risk, their elevation, their proximity to the river, their closeness to inundation areas, about return periods of different types of floods in the floodplain and the like. Taken together, these indicators inform us of the frequency of floods in floodplains and of the threat to the various elements at risk of being inundated. The indicators of the second category focus on general flood characteristics like duration, velocity, sedimentation load and inundation depth. Considered in concert they indicate the severity of inundation as well as its distribution in space and time. Summing up, exposure indicators confer specific information about hazardous threats to the various elements at risk (e.g., Alexander 1993, Heyman et al. 1991).

2.2.3. Susceptibility indicators

Susceptibility indicators measure how sensitively an element at risk behaves when it is confronted with some kind of hazard. Figure 1 relates susceptibility indicators to the affected social, economic and ecological systems or to individual units of these systems. Regarding social and economic systems, an important group of indicators refer to susceptibility in a narrow sense, measuring the absolute or relative impact of floods on individual elements at risk. For example, the impact of inundation depth and flood duration on buildings is frequently a major issue of damage analysis and research, attempting to identify building categories that feature similar susceptibilities. And this makes sense, because wooden houses are much more susceptible to floods than stone houses and buildings with only one storey usually experience greater (relative) damage than houses with several storeys. Susceptibility indicators in a broader meaning relate to system characteristics and include the social context of flood damage formation, especially the awareness and preparedness of affected people regarding the risk they live with (before the flood), their capability to cope with the hazard (during a flood), and to withstand its consequences and to recuperate (after the flood event). Accordingly, the three relevant sets of indicators mentioned in figure 1 refer to preparedness, coping and recovery capabilities and strategies of individuals and social systems.

A lot of research work has been carried out regarding the vulnerability of social systems in terms of their susceptibility in a broader sense, and many indicators have been proposed in this context. Firstly, awareness and preparedness indicators for individuals and communities reflect the awareness and preparedness of threatened people and communities for dealing with hazardous events, including, for example, the number of households protected against physical flood impacts by means of technical measures, the number of people with insurance against flood damages, the number of persons ready for action in disaster management, as well as the quality of flood protection measures and disaster management organisations (e.g, Green et al. 1994: 47ff.). Secondly, since the ability of individuals and social systems to cope with the impact of floods is often correlated to general socio-economic indicators, coping indicators embrace general information on age, structure, poverty, gender, race, education, social relations, institutional development, proportion of population with special needs (children, elderly) and the like (e.g., Blaikie et al. 1994, Watts/Bohle 1993, Hewitt 1997, Smith 2001). This category also includes indicators for technical systems, because the social impact of floods significantly relates to the susceptibility of basic infrastructure and lifelines, which support the population's supply of basic needs. Technical susceptibility indicators specify flood-specific weaknesses and the ability of socio-technical systems like drinking water supply, waste water treatment, communication systems and energy supply to withstand the consequences of flood events (Gasser and Snitofsky 1990, Platt 1990). Thirdly, social susceptibility in a broader sense also relates to the capability of the actors to overcome the consequences of the hazard and to re-establish previous conditions. Recovery indicators are meant to measure this aspect. Among others, indicators refer to the financial reserves of affected households and communities, the substitutability of lost items, the cohesion of social systems, and the external support provided by friends, the government and private donors. Furthermore, the long term flood impacts on the standard of living and the general health conditions can either be measured in physical units or in time units, reflecting the time required to achieve conditions which are comparable to the time before the hazardous event.[ii]

Although less research has been carried out on economic systems and their susceptibility to floods in a broader sense, several susceptibility indicators do exist regarding the impact of floods on economic units and systems like firms, sectors and economic production areas. Just as in the case of social systems, the relevant indicators refer to preparedness, coping and recovery abilities and strategies. Economic preparedness indicators report on the technical and social preparedness of economic actors and systems, among others, on flood insurance and on the ability to transfer production to other locations. Coping indicators deal with the strength of actors to cope with flood events (Parker et al. 1987, Green et al. 1994). Eventually, recovery indicators give information on long term impacts like productivity, competitiveness and bankruptcy and report on the time required to re-establish previous conditions.

While the frequent occurrences of floods and their vital significance for floodplain ecosystems is often referred to as a beneficial effect of floods, there are also negative ecological flood impacts. Especially if the flood water is polluted or if large sedimentation processes occur, ecological systems can be disrupted significantly (Haase 2003). Therefore, it is reasonable to talk about the flood susceptibility of ecological systems, too. Although it is not constructive to relate the susceptibility to individual biological units, it is sensible to derive susceptibility indicators in a broader sense as they relate to ecosystems as a whole. Such indicators can be derived from the debate concerning ecological resilience. Ecological resilience is a property of a system and refers to its ability to absorb external disturbances or changes and still persist (Holling 1973). In this context, indicators are important which refer to the amount of change or disruption that a system can absorb, to its capacity to be capable of self-organisation and adaptation (Carpenter et al. 2001) and to the rate at which it returns to equilibrium after a disturbance (Pimm 1984).

After having identified and quantified the most important indicators for elements at risk, exposure and susceptibility in a narrow and a broader sense, it is the task of vulnerability analysis to identify the most important relationships between expected flood damages and the exposure and susceptibility characteristics of the affected socio-economic and ecological systems. Typical results are shown in the right part of Figure 1, indicating the development of expected damage to an element at risk depending on susceptibility and exposure characteristics. Hence, the above mentioned broad definition of vulnerability can be made more explicit. Vulnerability can be defined by the characteristics of a system that describe its potential to be harmed. It can be expressed in terms of functional relationships between expected damages regarding all elements at risk and the susceptibility and exposure characteristics of the affected system, referring to the whole range of possible flood hazards.

2.3. Risk perception

The notion of risk perception refers to the intuitive risk judgements of individuals and social groups in the context of limited and uncertain information (Slovic 1987). These judgements vary between individuals due to different levels of information and uncertainty, due to different intuitive behaviour, and also due to specific power constellations and positions of interest. As a consequence, the individuals of a community may assess the risk of being flooded very differently, because they do not have the same information about the probability of flood hazard events in their region, about flood mitigation measures and their effectiveness, and they perhaps have a different historical background regarding the experience of living in a floodplain and of being flooded. Due to their specific perception of flood risk individuals, social groups and also public persons like mayors, politicians and employees in the public sector dealing with flood protection and disaster management may handle this issue very differently. Experts responsible for flood protection may try to maximise their scientific information on flood hazards and flood risk in order to optimise the effectiveness of flood protection measures. Politicians may be more interested in attracting additional inhabitants or enterprises into a floodplain region in order to strengthen the regional economic development. As a consequence, they may object to unattractive measures of flood risk management. And, finally, some individual inhabitants may feel that there is a degree of flood risk which they want to reduce by means of private measures. Others might be inclined to do nothing, either because they do not share this perception, or they believe that these measures will not pay, or they simply assume that flood protection is a public policy task. In face of the very diverse risk perceptions within society, a communication process on flood risk and flood risk perception should be encouraged as a basis for policy. If prevailing perceptions and value concepts become transparent and open to public debate, a common perception of communities may evolve and contribute to an increased acceptance of flood protection policies.

2.4. The relationship between flood damage and vulnerability

Flood damage analysis aims at quantifying flood damages for specific future scenarios with different flood events and flood policies in order to quantify the benefits of flood protection measures ex ante and, thereby, support policy decisions. In this context the concept of damage potential is crucial. The damage potential of a specific area represents the maximum possible amount of damage which may occur if the area becomes inundated. In these analyses vulnerability aspects must be considered in order to estimate the proportion of the damage potential which will finally materialise, i.e. to determine expected damages. In many instances, a vulnerability factor is derived for the most important vulnerability indicators having a substantial impact on the degree of damage produced during a flood event. In some vulnerability analyses, such a factor is derived from expert knowledge and empirical data on flood damages and then expressed on a scale between 0 (no loss at all) and 1 (total loss) in order to quantify the expected damage reduction for several categories of elements at risk (e.g., Elsner et al. 2003, Glade 2003). As will be outlined in more detail below, the most important vulnerability indicator for estimating damages in current flood damage analyses is the exposure indicator "inundation depth".

2.5. The relationship between risk perception and vulnerability

With regard to the social and economic features of vulnerability, the notion of risk perception is crucial, too. In this context, the concept of preparedness, which has already been discussed above in the context of social susceptibility indicators, plays a specific role. If (average) flood risk perception is low in a region – perhaps due to the fact that flood events rarely occur or the level of flood protection in terms of dykes and levees is high – many laymen, experts and politicians do not think that they could ever be affected by flooding in their area. As a consequence, they would probably not take any action to decrease the risk or to prepare for the occurrence of flooding. Even if they were warned in advance of an emerging flood hazard they would probably either not believe that this could really happen, or they would just not know what to do. Conversely, if people are well aware of a flood risk – perhaps because they experience a flood with varying severity time and again – they tend to be better informed and prepared (Baan and Klijn 2004). As a rule of thumb it can be stated that regions with low levels of flood risk perception and a low degree of preparedness for coping with flood events tend to experience flood damage levels above average – their vulnerability to flood events is usually high.[iii] Hence, there might exist a vulnerability factor with regard to risk perception and preparedness of communities and individuals.

3. STATE OF THE ART OF SOCIO-ECONOMIC FLOOD DAMAGE ANALYSIS AND EVALUATION

Traditionally, flood defence planning has focused on safety standards, such as dike design levels or reservoir volumes required to ensure pre-defined protection levels for the population and the economy. Protection of the community against floods with a frequency of 1250 years and more serves as a good example, as is the case with the flood protection law of the Netherlands (Baan and Klijn 2004). However, this approach neglects the amount of valuables protected by a defence system and, hence, disregards the efficiency of flood protection measures. While economic costs of alternative flood defence options are usually considered in the decision-making process, the benefits of flood protection in the form of prevented damages should be taken into account, too. The new paradigm for flood risk management (see, for example, Sayers et al. 2002 and Schanze in this issue) specifically includes the economic analysis of costs and benefits of flood protection and mitigation measures in the context of risk analysis. Here, not only the safety of a defence system and its associated costs and benefits are considered, but also the damages to be expected in case of its failure. As a consequence of the application of cost-benefit and risk analysis, safety standards could better be adjusted to the specific circumstances, because it could turn out that the costs of ensuring an overall safety standard considerably exceed the benefits in some areas.

Usually, there are two integral parts in the current state-of the-art ex ante estimation of flood damages.[iv] Firstly, the flood hazard needs to be determined by means of exposure indicators, using flood parameters like expected inundation area and depth, velocity and flood duration. Secondly, the expected damage needs to be estimated. For this, all valuable property located within the endangered area, i.e. the damage potential, needs to be quantified. The expected damage is then calculated by using depth-damage-functions, which show the total damage of the valuable property (e.g. buildings, cars, roads, etc) or its relatively damaged share as a function of inundation depth. Depending on whether the functions relate to the absolute damage or the damage share, they can be called absolute or relative depth-damage functions, respectively. Over the past decades, a great variety of different methods for the ex ante estimation of flood damages emerged. According to their scale and goal, these methods can be roughly divided into three categories: Macro-, meso- and micro-scale analyses (Gewalt et al. 1996). In the following a short overview is given over the most important state-of-the-art approaches of flood damage analysis.

3.1. First part of flood damage analysis: Determination of flood characteristics

The first part of flood damage evaluation, the determination of inundation area and depth, is necessary to get basic information about the flood hazard which generates flood damages. In this context, no clear distinction between macro-, meso- and micro-scale methods can be made – only that small-scale analyses tend to use more accurate methods. The methods vary considerably due to the character of the flooding – e.g., the simulation for storm surges is more complex than for river floods because of tidal dynamics – and with regard to the question whether the research area is protected by flood defence systems or not. The variety of methods ranges from the definition of flood plains by fixed contour lines for one or more scenarios (e.g., Ebenhöh et al. 1997; Klaus and Schmidtke 1990) to the calculation of water levels for floods with different frequencies (e.g., MURL 2000) to dynamic flooding simulations, which also take the extent of dike breaches, the flow volume and the velocity of the flooding event into account (e.g., Mai and von Liebermann 2002).

3.2. Second part of damage analysis: Estimation of damage potential and calculation of expected damages

The main differences between the three mentioned micro, meso and macro approaches relate to the spatial accuracy of damage potential analysis, to the differentiation of land use categories and to the damage functions used. Before some typical methods for the three approaches are outlined in the following, it has to be mentioned that most of the studies – regardless of whether they are performed for macro, meso or micro scale – primarily focus on the estimation of direct, tangible damages, which means damages to assets which can be expressed in monetary terms. Intangible and indirect damages have been rarely considered to date, due to methodological difficulties.

3.2.1. Macro-scale approaches

One typical example of macro-scale analyses is the study for the German Coasts (Ebenhöh et al. 1997; Behnen 2000), which is based on the Common Methodology of the Intergovermental Panel on Climate Change (IPCC 1991). Here, the calculation of damage potentials is carried out for the level of municipalities. The main data sources for this evaluation are official statistics. However, sometimes data are not accessible for this level of aggregation. While for example, the number of inhabitants is directly available from the municipality level statistics, other categories of valuables, such as residential capital or fixed assets, are only published for the state level. As a consequence, these categories of valuables have to be disaggregated to the municipality level by using the number of inhabitants or employees. Of course, such a procedure generates data with a low degree of accuracy. Furthermore, the spatial distribution of the damage potential within the municipalities is not differentiated, i.e. an equal distribution of the valuables over the whole area is assumed. This increases the degree of inaccuracy. However, if the aim of the study is just to estimate the approximate level of damage related to sea-level rise, it might be sensible to apply a macro approach.

158

3.2.2. Meso-scale approaches

Within meso-scale analyses[v], the damage potential is derived from aggregated data, too. Just as in the macro-scale approach, the data on valuables stem from official statistics at the municipality level. However, in order to enable a more realistic localisation of the valuables within the municipalities, each of the categories for the valuables is assigned to one or more corresponding land-use categories. For example, residential capital is assigned to residential areas, fixed assets and inventories of the manufacturing sector are assigned to industrial areas and livestock is assigned to grassland. This approach allows a differentiation between areas of high value concentration, such as urban areas and especially city centres on the one hand, and areas with very low damage potential like agricultural land or forests on the other hand.

Today digital land-use data like the digital landscape model from the German ATKIS (Official Topographic Cartographic Information System) is frequently used for this approach, which allows its spatial imple-mentation by means of the Geographic Information System (GIS). By intersecting maps of inundation area and damage potential in a GIS and relating them accordingly, the amount of valuables or people affected can be determined. The vulnerability factor of the valuables, i.e. the share that is expected to be damaged, is in most cases exclusively related to inundation depth. Hence, relative depth-damage functions are used to calculate the expected damages. They show the damaged share of the category of the valuable as a function of inundation depth (fig. 2). Depth-damage functions can be derived from estimations of expert assessors (synthetic data) and/or from empirical flood damage data (survey data).

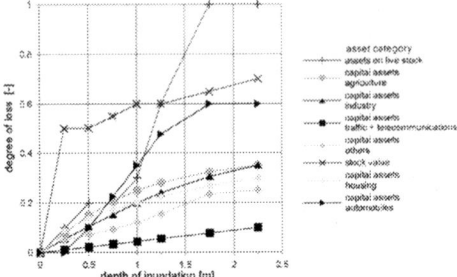

Source: Elsner et al. 2003

Figure 2. Depth-damage-functions for different asset categories (based on Klaus and Schmidtke 1990)

3.2.3. Micro-scale approaches

Within micro-scale analyses damage potentials and expected damages are evaluated on an object level, i.e. single valuables of one category, such as specific types of residential or non-residential properties, are differentiated.

Two different micro-scale approaches of damage calculation can be distinguished.

A micro-scale damage potential evaluation was used within the MERK-Project (Micro-scale Risk Evaluation for Coastal Lowlands; Reese et al. 2003), which was implemented for several cities and municipalities in the state of Schleswig-Holstein, Germany. In the context of this method the value of individual objects is considered. This means that, e.g., the total value of every single building in the research area is determined, using "normal construction costs" according to the official guideline for the assessment of property values. This approach requires a detailed site survey, whereby building characteristics such as age, construction design and type of usage are mapped. Just as in meso-scale analyses, the damaged portion of the valuable objects at risk is calculated according to relative depth-damage functions.

A different micro-scale approach was developed by the Flood Hazard Research Centre in the UK (Penning-Rowsell and Chatterton 1977; Penning-Rowsell et al. 2003).[vi] This method does not refer to the total value of objects, i.e. a damage potential analysis is not carried out. Instead, absolute depth-damage functions are used, which specify absolute damage amounts related to inundation depth. Since such absolute damage amounts vary strongly depending on the object or building regarded, a considerably differentiated set of damage functions is needed for this approach, as well as detailed information about building characteristics.

3.2.4. Intermediate approaches

The classification of methods mentioned above should not be interpreted too stringently. It aims at providing an outline of typical methods of flood damage analysis used for different spatial scales. Due to the great variety of damage studies, there are also many approaches with intermediate methods, which combine elements of all three types (see e.g., IKSR 2001, Gemmer 2004, DEFRA 2001, Bateman et al. 1991, ProAqua et al. 2001, MURL 2000, Meyer 2005, Kok et al. 2004).

In face of the great variety of methods of damage analysis, the choice of an appropriate method (or of a combination of elements of different approaches) does not only depend on the size of the area under consideration, but also on other factors like the availability of necessary data, time, manpower and/or money resources and not least on the goal of the respective study and the management level for which it should provide decision-making support. The latter factors determine the political demands regarding the accuracy of the results and, hence, will decide upon the application of micro-, meso- or macro-approaches for a given study region.

4. SHORTCOMINGS OF THE CURRENT STATE-OF-THE-ART DAMAGE ESTIMATION METHODS

Despite the fact that, from an economic perspective, the application of current state-of-the-art methods of flood damage analysis is a clear progression when compared to the safety standard approach, it must be considered as well that the state-of-the-art methods presented above are characterised by several deficiencies. Particularly, the complex interrelations of flood vulnerability analysis as described in section 3 are considered only in an extremely reduced sense, while existing socio-economic evaluation approaches are not taken into account. The five most important shortcomings are portrayed in the following.

(1) Current flood damage and vulnerability analyses have a dominant focus on tangible flood effects. Despite the fact that economic methods for the evaluation of intangibles have existed and have been discussed for many years in economics literature (e.g., Hanley and Spash 1993, Brent 1996), and economic studies on the evaluation of intangible health effects (e.g., Johnson et al. 2000, Sendi et al. 2002), loss of life (e.g. Landefeld and Seskin 1982) and environmental effects (e.g., Garrod and Willis 1999) are at hand, these methods are not (or, at most, very rarely) applied in the context of flood vulnerability analysis. A major reason for this deficiency might be that flood damages are often calculated by engineers or hydrologists with a business economics background. Therefore, economic methods regarding welfare effects of the whole economy might not be recognized. Another reason might be that the evaluation of human life in monetary terms is rejected by many people on ethical grounds. However, even if the monetisation of some of the so-called intangibles are controversial (even among economists), it is still widely accepted that effects on health and the environment can, at least partly, be quantified in monetary terms in order to approximate the respective welfare losses. Therefore, if appropriate methods are at hand to quantify intangibles, this should be done to improve the estimations of flood damage potential and expected damages.

(2) Indirect effects are also outside the scope of most analysts who are executing flood vulnerability and damage analyses. However, if the economic activity in a region is brought to a standstill, this does not only imply a loss of production and a decrease in supply of consumers within the affected region. It might also lead to severe consequences for other sectors within the economy, which are closely connected through intermediate products, trade, services like electricity and telecommunication and company relations. Especially if production processes for export goods are affected or the economic sectors hindered by floods are highly concentrated and/or specialised, there may be no possibility of shifting production to other national producers. As a consequence, production and sales might be lost to manufacturers in other nations, such that national value-added and exports decrease. While indirect effects in the form of production and sales losses in inundated regions are sometimes considered in flood vulnerability studies by means of average loss of

value-added or additional costs, effects outside the inundation area are usually neglected – often due to a lack of empirical data (Penning-Rowsell et al. 2003, ch. 5). However, analytical methods for estimating such indirect effects are available, especially in the form of economic input-output models. Pioneer work for estimating the structural economic effects of large scale inundation by means of input-output modelling has been executed in the Netherlands (van der Veen et al. 2003), and should increasingly be applied in the context of flood damage analysis.

(3) Regarding the vulnerability relationships between expected damage and different system characteristics, vulnerability factors are usually used or calculated for one exposure indicator only. Frequently, inundation depth is the main and only flood characteristic used to estimate expected flood damage by means of depth-damage curves. While it is known that other variables such as velocity, turbulence, flood duration as well as toxic or sedimentation load can have a significant impact on flood damages[vii], these variables are usually assumed to be strongly correlated with inundation depth – and therefore ignored in the analysis. Since the other variables are also difficult to measure or to estimate, inundation depth is still the major variable for calculating flood damage today (Smith 1998, 40f). Only a few authors have tried to include complementary exposure variables, such as flood duration, as secondary variables in the analysis and generated depth-damage curves with specific variants for different flood durations (see Figure 3) (Penning-Rowsell and Chatterton 1977, Penning-Rowsell et al. 2003). Accuracy of flood damage analysis could improve if such expanded depth-damage curves were to be developed and applied more frequently.

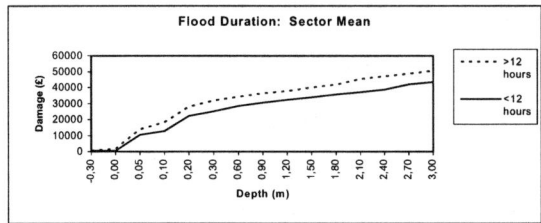

Source: Penning-Rowsell et al 2003: ch. 3
Figure 3. Depth-damage curve differentiated by flood duration

(4) Socio-economic susceptibility indicators in a broader sense are not considered, for the most part, in flood damage and vulnerability studies to differentiate and improve damage estimations. Factors such as individual and public preparedness before flood events, the quality of coping strategies during a flood and, closely linked to this, the perception of flood risks in the affected population are usually excluded from ex ante flood damage calculations. However, as evidence from the Elbe flood 2002 showed, individual preparedness in terms of technical measures in the buildings and flood-adapted usage of the lower storeys can reduce the damage by a range of 5-30% (DKKV 2002: 46-51). Therefore, susceptibility aspects should be considered more carefully in the context of flood damage analysis. One of the rare examples to include

162

socio-economic factors in flood damage analysis stems once again from the UK. In the Flood Hazard Research Centre, flood researchers developed an approach for estimating the impact of early flood warning lead time on damage. As can be seen in Figure 4 for different levels of inundation depth, an increase in warning time by more than two hours has the potential of reducing damage by more than 10% (Penning-Rowsell et al. 2003; ch. 3). This reveals that human efforts and coping strategies during the warning lead time of a flood have a clear impact on flood damage. However, these percentages are still low compared to the efforts and investments often undertaken to improve early flood warning systems. Differentiating these curves further for different types of coping strategies and risk perception patterns could generate more evidence regarding the significance of socio-economic susceptibility indicators in flood damage analysis.

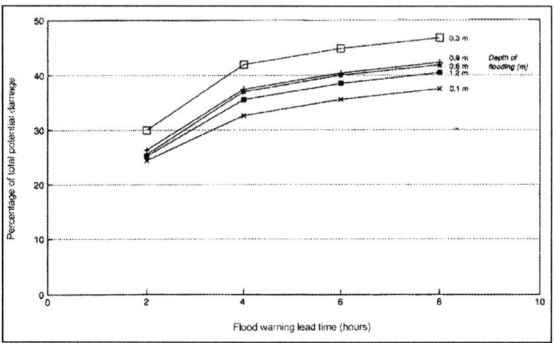

Source: Penning-Rowsell et al 2003: ch. 3

Figure 4. Impact of flood warning lead time on flood damage

(5) Last but not least, it should be emphasised that the final evaluation of flood damage should not be executed on the basis of monetary cost-benefit results alone. Even if new economic methods for estimating intangibles are applied, there will always remain a number of intangibles which cannot be monetised or which society does not accept in monetary terms, among them for example loss of life, loss of unique valuables like diaries, loss of cultural heritage and distribution effects of floods – to name just a few. Current state-of-the-art approaches of flood damage evaluation do not consider these effects, although empirical surveys have shown that people usually bemoan these intangible flood damages most (Green et al. 1994: pp. 52 ff, Hagemeier 2005: pp. 88 ff). Therefore, in order to take these effects into account in the evaluation of flood risk management strategies, multi-criteria methods should be developed and applied in the context of flood damage analysis and risk assessment.

These five shortcomings pose a substantial challenge for flood damage and flood vulnerability research. Diminishing or even eliminating them and improving the state-of-the-art in flood research and flood risk management accordingly would be a great success.

5. OUTLOOK

It is a challenge of flood research to find new and innovative approaches for overcoming the shortcomings of current flood damage and vulnerability analysis approaches and, thereby, to strengthen the overall approach of flood risk management with special regard to its socio-economic components. In the context of the Integrated Project FLOODsite, financed by the EU in the 6th framework programme, some of the shortcomings of flood vulnerability analysis are the object of research of a group of European social scientists. The research objectives are:

(1) Providing methodological guidelines for the monetary estimation of flood effects on human health and the environment:
(2) Providing methodological guidelines for the monetary estimation of indirect economic effects based on input-output modelling techniques;
(3) Advancing the development of functional vulnerability relationships between expected damage and flood characteristics besides inundation depth;
(4) Advancing the development of functional vulnerability relationships between expected damage and indicators of socio-economic susceptibility in a wider sense, focusing especially on risk perception, preparedness and coping indicators.
(5) Developing multi-criteria tools in order to include non-monetary intangible damage into the assessment framework of flood damage analysis.

Furthermore, in order to disseminate the knowledge gained from research, an overall guideline document on the state-of-the-art flood damage and vulnerability analysis approaches will be produced, including guidelines for innovative approaches for reducing current shortcomings. This document is meant to contribute to the harmonisation and improvement of flood vulnerability methods used, and to expand their application all over the EU, especially in countries where risk analysis and flood vulnerability analysis are uncommon methods today.[viii]

ACKNOWLEDGEMENTS

This research was funded by the European Commission under contract number GOCE-CT-2004-505420. Furthermore, we would like to express many thanks for instructive comments on an early draft to Annett Steinführer and Christian Kuhlicke.

References

Adger W.N. (2000) Social and ecological resilience: are they related? *Progress in Human Geography* 24 (3), 347-364.

Alexander D. (1993) *Natural disasters*. New York: Chapman & Hall.

Baan P.J.A. and Kljin F. (2004) Flood risk perception and implications for flood risk management in the Netherlands. *International Journal of River Basin Management* 2 (2), 113-122.

Bateman I., Bateman S., Brown D., Doktor P., Karas J.H.W., Maher A. and Turner R.K. (1991) Economic Appraisal of the Consequences of Climate-Induced Sea Level Rise: A case Study of East Anglia. Report to the Ministry of Agriculture, Fisheries and Food, University of East Anglia, Norwich.

Behnen T. (2000) Der beschleunigte Meeresspiegelanstieg und seine sozioökonomischen Folgen: Eine Untersuchung der Ursachen, methodischen Ansätze und Konsequenzen unter besonderer Berücksichtigung Deutschlands, *Hannoversche Geographische Arbeiten* 54, Hannover.

Beyene M. (1992) Ein Informationssystem für die Abschätzung von Hochwasserschadens-potentialen. Mitteilungen des Instituts für Wasserwirtschaft der RWTH Aachen. Aachen.

Blaikie P., Cannon T., Davis I. and Wisner B. (1994) *At risk -- natural hazards, people's vulnerability and disasters*. London.

Brent R. J. (1996) *Applied Cost-Benefit Analysis*. Cheltenham.

BWK (Bund der Ingenieure für Wasserbau, Abfallwirtschaft und Kulturbau) (eds.) (2001) Hochwasserschadenspotenziale. Bericht 1/2001, Düsseldorf.

Carpenter S., Walker B., Anderies J.M. and Abel N. (2001) From metaphor to measurement: resilience of what for what? *Ecosystems* 4, 765-781.

Cutter S.L. (1996) Vulnerability to environmental hazards. *Progress in Human Geography* 20 (4), 529-539.

DEFRA (Department for Environment, Food and Rural Affairs) (2001) National appraisal of assets at risk from flooding and coastal erosion, including the potential impact of climate change. Final Report.

Ebenhöh W., Sterr H. and Simmering F. (1997) Potentielle Gefährdung und Vulnerabilität der deutschen Nord- und Ostseeküste bei fortschreitendem Klimawandel. Case Study in Anlehnung an die Common Methodology der IPCC Coastal Zone Management Subgroup, Oldenburg.

Elsner A., Mai S., Meyer V. and Zimmermann C. (2003) Integration of the flood risk in coastal hinterland management. Proc. of the Int. Conf. CoastGis, Genua, Italy, 2003.

Garrod G. and Willis K.G. (1999) *Economic Valuation of the Environment*. Edward Elgar, Cheltenham, UK und Northhampton, MA, USA.

Gasser J. and Snitofsky E. (1990) Vulnerability analyses plan for wastewater emergencies. *American City and County* 105, 81-82.

Gemmer M. (2004) Decision support for flood risk management at the Yangtze river by GIS/RS-based flood damage estimation. Dissertation at University of Giessen.

Gewalt M., Klaus J., Peerbolte E.B., Pflügner W., Schmidtke R.F. and Verhage L. (1996) EUROflood – Technical Annex 8. Economic Assessment of Flood Hazards. Regional Scale Analysis-Decision Support System (RSA-DSS), München.

Glade T. (2003) Vulnerability Assessment in Landslide Risk Analysis. *Die Erde – Beitrag zur Erdsystemforschung* 134 (2), 123-146.

Green C., van der Veen A., Wierstra E. and Penning-Rowsell E. (1994) Vulnerability refined: analysing full flood impacts. Penning-Rowsell E. and Fordham M. (eds.) *Floods across Europe -- Flood hazard assessment, modelling and management*. Middlesex University Press, London.

Haase D. (2003) Holocene floodplains and their distribution in urban areas – functionality indicators for their retention potentials. *Landscape and Urban Planning* 66, 5-18.

Hanley N. and Spash C.L. (1993) *Cost-Benefit Analysis and the Environment*. Edward Elgar, Brookfield.

Hamann M. and Klug H. (1998) Wertermittlung für die potentiell sturmflutgefährdeten Gebiete an den Küsten Schleswig-Holsteins. *Vechtaer Studien zur Angewandten Geographie und Regionalwissenschaft* 20, 63-70.

Hewitt K. (1997) Regions of risk. A geographical introduction to disasters. Harlow.

Heyman B.N., Davis C. and Krumpe P.F. (1991) An assessment of world wide disaster vulnerability. *Disaster Management* 4, 3-36.

Holling C.S. (1973) Resilience and stability of ecological systems. *Annual Review of Ecological Systems* 4, 1-23.

IKSR (Eds.) (2001) Atlas der Überschwemmungsgefährdung und möglichen Schäden bei Extremhochwasser am Rhein. Online publication: www.iksr.org/rheinatlas/print-version/IKSR_Atlas_II.pdf.

IPCC (Intergovermental Panel on Climate Change) (1991) Assessment of the Vulnerability of Coastal Areas to Sea Level Rise - A Common Methodology. Report of the Coastal Zone Management Subgroup of IPCC Working Group III, Den Haag.

Johnson F.R., Banzhaf M.R. and Desvousges, W.H. (2000) Willingness to pay for improved respiratory and cardiovascular health: a multiple-format, stated-preference approach. *Human Health* 9 (4), 295-317.

Kiese M. and Leineweber B. (2001) Risiko einer Küstenregion bei Klimaänderung. Ökonomische Bewertung und räumliche Modellierung des Schadenspotentials in der Unterweserregion. *Hannoversche Geographische Arbeitsmaterialien* 25. Hannover.

Klaus J., Pflügner W., Schmidtke R.F., Wind H. and Green C. (1994) Models for Flood Hazard Assessment and Management. Penning-Rowsell E. C. and Fordham M. (eds.) *Floods across Europe. Hazard assessment, modelling and management*. London, 67-106.

Klaus J. and Schmidtke R.F. (1990) Bewertungsgutachten für Deichbauvorhaben an der Festlandsküste – Modellgebiet Wesermarsch. Untersuchungsbericht an den Bundes-minister für Ernährung, Landwirtschaft und Forsten. Bonn.

Klein R.J.T., Nicholls R.J. and Thomalla F. (2003) Resilience to natural hazards: how useful is this concept? *Environmental Hazards* 5, 35-45.

Knogge T. and Wrobel M. (2000) Klimasensitive Wirtschaftsbereiche: Ergebnisse eines Verfahrens zur Identifikation eines regionalen Clusters am Beispiel der Unterweserregion. *Bremer Diskussionspapiere zur ökonomischen Klimafolgenforschung* 7.

Kok M., Huizinga H.J., Vrouwenfelder A.C.W.M. and Barendregt A. (2004) Standard Method 2004: Damage and Casualties caused by Flooding. Client: Highway and Hydraulic Engineering Department, Delft.

Kron W. and Thumerer T. (2002) Water-related disasters: Loss trends and possible countermeasures from a (re-)insurers point of view. Munich Reinsurance Company, Germany (http://www.mitch-ec.net/workshop3/Papers/paper_thumerer.pdf).

Landefeld J.S. and Seskin E.P. (1982) The economic value of life: linking theory to practice. *American Journal of Public Health* 72 (6), 555-566.

Mai S. and von Liebermann N. (2002) RISK – Risikoinformationssystem Küste. *Jahrbuch der Hafenbautechnischen Gesellschaft* 53, 44-56.

Meyer V. and Mai S. (2003) Verfahren zur Berechnung der Schäden nach Deichbruch an der deutschen Nordseeküste. Kelletat D. (eds.) Neue Ergebnisse der Küsten- und Meeresforschung. Tagungsband der 21. Jahrestagung des Arbeitskreises "Geographie der Meere und Küsten" (AMK) 2003 in Essen. *Essener Geographische Arbeiten* 35, 169-178.

166

Meyer V. (2005) Methoden der Sturmflut-Schadenpotentialanalyse an der deutschen Nordseeküste. Dissertation at University of Hannover, UFZ-Dissertation 3/2005.

Mitchell J.K. (1989) Hazards research. Gaile, G.L. and Willmot, C.J. (eds) *Geography in America*. Columbus, OH, Merrill, 410-424.

MURL (Ministerium für Umwelt, Raumordnung und Landwirtschaft des Landes Nordrhein-Westfalen) (eds.) (2000) Hochwasserschadenspotentiale am Rhein in Nordrhein-Westfalen. Abschlussbericht, Düsseldorf.

OSAM GmbH Rostock (1995) Analyse zum Hochwasserschadenspotential an der Ostsee- und Boddenküste in Mecklenburg-Vorpommern. Teil 1: Zusammenfassung aller Landkreise und kreisfreien Städte. Teilbereich: Statistik. Analyse im Auftrag des Staatlichen Amtes für Umwelt und Natur Rostock, Rostock.

Parker D.J., Green C.H. and Thompson P.M. (1987) Urban flood protection benefits: a project appraisal guide (The Red Manual). Aldershot, UK, Gower Technical Press.

Penning-Rowsell E., Johnson C., Tunstall S., Tapsell S., Morris J., Chatterton J., Coker A. and Green C. (2003) The Benefits of flood and coastal defence: techniques and data for 2003. Flood Hazard Research Centre, Middlesex University.

Penning-Rowsell E.C. and Chatterton J.B. (1977) The benefits of flood alleviation: a manual of assessment techniques (The blue manual). Aldershot, UK, Gower Technical Press.

Pimm S.L. (1984) The complexity and stability of ecosystems. *Nature* 307, 321-326.

Platt R. (1990) Lifelines: An emergency management priority for the United States in the 1990s. *Disasters* 15, 172-176.

ProAqua, PlanEVAL and RWTH (2001) Integriertes Donau-Programm: Risikoanalyse Donau. Studie zur Hochwasserschadensminderung an der baden-württembergischen Donau. 1. Abschnitt: Ulm bis Ertingen-Binzwangen. Auftraggeber: Gewässerdirektion Donau/Bodensee.

Reese S., Markau H.-J. and Sterr H. (2003) MERK – Mikroskalige Evaluation der Risiken in überflutungsgefährdeten Küstenniederungen. Abschlussbericht, Kiel.

Sayers P.B., Hall J.W. and Meadowcroft I.C. (2002) Towards risk-based flood hazard management in the UK. Proc. of ICE, *Civil Engineering* 150 (May 2002), 36-42.

Sendi P., Gafni A. and Birch S. (2002) Opportunity costs and uncertainty in the economic evaluation of health care interventions. *Human Health* 11 (1), 23-31.

Slovic P. (1987) Perception of risk. *Science* 236, 280-285.

Smith K. (2001) *Assessing risk and reducing disaster*. London.

Smith K. and Ward R. (1998) *Floods – Physical processes and human impacts*. Chichester.

Tobin G.A. (1999) Sustainability and community resilience: the holy gray of hazards planning? *Environmental Hazards* 1, 13-25.

Veen A. van der, Steenge A.E., Bockarjova M. and Logtmeijer C.J.J. (2003) Structural economic effects of large scale inundation: a simulation of the Krimpen dike breakage. EUR Report 20997 EN, Office for Official Publications of the European Communities, European Commission, Brussels.

Watts M.J. and Bohle H.G. (1993) The space of vulnerability: the causal structure of hunger and famine. *Progress in Human Geography* 17, 43-67.

Weichselgartner J. (2001) Naturgefahren als soziale Konstruktion. Dissertation at University of Bonn, Faculty of Mathematics and Natural Sciences.

[i] The notion of vulnerability is used very differently throughout the literature. Three schools of thought of vulnerability definitions can be differentiated. The first one focuses on exposure to biophysical hazards, including the analysis of distribution of hazardous conditions, human occupancy of hazardous zones, degree of loss due to hazardous events and the analysis of characteristics and impacts of hazardous events (e.g., Heyman et al. 1991, Alexander 1993). The second school of thought looks to the social context of hazards and relates (social) vulnerability to coping responses of communities, including societal resistance and resilience to hazards (e.g., Blaikie et al 1994, Watts and Bohle 1993). The third school combines both approaches and defines vulnerability as a hazard of place which encompasses biophysical risks as well as social response and action. (Cutter 1996, Weichselgartner 2001: 169 ff). The third school is increasingly gaining in significance in the scientific community in recent years. This article also builds upon the arguments of the third school of thought.

[ii] It should be mentioned that there also exists a discourse on natural hazards and social resilience, which is closely related to the social vulnerability debate (e.g., Tobin 1999, Adger 2000, Carpenter et al. 2001, Klein et al. 2003). Social resilience can be defined as the ability of groups or communities to deal with external stress and it can, therefore, be understood as an antonym for social vulnerability (Adger 2000). The term social resilience is closely connected to the term ecological resilience, which will be defined below.

[iii] One German example to illustrate this rule of thumb: In the Rhine River basin two major flood events of comparable size occurred in 1993 and 1995. While people were less aware of the flood risk in 1993, their experience of the 1993 flood increased their awareness and preparedness. As a consequence, the amount of damage was only half in 1995 compared to 1993 (Kron and Thumerer 2002).

[iv] The difference between ex-post and ex-ante estimation of damages is important. Ex post estimations are executed after a flood in order to know the actual amount of damage to society and to compensate flood victims. Usually, these calculations are very detailed and object-specific. On the contrary, in order to assess different flood protection measures and their effects in the future, flood damages must be estimated ex ante. These calculations refer to expert knowledge and empirical data of actual ex post flood damages, but they use standardised functions to estimate future damages on a lower degree of accuracy.

[v] The meso-scale approach was originally developed by Klaus and Schmidtke 1990 (see also Klaus et al. 1994) within their case study for the Wesermarsch district at the German North Sea Coast. Since then, several further studies for other German regions or states were carried out, adopting, varying and improving this approach (OSAM 1995, Hamann and Klug 1998, Colijn et al. 2000, Knogge and Wrobel 2000, MURL 2000, Kiese and Leineweber 2001, Meyer and Mai 2003)

[vi] For the adaptation of this approach to Germany see Beyene (1992), BWK (2001)

[vii] To illustrate this aspect: turbulence and velocity are important variables determining the formation of road, rail track and pylon damage. The pure incidence of inundation does not lead to major damage regarding these elements at risk. Due to a lack of information and a lack of correlation to inundation depth, these damage types are usually not included in flood damage and vulnerability studies. Furthermore, experiences from the Elbe flood 2002 showed that actual damage to buildings and household contents were multiplied if toxic or sedimentation loads were involved (DKKV 2002, p. 50).

[viii] After being completed, the guidelines will be available at the FLOOD*site* web-page: http://www.floodsite.net.

Chapter 14

FLOOD RISK ASSESSMENT IN THE NETHERLANDS WITH FOCUS ON THE EXPECTED DAMAGES AND LOSS OF LIFE

ALEX ROOS, BAS JONKMAN
Ministry of Transport, Public works and Water Management,
The Netherlands

Keywords: Risk of flooding in the Netherlands, probability of flooding, assessment of damage and casualties.

1. INTRODUCTION AND OVERVIEW OF DEVELOPMENTS IN FLOOD PROTECTION

1.1. Introduction

Large parts of the Netherlands lie below sea level and are threatened by river floods. The flood depths in some areas can therefore become higher than 7 meters. Without the protection of dunes, dikes and hydraulic structures more than half of the country would be almost permanently flooded as is shown in Figure 1. Therefore, flood protection has always received much attention. There is always the possibility of flooding. But how serious is this danger? It is difficult to say. Especially shortly after a (near) disaster the situation is perceived as unsafe. Dunes and water defences protect the country, yet never 100%. There is no such thing as absolute safety against flooding. The question is which risks are acceptable and which ones are not. This is an ever-recurring socio-political consideration, which is fed by developments in the state of knowledge.

In the last decade of the 20th century methods have been developed to determine the probability of flooding and its consequences. The outcomes of this research offer new insights and moreover new possibilities to carry out a cost-benefit analysis for various flood protection strategies.

J. Schanze et al. (eds.), Flood Risk Management:
Hazards, Vulnerability and Mitigation Measures, 169–183.
© 2006 *Springer.*

Figure 1. The Netherlands without flood protection (the dark area can be flooded due to influence from the sea)

1.2. History

Due to its location, the Netherlands is always threatened by floods. Life in the delta of the Rhine and Meuse involves risks, but has also enabled the Netherlands to develop into one of the main gates of Europe. In the past river floods provided fertile soil and clay for brickworks, but also negative effects occurred, such as the loss of goods and chattel and the danger of drowning. As welfare increased and population density grew, more and better protection systems were built to prevent flooding. Since the Middle Ages more and more dikes, quays and hydraulic structures have been constructed. Whether the protection against the water is sufficient, is an all-time question; one that is asked at this moment and one that will be asked constantly in the future.

Since the danger of flooding is difficult to determine in advance, politics and society usually adopted a reactive position until recently. An 'almost' flood should not repeat itself. Until 1953 dikes were constructed to withhold the highest known water level. In 1953 a flood from the North sea occurred in the south of the Netherlands, killing over 1800 people and causing the disruption of a large part of the Netherlands. This flood disaster resulted in major investments to improve the water defences, based on a more pro-active base. After the 1953 flood the Delta Committee was installed to investigate the possibilities for a new safety approach. Safety was not based on the highest occurred level anymore, but on a rough cost-benefit analysis. In an econometric analysis the optimal safety level was determined for the largest flood prone area, Central Holland. This work laid the foundations for the new safety approach, in which dikes are dimensioned based on a design water level with a certain probability of occurrence.

The Deltaworks, which were constructed to protect the Southwest part of the country against inundation from sea, were given priority over protection against river floods. After the completion of the Deltaworks, the strengthening of the river dikes began at full speed. As the last big river flood dated back to 1926, there was strong opposition from environmentalists who where against the strengthening programme. The strengthening of river dikes resulted in loss of ecological areas, landscape and sites of cultural value. In 1993 the Government and Parliament agreed upon a new approach, saving the landscape, nature areas and places of cultural value. The river floods in 1993 and 1995 once again drew attention to the risks of life in a delta; afterwards the water defences were reinforced at an accelerated rate. In 2001 most water defences were at strength, and in accordance with the safety standards referred to in the Flood Defence Act (1996). 53 so-called dike ring areas are distinguished, ie. areas protected against floods by a series of water defences (dikes, dunes, hydraulic structures) and / or high grounds. The safety standards for these dike rings are based on the probability of exceedance of a design water level. Design and safety evaluation are based on these design water levels. For the coastal areas design water levels (see above) have been chosen with frequencies between 1/4000 [1/year] and 1/10.000 [1/year]. For the Dutch river area the safety standards were set between 1/1.250 [1/year] and 1/2.000 [1/year]. These safety standards for the various dike rings are shown in Figure 2.

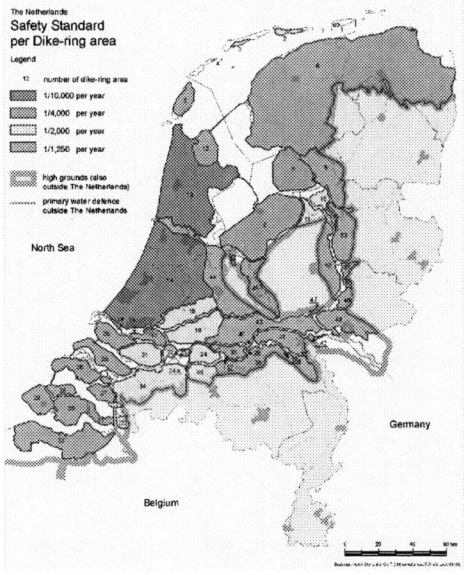

Figure 2. Overview of protection standards for dike rings given in the 'water defence act'

172

2. PRESENT SITUATION

2.1. Safety assessment 2001

As laid down in the Flood Defence Act of 1996, the safety of all primary flood defences must be assessed every five years. The first assessment period was between 1996 and 2001. For that purpose, the guideline for assessment of flood defences was developed by the Dutch Rijkswaterstaat. Also the hydraulic boundary conditions were laid down in the "Hydraulic boundary conditions for flood defences 1996". For every section of a flood defence the normative water level and wave conditions were described. Based on the hydraulic boundary conditions, combined with data (geometry, geotechnical data) of the flood defence, the safety assessment was carried out. An example of the results of an assessment for a dike ring area is shown in Figure 3.

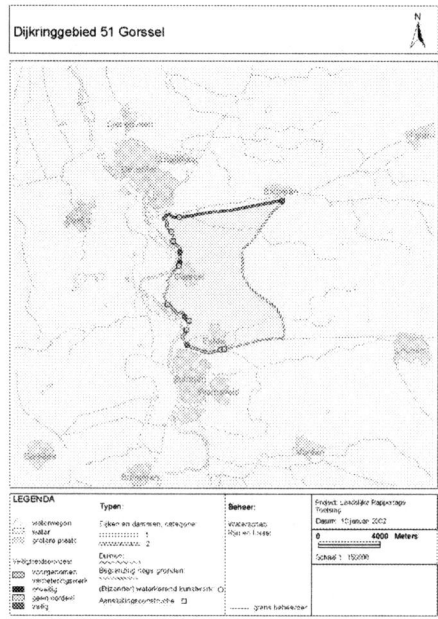

Figure 3. Result of first Safety Assessment 2001, for one dike ring area

In the Netherlands there is a total amount of 3565 kilometre of primary flood defences. 90 % of those defences are managed by water boards, the rest of the water defences are managed by Rijkswaterstaat or communicies. The first safety assessment resulted in a score of "safe" for 42 % of the flood defences. ½ % was unsafe and for 39 % no definitive judgement could be made. For the remaining 19 % improvement measures were already prepared or carried out.

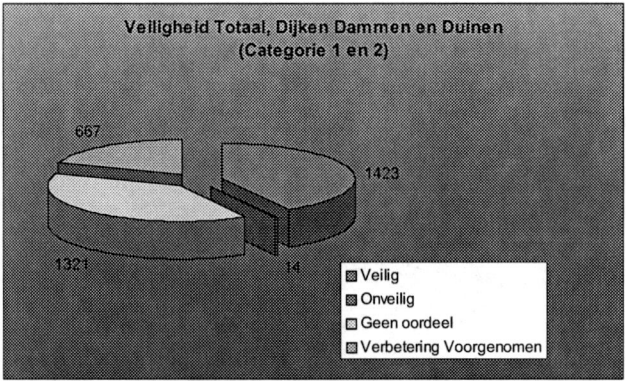

Figure 4. Result of first Safety Assessment 1996-2001

2.2. Developments in coping with floods in the Netherlands

While working in the present situation, the development of future improvements in assessment models or safety philosophy is continuously proceeding. While damage protection in the Netherlands traditionally aimed at reduction through improved dike construction, nowadays new political movement can be seen that searches for measures to prevent flooding *without* raising the dikes along the rivers. For example by giving the river more space. This 'Room for the Rivers' concept is a widespread in the Netherlands now as a possible alternative for dike strengthening. In the coming document of the national policy on spatial planning in the Netherlands it is expected that room for rivers will be described. Another 'hot item' are the so-called 'Emergency Retention Areas': Areas that can be inundated in a controlled way to prevent uncontrolled flooding of other areas. They will be used when a discharge occurs that exceeds the design discharge. These Emergency Retention Areas are, though controversial, a serious item in today's Dutch political discussion. Also more attention can be found for evacuation planning, early warning systems, insurance of flood damage and for the link with spatial planning.

The work of the Delta Committee was conducted in 1958. Based on an economical optimisation taking into account the probability of flooding and the possible damage, an optimal safety level was derived. This exercise can be illustrated by the upper circle in Figure 5. The lower circle describes the five yearly safety assessment of flood defences as laid down in the flood defence act. Because of the economical growth since 1958, the safety level should be re-evaluated.

Together with expected developments of the rise of the sea level, higher river discharges and soil subsidence these factors require a pro-active policy, in which the increase of the interests and investments to be protected will have to be taken into consideration. Knowledge of water and water defences is indispensable when considering the desired protection level against flooding.

174

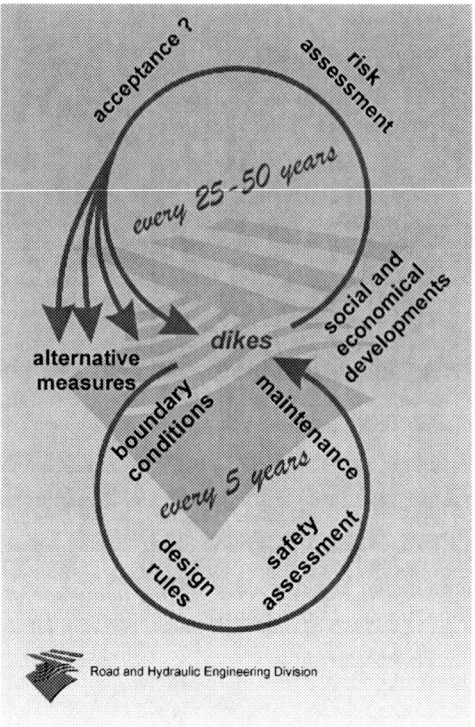

Figure 5. Result of first Safety Assessment 1996-2001

2.3. Evaluation of the current safety philosophy

The Technical Advisory Committee for Water Defences has developed a method to determine the probabilities of flooding and has successfully tested its application for four dike ring areas (TAW 2000). A dike ring area is a flood prone area protected by dikes, dunes, structures and / or higher grounds. Based on this study the Ministry of Public Transport, Public Works and Water Management carries out a project in which the probabilities of flooding are calculated for all 53 dike ring areas.

The transition from an individual dike section to a dike ring approach: the strength of a dike ring (consisting of dikes, engineering structures and dunes) can be calculated as a whole. The method will considerably increase understanding of possible weak links in the protection system. It takes equal account of various failure mechanisms of a dike (ring). This is different from the current approach, in which the safety analysis of the dike is dominated by the mechanism of overtopping and overflow of water. Moreover, it takes all uncertainties into account in a systematic and verifiable way, when calculating the probabilities of flooding. In the current approach uncertainties are for the greater part discounted afterwards by additional safety margins.

Knowing this probability of flooding gives the opportunity to use a risk-assessment (or cost-benefit analysis) to determine, whether the current - or in future expected - flood risks are acceptable. When evaluating the acceptability of the probability of flooding in an area, the potential damage caused by floods and danger for the population are key-information. In the next years the Ministry wants to portray the damage as a result of a flood together with other parties involved. At that moment it will also be possible to calculate the costs and benefits of the entire range of measures. These measures might include research (inspection and testing, study and research), reinforcement and elevation of the water defences, 'room for the rivers', retention areas as well as restriction of flood consequences by means of spatial planning or technical and administrative measures.

3. METHODS FOR ASSESSING THE RISK OF FLOODING

The method for evaluating the risk of flooding consists of four steps. First the risk of flooding of a dike ring area is calculated. The next step is calculating the expected damage given different flooding scenarios. The last step consists of estimating the number of casualties, in which evacuation can be taken into account. With these consecutive steps an analysis of cost and benefits of flood defences can be carried out. Based on the cost-benefit ratio, together with understanding of the number of casualties in different scenarios, a societal and political discussion about the desired safety level can be implemented. In the next sections the consecutive steps of modelling of flooding risks are described.

3.1. Calculation of the probability of flooding of a dike ring area

First, the flood defences around a dike ring area are divided in a number of sections. Each section represents a dune, a structure or a dike. Each section can fail as the result of different failure mechanisms, such as overtopping, piping or geotechnical failure. Structures can also fail because of human error.

In order to calculate the probability of failure of a section of a dike, a computer program called PC-Ring was developed in the Netherlands (Steenbergen and Vrouwenvelder 2003). The program is based on the First Order Reliability Method (FORM, Hasofer and Lind 1974), where the different mechanisms are combined by the Hohenbichler-Rackwitz method (Hohenbichler and Rackwitz 1983).

The method was applied to four dike ring areas in 2000 (TAW 2000). A comparison between the current safety standard and the probability of flooding could me made. Also weak spots in the dike ring area were found, and a cost-benefit analysis was carried out. Also it became clear that for a number of structural flood defences, the amount of available data was insufficient. Also the assessment method for these structures was inadequate.

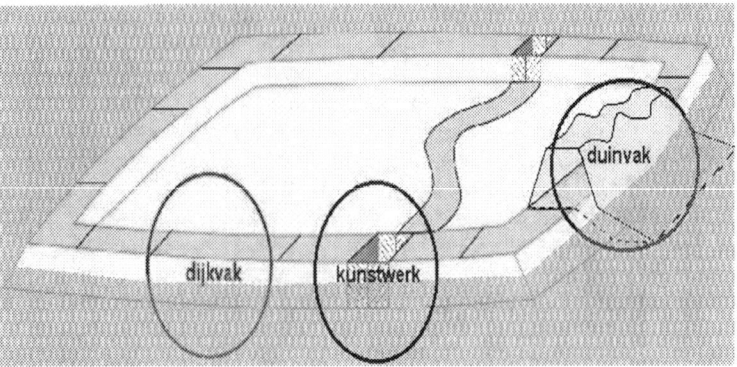

Figure 6. Dividing a dike ring in sections

In the present project "FLORIS, Flood Risks and Safety in the Netherlands" the safety of a larger number of dike ring areas are investigated. Also some improvements in the assessment method for the risk of flooding are carried out. In 2005 the results of this project are presented.

3.2. Calculation of damage

Starting point for the calculation of damage is the choice of a flooding scenario. In this scenario the location of the dike breach and the hydraulic boundary conditions during the breach are chosen. For each scenario the damage is calculated. By combining the different scenarios with the probability of flooding, the expected damage follows by:

$$R = E(S) = \sum_{i=1}^{all\ scenarios} S_i \cdot p_{F_i}$$

in which
R = the risk, the expected damage
S_i = the damage in scenario i
P_{Fi} = the probability of flooding for scenario i

For the following description, it is assumed that the flooding scenario is chosen. With the use of a hydraulic model, the flow and rise rates of the water in the dike ring, given a specific breach and flood scenario, are known, and the resulting damage can be estimated.

In order to do that, relations were proposed between depth, flow rate and rise rate on the one hand and damage on the other, for all valuable artefacts in a dike ring, e.g. different types of residential buildings (farms, small houses, apartment buildings, etc.), public buildings (schools, stadiums, gyms, municipal buildings, churches, etc.), factories and industrial buildings, cars, agricultural equipment, crops and live stock, and so on (Vrisou van Eck et. al. 1999). The relations were based on expert opinion and on damage claims in 1993 and 1995, when both Maas and Rhine rivers had relatively extreme discharges (100 year return discharge). No dikes failed, but the upstream part of the river Maas has no dikes and some damage was suffered there.

The expected damage for a particular artefact is calculated with:

$$S = f(d) * S_{max}$$

in which
S = damage for a certain category (agriculture, housing, etc)
f(d) = damage factor
S_{max} = maximum amount of damage for a certain category

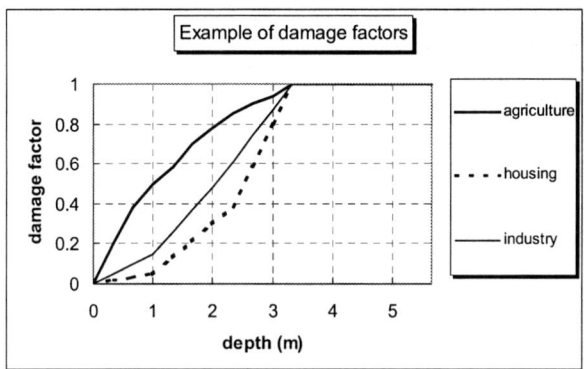

Figure 7. An example of damage functions

The damage factors represent the percentage of the value of the aspect. Until now the damage factors are solely dependend of the maximum water level. Damage types are expressed as monetary (man made capital, interruptions of production) and non-monetary damage (human lives). Until now most attention has been paid to damage-assessment for floodplains, and the nature of the threat being river-flooding.

These relations were combined in yet another computer model (Vrisou van Eck et. al. 2001). Besides these water-damage relations, the assets in the polder have to be input to the model. Different national databases, maintained by different Ministries, where linked to the model. The output consists of the (average) damage for each grid cell and can be presented by GIS (Graphical Information Systems).

The method for assessing the damage on flooding was applied to all the dike ring areas in the Netherlands (DWW 2003). For that cause, a simplified (conservative) flooding scenario was chosen. The calculated amount of damage varies from 300-400 billion euros for the central part of the Netherlands to several millions of euros for other dike ring areas (for example 200 million euros for Terschelling, an island of the Frisian coast).

178

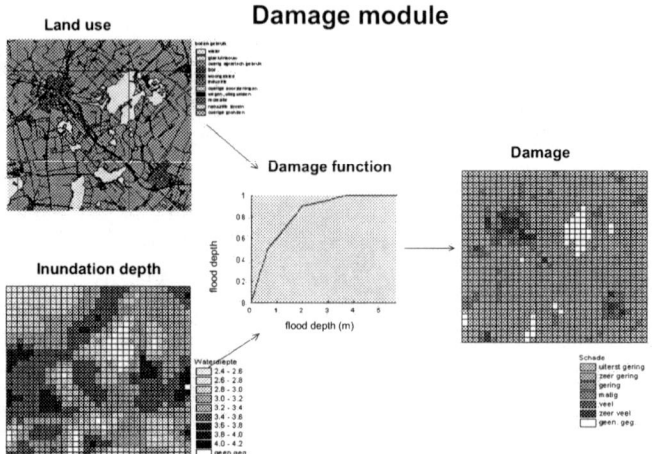

Figure 8. Schematic overview of damage module

3.3. Calculation of casualties

Development of a flood mortality function is based on the 1953 flood disaster in the Netherlands. During this flood a storm surge from the North Sea inundated large parts of the southwest of the Netherlands. This event caused enormous damage and shock. Moreover, 1836 persons were killed due to this disaster. Based on the available information on fatalities it was investigated whether a relation between hydraulic flood characteristics and flood mortality could be found. In this case the effects of evacuation are neglected: the disaster struck unexpectedly at night, no evacuation could be carried out. The loss of life can thus be directly related to the presence of persons at different locations and the flooding circumstances.

Based on the collected data from the 1953 flood the following approach is proposed to relate flood mortality (i.e. the fraction of inhabitants in an area that lose their life in the flood) to hydraulic characteristics of the flood. In recent studies three categories of flood deaths are distinguished:

- Drowning persons due to rapidly rising water
- Drowning persons due to high flow velocities
- Deaths due to other causes, such as hypothermia, heart attacks, shock, failed rescue, etc.

In this paper we will only address the first category.

When the water rises rapidly, dangerous situations may occur. People will not be able to reach high grounds or even to reach the higher floors of buildings. It is expected that especially the combination of water depth and rate of rising causes the danger, since dangerous situations will especially occur in larger water depths. The fatalities caused by rapid increase in water depth during the 1953 flood are shown in Figure 9. From this Figure it can be seen that mortality increases with water level. The following function is derived:

$$f(h)_{rise} = 9.18 \cdot 10^{-4} \cdot e^{1.52 \cdot h} \qquad f(h)_{rise} \leq 1$$

with f(h)rise is the fraction of inhabitants killed by rapidly rising water levels and h is the water depth.

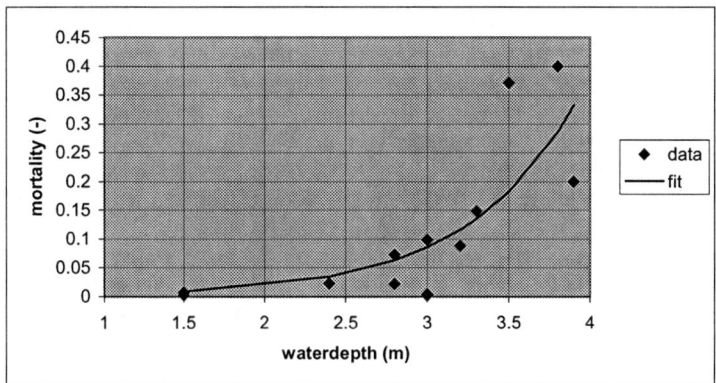

Figure 9. Function for estimation of mortality for rapidly rising water

Based on the known values of the rate of water level rise it is assumed that this function should be applied if the water rises with 1 m/hr or more. A specific problem is the extrapolation of this function for larger water depths. No data is available for water depths of more than 3.9 m. As the proposed function is very steep at the upper end, extrapolation may result in unrealistic numbers of deaths. For instance, at about a water depth of about 5.3 m, the maximum fraction of 100 % deaths is reached. In further research the elapse of this function for larger water depth is to be assessed. Now it is assumed that the 100 % value applies to all water depths of 5.3 m or more.

The method is applied a several times in the recent years. A case study carried out by Delft Hydraulics calculated an estimate of 80,000 casualties for a flooding near Rotterdam. Other studies found ranges varying from 100,000-200,000 for the central part of Holland (a large, low lying area which is threatened by storm surges) to 5,000-10,000 casualties for the Betuwe, a polder along the Rhine.

3.4. Estimation of the time needed for evacuation

Application of the above described relations between flooding characteristics and number of fatalities is likely to result in an overestimation of the number of fatalities as possibilities for evacuation or escape are not accounted for. For instance, when flow velocities and water depths are large, a large number of inhabitants is expected to die. However, in reality the number of fatalities will differ depending on whether these large depths and strong flows occur immediately after failure of the dike, or several hours or days later. To estimate the percentage of the population that is able to evacuate or escape before the flood water reaches their houses, information is needed about the time required time for an evacuation or unorganised escape.

Evacuation has been studied in great detail in the Poldevac project. In this project, a detailed transport model was developed and applied to simulate evacuation. Delays and risk for traffic jams due to the presence of cross roads and traffic lights were accounted for. Ideally such a transport model should be available for the entire country. Only with this type of model it is possible to produce accurate estimates of requested evacuation times and of the time needed for people living in different locations to reach a safe place. Unfortunately this model is not available for the case study area. Also, this type of model requires an extensive amount of input data and is therefore not suitable for a more generic use for other flood prone areas.

Recently a more simple conceptual method was developed to simulate an evacuation of a flood prone area in the Netherlands. This model mainly considers preventive evacuation before the beginning of the flood. A preventive evacuation consists of three stages: the decision making, initiation of the evacuation, and the evacuation itself. The time needed for each phase depends on the availability of an evacuation plan (how well are inhabitants and local authorities prepared), the number of people to be evacuated and the available infrastructure. An example of an evacuation function is shown in Figure 10.

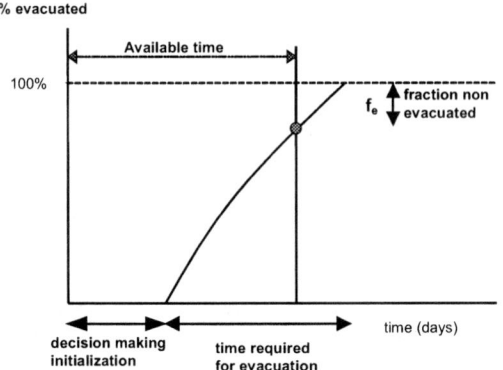

Figure 10. Evacuation curve

The available time for evacuation depends on the predictability of the water levels at sea or in the river and the failure mechanism. While extreme river discharges in the Netherlands can be predicted up to several days ahead, extreme sea water levels have a much shorter prediction time (6 – 10 hours). Failure of a dike is relatively easy to predict in case of overtopping, but is much more difficult to foresee in case of the failure mechanism piping. In case of evacuation or escape after failure of the dike, the available time only depends on the travel time of the flood wave to a certain location within the flooded area. The time needed for decision making and initialization in that case equals the time needed to warn people and for the people to prepare themselves for departure. The time required for evacuation depends on the capacity of the infrastructure. In the case of evacuation after failure of the river dike this will mainly be the capacity of the roads as the railway system is expected to be dysfunctional. Depending on the available time and the requested time, a certain percentage of the inhabitants will be able to escape in time. The percentage that is not able to escape is indicated by 'fe' in Figure 10. These persons will be exposed to the flooding conditions and run a risk of drowning. Since high rise buildings provide shelter places during a flood, it is assumed that persons present in high rise buildings are safe and can be considered as 'evacuated'.

To estimate the number of fatalities caused by a flood event, the relationships between flood characteristics and fatalities and the knowledge on evacuation times need to be integrated into a so-called loss of life model. As most information with respect to the flooding characteristics, locations of houses and number of inhabitants varies spatially, it is preferred to develop such a model in a Geographic Information System (GIS).

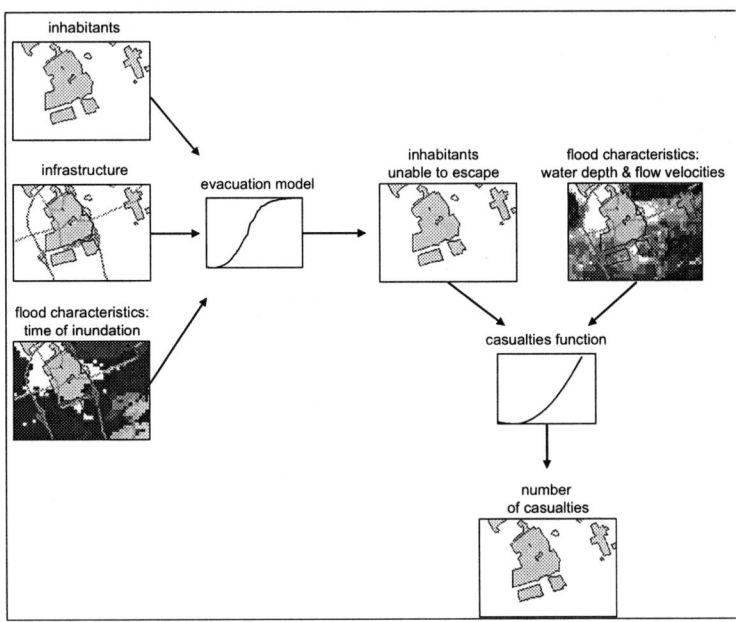

Figure 11. Schematic overview of evacuation module

Figure 11 shows the overall framework of the loss of life model. Information on the number of inhabitants, the capacity of the available infrastructure and the time of inundation are combined in the evacuation model to estimate the number of people that are unable to escape. This, together with information about the water depth and flow velocities is used as input for the flood mortality functions. The outcome consists of a map showing the number of fatalities at different locations within the flooded area. GIS maps indicating the number of inhabitants were derived from the damage and loss of life model developed by Rijkswaterstaat, DWW (Vrisou van Eck et al. 2000). Characteristics of the flood, such as time of inundation, water depth and flow velocities are based on hydraulic computations with the SOBEK model. The evacuation model was developed in a GIS-package.

4.   FUTURE CHALLENGES

The understanding of the urgency of a re-evaluation of the safety standards in the Netherlands is growing. First results comparing the risk of flooding with other risk sources (airports etc.) are available, and it becomes clear the risk of flooding is much larger then other risk sources.

In the near future the societal discussion about flooding risks will be take place. This discussion will demand the availability of robust and complete methods. Also other discussions, like for instance about the level of acceptable risk take place. The process of the societal discussion is not clear yet.

After the new safety standards are set, a complete new approach for assessing safety must be introduced. Questions how to maintain the safety must be solved before.

In the mean time, research will be conducted in some fields, for example:

• Research on strength of flood defences, wave interaction in estuaries
• Improvement of crisis management (warning systems, preparation on flooding, evacuation) and effect on flooding risk
• International cooperation (mainly rivers) with the focus on reducing the normative discharges

Further development takes place on the

• method for probability of failure,
• casualty function and
• evacuation model.

References

DWW (2003) Schade na grootschalige overstromingen (in Dutch).

Hasofer A.M. and Lind N. (1974) An exact and invariant first order reliability format. Proceedings of the ASCE, Journal of Engineering Mechanics Division.

Hohenbichler M. and Rackwitz R. (1983) *First order concepts in system reliability: Structural Safety 1*. Elseviers Science Publishers B.V., Amsterdam.

Steenbergen H.M.G.M. and Vrouwenvelder A.C.W.M. (2003) Theory manual PC-Ring, version 3.0, Part A: description of mechanisms, Part B: Statistical models, Part C: Numerical Techniques (all in Dutch). TNO-Bouw reports 2003-CI-R0020, 21 and 22.

Technical Advisory Committee on Water Defences (TAW) (2004) Towards a new safety approach, A calculation method for probabilities of flooding.

Vrisou van Eck N., Kok M. and Vrouwenvelder A.C.W.M. (1999) Standard assessment of damage and victims caused by floods, Part 1: Standard method, Part 2: Background (in Dutch). HKV Lijn in water and TNO-Bouw.

Vrisou van Eck N., Huizinga H.J. and Dijkman M. (2001). HIS Damage and victims module – user manual (in Dutch). PR236.40 HKV Lijn in water and Geodan Geodesie.

PART 5

MITIGATION MEASURES

Chapter 15

FLOOD PROTECTION IN THE TISZA RIVER BASIN

ZOLTAN BALINT[1] AND SÁNDOR TÓTH[2]
[1]Upper-Tisza Environmental and Water Directorate Nyiregyhaza,
Hungary,
[2]National Environmental and Water Directorate, Budapest,
Hungary

Keywords: Transboundary river basin, hydrological measurements, vulnerability, structural
and non-structural measures, flood warning, flood management.

1.   INTRODUCTION

The Tisza River Basin is shared by five nations: Ukraine, Romania, Slovakia, Hungary and Serbia-Montenegro. The river itself is the frontier along several kilometers between Ukraine and Romania, Ukraine and Hungary and between Slovakia and Hungary. All blessings and all disasters a river can bring are also shared by the five nations. For people living close to the river, it is their source of survival, their friend, their partner, and their enemy. In local languages she is a female. They love her, they fight with her, but they cannot live without her.

Flood hazard in Hungary and particularly in the Tisza Valley is high. 22% of the total area of the country is flood plain (Szlavik 2003). This is the highest value in Europe, the only situation comparable with it is that of the Netherlands. And three quarters of that floodplain is located in the Tisza Valley. Therefore flood protection in the Tisza River Basin is a very important national and international task for all the riparian nations.

2.   NATURAL CHARACTERISTICS OF THE RIVER BASIN

River Tisza is one of the largest tributaries of the Danube with a total catchment area of 157,000 km^3. From North, East and South-East it is embraced by the Carpathian mountains, from the West it is bordered by the watershed of the Danube (Figure 1).

J. Schanze et al. (eds.), Flood Risk Management:
Hazards, Vulnerability and Mitigation Measures, 185–197.
© 2006 Springer.

186

The origin of the name Tisza/Tisa is not clear. Some linguists think that it is of an Indo-European origin, and means muddy, silty. Romans called it Parsius, Pthirus, Pathissus, Tigas, Tisianus, and in the middle ages it was called Tisia or Tysia (Balint, Bodnar and Konecsny 2000).

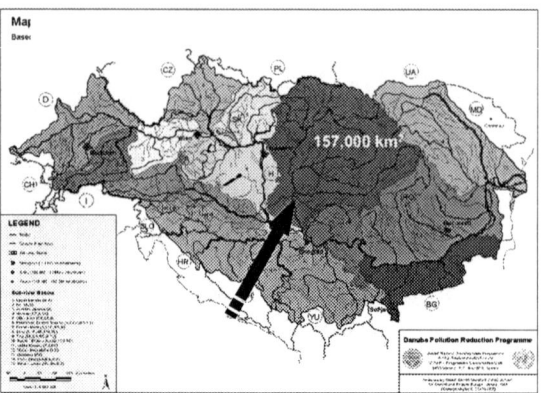

Figure 1. Tisza River basin within the Danube river basin

The river can be divided into three sections (Vagas 1982). The Upper-Tisza is the section between the sources and the confluence with River Szamos, the Middle-Tisza is between the Szamos and Maros rivers, and the Lower-Tisza is from the Maros confluence to the estuary into the Danube (Figure 2).

The source of the river is in Ukarine, in Transcarpathia. The mountainous part of the basin can be characterized by several equal size tributaries, each contributing to the creation of a flood wave. The highest point is the Pietros peak with 2305m, located in the Radnai Mountains (Balint 2001).

The variation of the mean annual temperature is fairly high, its range extends from 2 to 11°C. In the smaller valleys in the mountains inverse temperature gradient is frequently observed. Precipitation also varies significantly in time and in space. The physical and statistical rules of the formation and progress of floods in the Carpathian Basin are rather complicated. Most of the flood waves, about 94%, are caused by macro- or mezoscale cyclonal systems (perhaps coinciding with snow melt), while only 6% is caused by rapid snow melt, triggered by zonal, warm advection. The Carpathian Mountains present remarkable orographic effects. About 80% of the air masses carrying flood-forming precipitation arrive from a South-West direction, form the Atlantic Ocean and the Mediterranean Sea. The highest annual rainfall, 1720 mm was measured in the valley of the Sopurka Creek in the Sidovec Mountain. Generally the annual precipitation is higher than 1200 mm in the higher mountains, 800-1000 mm in lower mountains and less then 600 mm on plain lands.

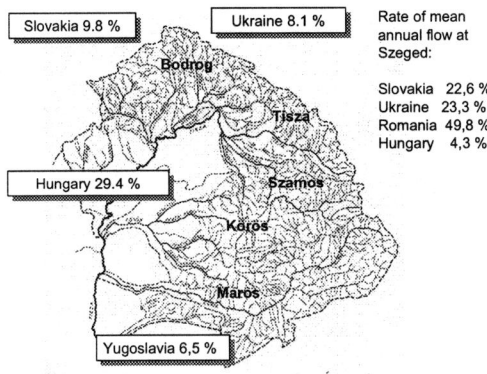

Figure 2. Tisza River basin with the system of water courses

Variation of the precipitation within the year determines formation of floods. Experience shows that in the Tisza River Basin floods can occur principally any month of the year. However, the bigger floods happen most frequently in spring time and the equally dangerous icy floods in winter time. Water level can rise as much as 7 m in 24 hours at the UKR-HU border.

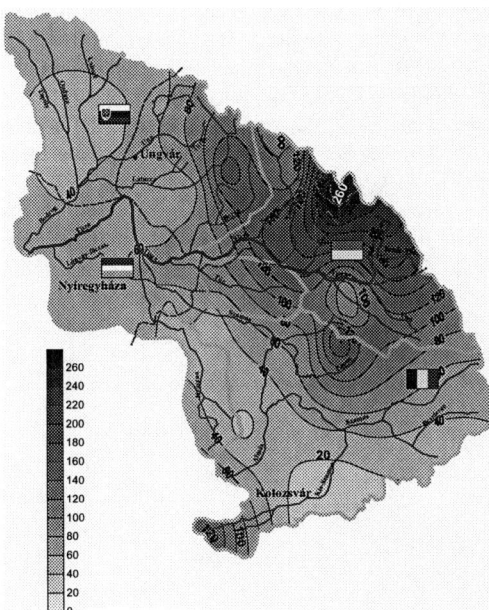

Figure 3. The areal distribution of precipitation causing the extreme flood in March 2001

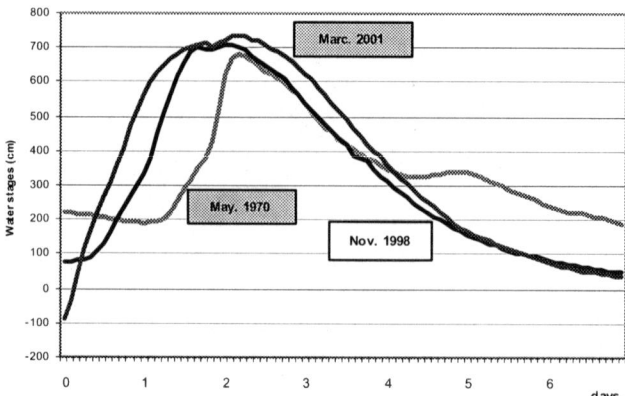

Figure 4. Hydrographs of the highest floods in the past years on the Ukrainian-Hungarian border section

On the Upper-Tisza the cause of the most dangerous floods has been, almost without exception, intensive rainfall with the combination of snowmelt or at least frozen soil conditions.

In March 2001 the progress of a South-Westerly cyclone was stopped by the mountains at the source of River Tisza (a typical flood-forming situation), resulting in the release of heavy rainfall. As seen in Figure 3, at the worst places up to 300 mm rain fell in 3 days time, which equals to the rainfall sum of 2-3 spring months in the region. As the rain fell on snow with an average depth and frozen soil, accumulation was very fast. Flood rise was very intensive, with a maximum intensity of 48 cm /hour at the Tivadar section of the river, close to the Ukrainian-Hungarian border. At the same station more then 11 m rise in water stage was measured in less then 48 hours (Figure 4).

Formation of the floods and the danger they carry are different in the different sections of the river. We have already seen the characteristics of the Upper-Tisza. The middle and lower sections face different problems. All the flood waves created in the mountains run sometimes simultaneously down into the main river, where coinciding with each other they form extremely long flood waves. The long, constant load on the dikes saturates them, causing underground seepage, boils and leakage (Figure 5).

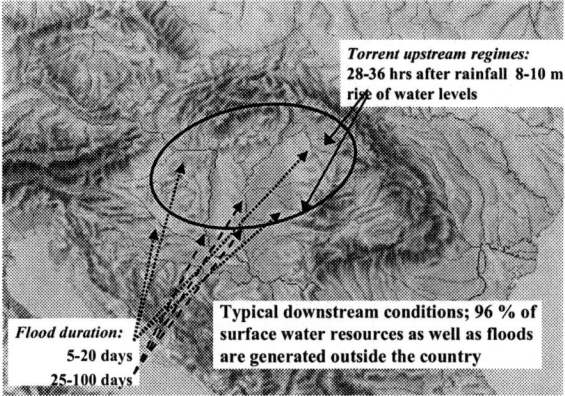

Figure 5. Flood characteristics

Flood peaks in the last century have been on the rise. (Figure 6). Further research is needed to make clear if this goes together with the rise of the flood discharges or water level keeps on rising even with unchanged flow. Research results carried out on the Upper-Tisza (ref. 2) show that a 20% rise in flood discharges can be detected in the last 50 years with unchanged annual and monthly rainfall patterns. Downstream flood managers emphasise the phenomenon that flood levels have risen even with unchanged flood discharges.

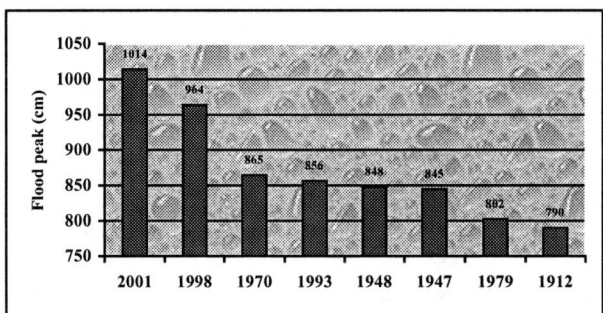

Figure 6. Rise of flood peaks at Tisza-Tivadar

3.   HISTORY OF FLOOD CONTROL DEVELOPMENT

One and a half centuries ago the Tisza Valley was a large, very flat, swampy territory. Much of its area was inundated either temporarily or permanently, as shown in Figure 7 (Vagas 1982).

190

Although lifestyle in that era is sometimes painted romantic from our pleasant air-conditioned offices, in practice the inundations hindered any development, a fundamental necessity for the survival of the communities.

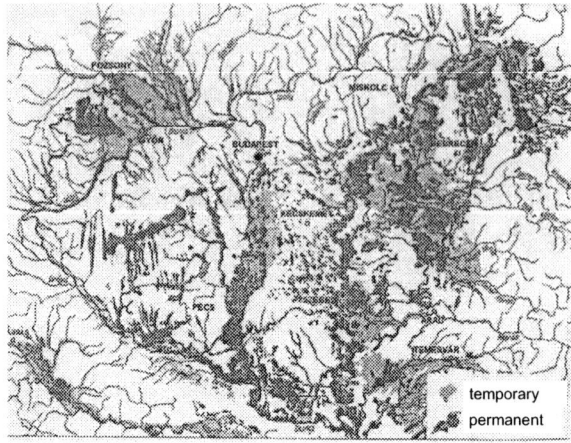

Figure 7. Inundations in the Tisza valley before the river regulation works

The flow in River Tisza needed some kind of control, some regulatory interactions. Two famous names highlight the start of the development. One is count Istvan Szechenyi, a wise politician-economist-writer, the other is Pal Vasarhelyi, an internationally recognized civil engineer.

Although the practical work concentrated on engineering regulatory works, we want to emphasise here that the basic aim was not that of implementing an engineering idea but the development driven by the need of the communities. The huge nation-wide programme was launched in 1846, terminated 97 years later, in 1937, by building the embankment along the Borsod open floodplain. A brief summary of the works is given in Figure 8 as below.

Ing. Pál Vásárhelyi
1796 - 1846

- Vásárhelyi's basic regulation aim was **to serve the interests** of **navigation** and **flood control** simultaneously.
- The substance of the concept can be summarized as follows:

  ✓cutting 102 overdeveloped bends across, thus shortening the river by 452 km (over 37 %) between Tiszaujlak *(Vilok)* and the mouth;
  ✓erecting embankments (flood levees) on both sides of the regulated river, spaced at 550-1900 m distance, depending on local conditions.

Figure 8. Brief history of river training and flood alleviation in the Tisza River Basin

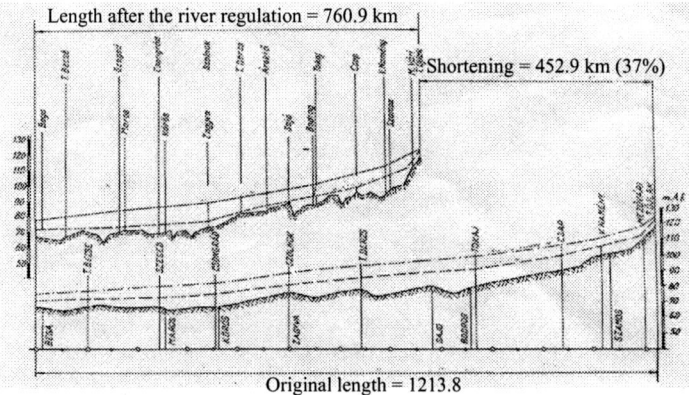

Figure 9. Impact of the river regulation on the river length and stream profile

Figure 9 demonstrates the effect of the enormous program implemented most of the time by manual labour (Vagas 1982). Building dikes by wheelbarrows was a challenging job. And, naturally, it could not be built very high and very strong. Therefore dike construction in the Tisza River Basin can be considered as a permanent part of flood control. It started with a small cross-section and new and new layers were put on top of the other. Hence the onion-type structure shown in Figure 10 (Vagas 1982).

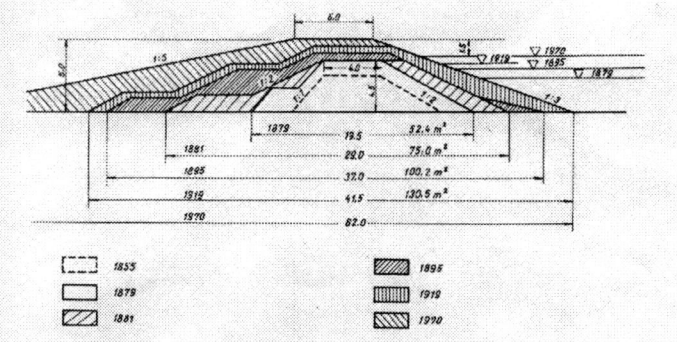

Figure 10. Typical cross-sections of the dikes in 1855, 1879, 1881, 1895, 1919 and after 1970

4. FROM RIVER REGULATION TO INTEGRATED FLOOD MANAGEMENT

In spite of the great effort and financial resources put into the development of flood defence structures, recent analysis shows that
- Only 60 % of the defence structure has been developed to meet design flood criteria (to withstand 1% probability flood)
- Flood conveyance capacity of the flood bed has reduced

192

- Four major floods occurred between 1998 and 2001.

Analysis of the history, recent events and trends lead us to some conclusions that are becoming very well accepted principles in flood related development nowadays.

- Flood peaks and flood volumes are constantly growing as a consequence of natural and man-made effects.
- Flood dikes are already very high, carrying risks of higher damage in case of structural failure.
- Part of the water that runs off the surface of the catchment at an ever increasing rate is actually missing from agriculture and from the ecosystem.
- Flood management is not a technical problem any more. It has to cover many aspects of legal, institutional, communication, emergency management, environmental, monitoring, land development etc. issues as well.

To satisfy the requirements of our modern society we have to use an integrated approach, which is even more demanded in case of a transboundary river like the Tisza. Several national and international projects already use this approach. Figure 11 demonstrates many components of the integrated flood management as worked out in the Joint Ukraine-NATO Project on Flood Preparedness and Response in the Carpathian Region in harmony with the recommendations of the International Strategy for Disaster Reduction, and that of the EU Best Practices on Flood Prevention, Protection and Mitigation.

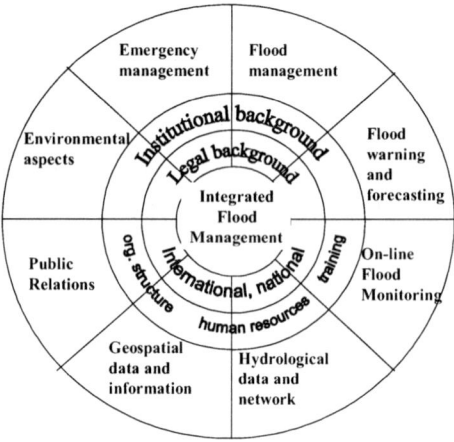

Figure 11. Integrated flood management

I would like to draw the attention to the results of this successful, joint Ukraine-NATO project (Balint 2002), because in many ways it was ahead of many other projects, even of those that were built on it.

- It established excellent co-operation between experts from ten NATO and non-NATO countries,
- The only unit area considered was that of the undivided river basin
- All components in Figure 11 were analysed

- It made truly trans-boundary, interdisciplinary, integrated recommendations. One of these was the establishment of four regional flood centres, one in each country, physically interconnected and sharing the international hydrologic data management system (Figure 12) besides having national data bases too.

Integrated flood management has to build on structural and non-structural measures. One successful example of trans-boundary non-structural development is the joint Hungarian-Ukrainian on-line remote sensing system established from a Hungarian government grant in two phases between 1998 and 2003 (Figure 13). The river and meteorological stations established in the project communicate via UHF radios with two data collection centres, one in Nyiregyhaza, Hungary and one in Uzhgorod, Ukraine. The two computer centres are interconnected by a microwave data and voice transmission system in such a way that the two sub-systems form a unit which seems for the user just one system. A precondition of that is, that two centres use exactly the same software.

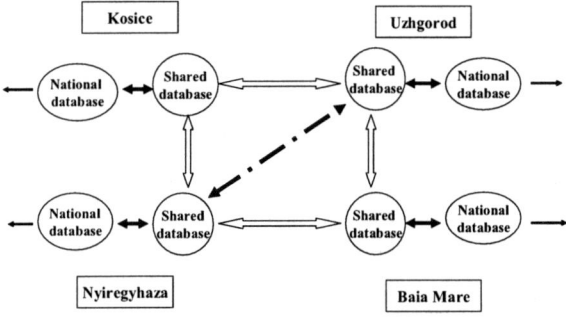

Figure 12. Proposed regional flood centres

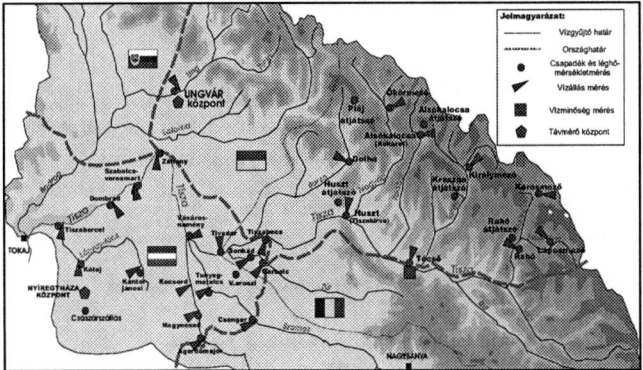

Figure 13. The stations of the joint Hungarian-Ukrainian on-line monitoring system

The most complex development program ever started in the Tisza River Basin is the so called "Vasarhelyi Plan". The plan is considered the

194

continuation and functional extension of the original program of controlling dangerous floods of the river (Figure 14).

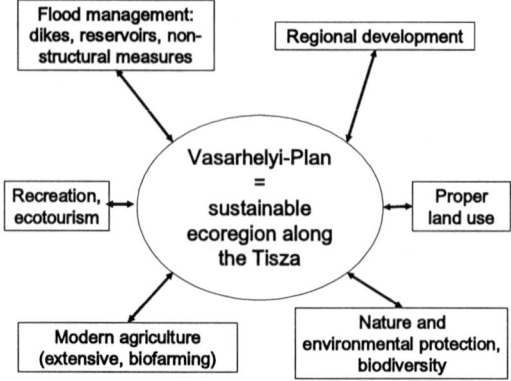

Figure 14. The Vasarhelyi Plan

The main components of the multi-criteria selection system, the main technical features and expected results are summarised in Figure 15.

Result of multicriteria selections

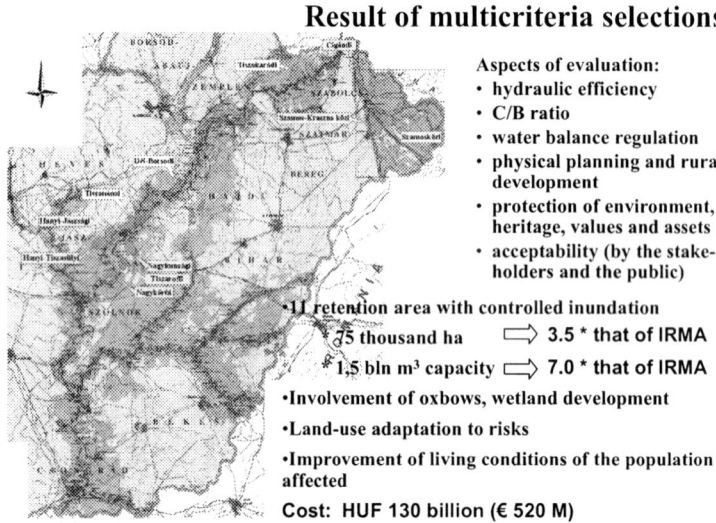

Aspects of evaluation:
- hydraulic efficiency
- C/B ratio
- water balance regulation
- physical planning and rural development
- protection of environment, heritage, values and assets
- acceptability (by the stake-holders and the public)

•11 retention area with controlled inundation
75 thousand ha ⇒ 3.5 * that of IRMA
1,5 bln m³ capacity ⇒ 7.0 * that of IRMA
•Involvement of oxbows, wetland development
•Land-use adaptation to risks
•Improvement of living conditions of the population affected
Cost: HUF 130 billion (€ 520 M)

Result: - Flood crest reduction by ~ 1,0 m; - protection against 1 in 1000 yrs floods

Figure 15. Technical features and expected results of the Vasarhelyi Plan

5. RISKS AND UNCERTAINTIES AT ALL LEVELS

Having said all the above, the questions arise: Have we found the ultimate, good solution in the integrated approach? Are there no more problems, uncertainties and risks? Can we just sit back and watch the final solutions being implemented? Our answer is: No way. We are just at the beginning of understanding risks and uncertainties. The whole complex natural-social-economic phenomenon is full of uncertainties and risks. Even to group them is not so easy.

For further thinking the terminology used by the ISDR (International Strategy for Disaster Reduction, 10) is adopted.

> **Hazard:** A potentially damaging physical event or phenomenon that can harm people and their welfare.
>
> **Risk:** The probability of harmful consequences or the expected loss of lives, …, resulting from interactions between hazards and vulnerable conditions.
>
> **Vulnerability**: The degree to which a socio-economic system is susceptible to the impact of hazards.

In the case of the Tisza River the hazard is the flood that we hardly can influence. Contrary to this, vulnerability depends a lot on the actions and the way of life of people and communities. The interaction between the flood and the society goes through several steps and several factors that influence events and progress of the phenomenon. We call them risk factors and uncertainties. These factors represent the stochastic nature of most of the phenomena we deal with, the uncertainties of data, parameters and information, the limitations in our knowledge and understanding, the spatial and time variability of data and phenomena that cannot be taken into account. A first collection of various factors is presented in Figure 16. Grouping could be done, of course, in many other ways, however, we think that the presented one is suitable for starting a collective thinking process. Within the limitation of this paper it is only possible to pick out a few of the factors as examples.

a) A lot of scientific work has been put into determining the most accurate design flood level, using various probability methods. However, we have to revise the levels continuously, because the input time series used as a basis for the calculations keep on changing. This fact brings significant uncertainties into the decisions and contributes severely to the risk itself.

b) Many hydrological calculations are based on mathematical statistical principles that require that data series should be homogeneous and independent. Nature does not know and does not respect this principle, and at its own will produces time series with periodicity and interdependent data. Again, the input into our system has significant approximations and uncertainties.

c) Human factors in our analysis are not more reliable than nature. We have to admit, that "scientific" statements are sometimes unfounded and distorted by fashionable common ways of thinking, sometimes even manipulated by the media. This is not a new phenomenon, we can trace it back to many centuries.

d) Human behavior is sometimes not predictable, many times because of lack of information or misinformation. This is particularly true in disaster situations. Therefore the magnitude of the risk depends a lot in many cases on socio-economic factors.

e) It is generally appreciated that catchment parameters are uncertain, because they cannot be measured with the required frequency and spatial distribution. However, we tend to hold basic hydrological measurements (rainfall, discharge, temperature) reliable. Since we have improved our water stage sensors, we had to realize that this most reliable parameter also has significant uncertainties. Sometimes the uncertainty in the input water level is bigger than the required accuracy of the forecast.

Table 1. Factors and features for flood risk management

	Risk factors and uncertainties	Vulnerability	Mitigation measures
Pre-flood preparedness	Design flood level Data series Applicability of methods Meteorological conditions Biased scientific statements Dike crest under design flood level Dike structure Interconnected systems Lack of DEM	Settlements under design flood level Unauthorized constructions Not proper building material Not adequate foundation Not proper furniture	Further research Dike construction Floodway clearance Flood reservoirs Confinement plans Surveys, GIS
Operational flood management	Rainfall forecast Rainfall, water stage, discharge measurement Lack of data Time delays Catchment and river bed parameters Lack of flood warnings and forecasts	Villages located high in the mountains Bad forecast communication Little time left because of rapid progress Reduced retention Lack of awareness	Better communication systems On-line remote sensing, radar, satellite mages Flood forecasting systems Confinement plans Emergency management
Socio-economic factors	Community response Responsible authorities Political credit (in rehabilitation) Condition of confinement structures Social comprehension Divided river basins National interests Legal-institutional structures	Low income families Little investment No other place to live Lack of insurance Passing on flood problems to downstream areas	Integrated actions Enhanced flood awareness, Complex systems Joint international monitoring and forecasting Develop international legal and institutional structures

We could continue analyzing the factors one by one, but we have to leave this exercise to the reader. Generally we placed the factors into three groups according to the functions where they play the most important roles. (Overlapping is sometimes unavoidable). Group 1 comprises factors to be dealt with mostly in pre-flood preparedness tasks, group 2 lists factors we have to face during operational flood control, and the third group contains the socio-economic factors.

6. SUMMARY

Instead of trying to summarize the very complex and complicated tasks of flood management in the Tisza Valley, let us quote the question of one of our journalist friends, directed to us at a ceremonial opening of a new section of our dike system:

Is this dike able to withstand the 100-year flood, that comes to us every second year?

His question reflects many of our problems and tasks in connection with flood risk management in the Tisza River Basin.

References

Balint Z. (1980) Correction of non-homogeneous time series in preparation of hydrological forecasting. IAHS-AISH Publ. 129, Oxford Symposium.

Balint Z. (ed.) (2001) Flood forecasting on the Upper-Tisza (in Hungarian). FETIVIZIG, Nyiregyhaza, Study Report.

Balint Z. (2002) Flood monitoring and forecasting. Final Report on the Joint Ukrainian-NATO Project on Flood Preparedness and Response in the Carpathian Region, Brussels.

Balint Z., Bodnar G. and Konecsny K. (2000) Effects of the forest cover on formation of the floods (in Hungarian). FETIVIZIG, Nyiregyhaza, Study Report (in Hungarian).

Koncsos L. (2001) Flood forecasting by hydrodynamic models (in Hungarian). Flood forecasting on the Upper-Tisza, FETIVIZIG, Nyiregyhaza.

Somlyódi L. (2000) Strategic pillars in the Hungarian water management (in Hungarian). Somlyódi L. Strategic questions in the Hungarian water management. MTA Publications, Budapest.

Szlavik L. (2000) Flood management (in Hungarian). Somlyódi L. Strategic questions in the Hungarian water management. MTA Publications, Budapest.

Szlavik L. (2003) The 1998 Flood (in Hungarian). Vizugyi Kozlemenyek, Budapest.

The United Nations World Water Development Report, Part IV. (2003).

Vagas I. (1982) Floods of River Tisza (in Hungarian). VIZDOK, Budapest.

Chapter 16

MANAGEMENT STRATEGIES FOR FLASH FLOODS IN SAXONY, GERMANY

ELIMAR PRECHT[1], HANS CHRISTIAN AMMENTORP[2], OLE LARSEN[1]
[1]DHI Wasser & Umwelt, Syke, Germany
[2]DHI Vand & Miljø, Hørsholm, Denmark

Keywords: Flash floods, MIKE 11, mitigation options.

1. INTRODUCTION

A devastating flood hit the German State of Saxony in August 2002. Heavy rainfall, exceeding 300 mm in 24 hours at some locations, led to severe flooding along the rivers and streams and caused considerable damage, particularly in the urban areas.

Figure 1. The town of Pockau was badly hit by the flood, along both the Flöha and (here) the Schwarze Pockau River. Pictures by Hr. Sacher

Several projects were launched in the wake of the flood in order to limit the damage of future flooding. DHI Water & Environment was awarded a project to propose and analyse potential flood mitigation measures along the upper Flöha River and tributaries. The Flöha River

J. Schanze et al. (eds.), Flood Risk Management:
Hazards, Vulnerability and Mitigation Measures, 199–205.
© 2006 Springer.

itself is a tributary of the Mulde River, a tributary of the Elbe. The upper catchment area of the Flöha lies in the mountainous area of the Erzgebirge. The analysis and conclusions of this project are briefly described in the following.

2. MITIGATING FLASH FLOODS

Local flood protection in the form of walls or dikes is often a cost-effective way of reducing the flood damage in urban areas. It has, however, some disadvantages. High walls may block the view of the river valley from near-by houses and thereby reduce their value. Flooding downstream may be increased, as the natural flood retention in the protected area is eliminated. The local protection may further create a false sense of security in the area as all the minor floods, which used to occur, are avoided. The unprepared community may therefore be hit particularly hard by the first flood which exceeds the design criteria of the protection walls, especially if the development of land use has taken place without consideration of the remaining flood risk. Generally, the need for constructing local flood protection should be limited, if possible, by seeking additional measures of flood mitigation.

The floods in the upper Flöha River catchment are flashy in character: The water rises within a few hours from a heavy rainfall event, but returns quickly to normal levels after the rain has stopped. Little time is available for taking any preventive action and warning the population.

Flood forecasting systems suitable for areas of this type do exist today, however. They operate by automatically collecting real-time data and short-term weather forecasts, performing fast model simulations of the runoff and river flow, and issuing warnings automatically by e.g. telephone or sirens. Even a few hours warning has been shown to considerably reduce the losses of lives and property. An automatic flood forecasting and warning system has therefore been recommended for this area, but the likeliness of implementation is at present, however, doubtful.

Would it be possible to automatically retain some of the water during heavy rain in the upstream catchment areas? Optimally, floodwater retention should cut off the peak of the runoff hydrograph and thereby reduce the maximum water level as much as possible in vulnerable areas downstream. This could be achieved by allowing water to spill over a lowered embankment into a polder, parallel to the river. Unfortunately, the topography in the Flöha catchment is not suitable for this type of retention.

An alternative solution was found, however. It is possible at several locations to construct flood retention dams, which have a culvert at the level of the river. Water will pass undisturbed through the culvert during normal and low flow, but be temporarily retained during high flow, limited by the capacity of the culvert.

3. THE MIKE 11 MODEL

A river and catchment model was developed in order to test the feasibility of this approach and other potential measures. The selected modelling system was MIKE 11, which has been applied successfully to this type of analysis before (see e.g. Biza et al. 2001). Rainfall-runoff models for all sub-catchments in the area were established along with a hydrodynamic model of the main rivers. The models were calibrated to reproduce observed flow and water levels, cf. the graph below, and subsequently applied to test various flood mitigation options.

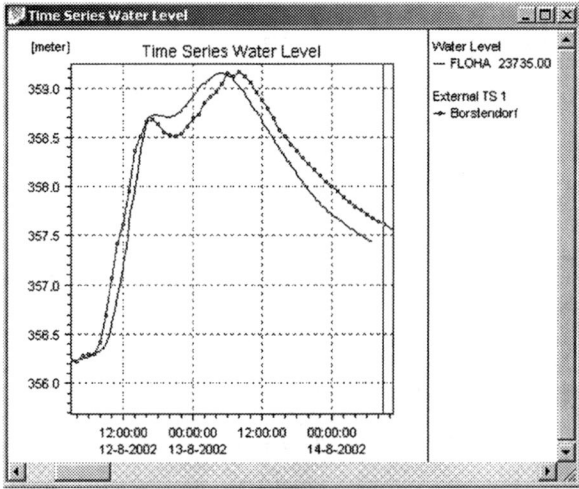

Figure 2. Model calibration at Borstendorf for the 2002 event, showing the simulated water level (black line) and the observed water level (blue line with markers)

Markers on buildings and trees from the 2002 flood showed that the water level had been exceptionally high at a few locations in urban areas. The model was applied to demonstrate that the temporary blockage of bridges by floating trees and debris was a likely cause of these high levels. A recommended flood mitigation measure is therefore to maintain the river bank areas, so that the amount of debris is limited.

202

Figure 3.   Trees and debris, washed down the river by the flood, caused temporary blockage of some bridges. Photos by Hr. Diener (left), Weser Fenster (right) (Druckerei Olbernhau 2002;with kind permission)

One of the tested scenarios was a change of land use from agriculture to forest. This was shown to have a negligible effect in this area, which has considerable forest areas at present.

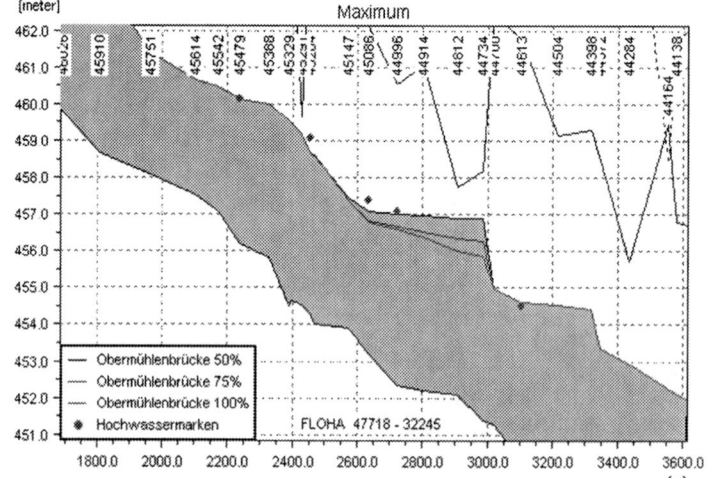

Figure 4.   Longitudinal profile of the Flöha River. Model simulations of different degrees of blockage of the bridge "Obermühlenbrücke" show that this is a likely explanation for the high water levels observed upstream the bridge

4.   THE PROPOSED SOLUTION

Several locations were found suitable for flood retention, see the map below.

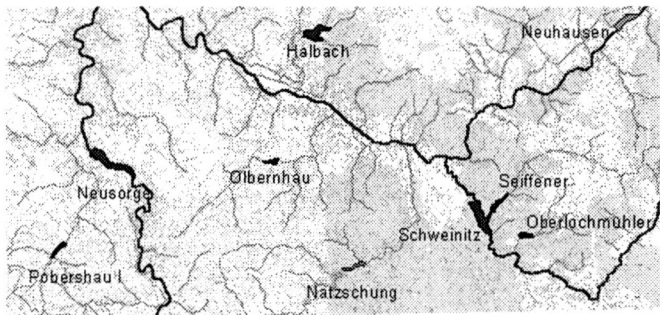

Figure 5.   Possible locations of flood-peak retention dams

The hydrological model was applied to generate dynamic flood events corresponding to different return periods, so that the mitigation schemes may be tested for realistic events of known probability. The impact of a flood-peak retention dam is illustrated in

Figure 6 and Figure 7 for a single river branch.

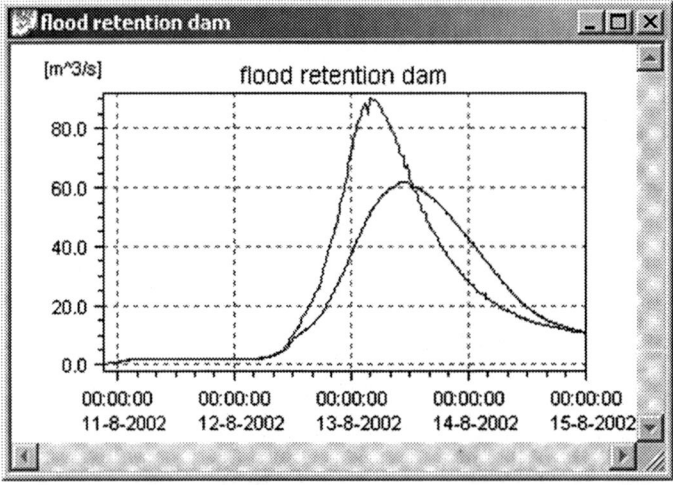

Figure 6.   The simulated discharge hydrograph before (black) and after (blue) the construction of a flood peak retention dam just upstream this location

204

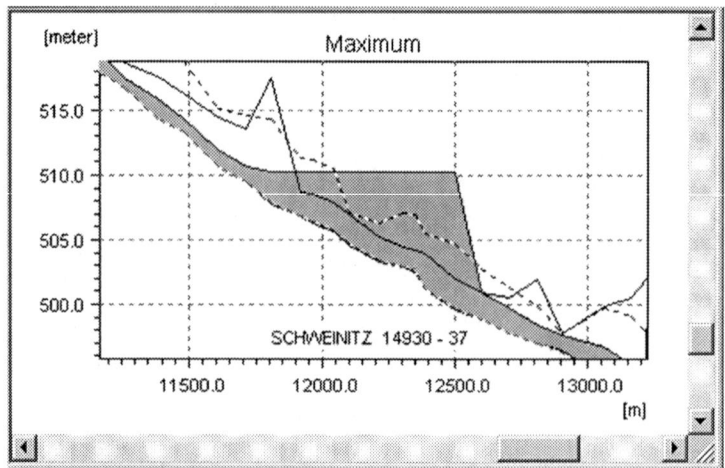

Figure 7.   Simulated maximum water level along the Schweinitz River at the location of a flood-peak retention dam

Different targets were set for the flood protection of different areas, depending on the land use. Urban areas should be protected against a 100-year flood, for example, while more frequent flooding would be acceptable in rural areas. The required protection was found to be feasible with a combination of flood-peak retention dams and local, low walls in the towns. Figure 8 shows the flooding in the town of Olbernhau before and after construction of the proposed measures.

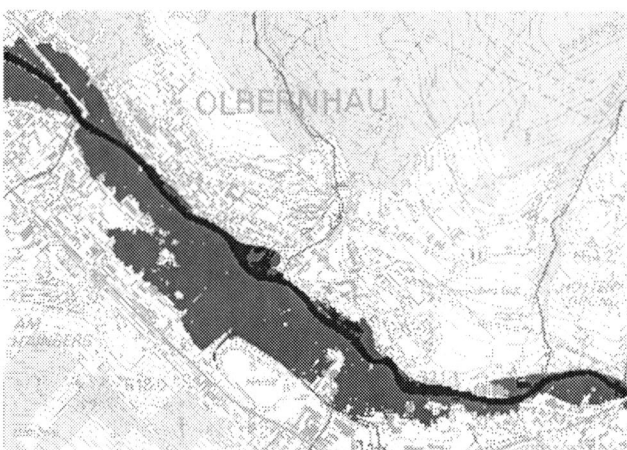

Figure 8. Flooding in Olbernhau town for a 100-year flood before (blue) and after (purple) implementation of the proposed mitigation scheme

The flood damage was assessed for a range of flood events with known probability using the generated flood maps. A benefit-cost analysis showed that the direct damage during the life-time of the proposed flood mitigation measures would exceed the investment.

5. CONCLUSION

In the wake of the august 2002 flood, the state of Saxony has launched a long range of activities to protect its citizens against future flooding. This follows a global trend, as the number of severe flood disasters has increased dramatically over the recent decades, see e.g. Kundzewicz (2002). This has prompted authorities around the world to strengthen the flood defences and preparedness at many locations. Flash flood prone areas have often been neglected, however, due to the conceived difficulties in handling this type of flood.

Recent technological development has changed this situation. Forecasting of flash floods is now feasible through automatic systems, which continuously predict the runoff using telemetry, weather radar, and advanced mathematical models. These models are also applied to test various flood mitigation options, as illustrated in this paper. It is important for these applications that the selected models are capable of describing all elements of the flooding correctly. This includes the rainfall-runoff process and the dynamics of river flow in flood situations, where structures in the river and on flood plains often play a significant role in the spreading of inundation.

References

Biza P., Gimun V., Knap R., Ammentorp H.C., Smith G. and Ohly T. (2001) The use of a GIS-based software tool for cost-benefit analysis of flood mitigation. 4th DHI Software Conference, Elsinore, June 6-8.

Druckerei Olbernhau (2002) 2 Tage im August. Olbernhau.

Kundzewicz Z.W. (2002) Non-structural Flood Protection and Sustainability. *Water International* 27 (1), 3-313.

Chapter 17

RED RIVER FLOODING
Mitigation Planning in an International River Basin

DAVID LEMARQUAND
Public Safety and Emergency Preparedness, Ottawa, Canada

Keywords: International river, floodplain management, river basin cooperation and
organisations, mitigation, Red River, disaster diplomacy, flooding.

1. INTRODUCTION

The Red River flood of 1997 in the eastern part of the Prairie-Great Plains region of North America was the worst on record in many parts of the basin and caused widespread damage in Minnesota and North Dakota in the United States and in Manitoba in Canada. In the flat, wide floodplain, the floodwaters spread up to 40 kilometres wide. It flooded the major regional centre of Grand Forks/East Grant Forks, numerous smaller communities as well as coming precipitously close to flooding the city of Winnipeg.

The Red River flows north from the United States into Canada. As an immediate response to this international flooding disaster, President Bill Clinton and Prime Minister Jean Chrétien asked the International Joint Commission (IJC) to study the causes of the flooding and to make recommendations on means to reduce, mitigate and prevent harm from future flooding in the Red River basin.

This paper looks at some the constraints and opportunities created for river basin planning and floodplain management in an international river basin following a major flood. The paper focuses on mitigation. Other elements of the disaster recovery cycle – preparedness, response, recovery – are of equal importance, but not addressed. Mitigation refers to actions to reduce flood risks, or limit damages should there be a flood.

J. Schanze et al. (eds.), Flood Risk Management:
Hazards, Vulnerability and Mitigation Measures, 207–218.
© 2006 *Springer.*

2. THE RED RIVER BASIN AND THE FLOOD OF 1997

The Red River, or Red River of the North in the United States, flows from the United States into Canada. From its headwaters in Minnesota and South Dakota it flows north across the Canada-U.S. international boundary to its outlet at Lake Winnipeg in Manitoba. Its course is through the flat, fertile bed of ancient glacial Lake Agassiz, formed from retreating glaciers. For the 877 kilometres (km) from Wahpeton, North Dakota to Lake Winnipeg the river drops just 75 meters (m). Near the boundary, the slope averages four centimetres (cm) per km. The flat central portion of the basin is about 100 km across. When the river overtops its banks, there is little to hold the waters back, creating what has been called the "Red Sea."

The basin occupies portions of eastern North Dakota, northwestern Minnesota, southern Manitoba, and a small portion of northeastern South Dakota. It covers 116,500 km^2, excluding the Assiniboine River basin, which joins the Red River at Winnipeg. Close to 90 percent of the basin is in the United States.

The river forms the boundary between North Dakota and Minnesota and has 25 tributary basins. Four of the sub-basins are inter-jurisdictional, including two that cross the border.

Flooding in the Red River differs from many other types of basins. For example:

- Spring snow melt floods can be influenced by weather conditions occurring in autumn and winter, months in advance of the flood,
- The long advance period enables good flood forecasting and flood preparation,
- The flat terrain makes the basin difficult to model hydrologically and hydraulically,
- Floods rise and fall slowly, meaning that defences must be robust, and evacuation periods are long,
- The wide, flat floodplains allow relatively small increases in flood elevation to flood vast areas,
- Roads, culverts, bridges and other physical structures can have large impacts on local flood hydraulics (Slmcleod and Wilkes 1999).

The 1997 flood originated with a wetter than normal fall and record snowfalls during the winter. After the start of an early thaw in the south, the basin experienced a major blizzard of up to 48 cm snow accumulation. Tributary flooding peaked at the same time as the main channel. Flooding persisted from early April through mid-May.

In the United States, the flood exceeded all previous historical and recorded floods. The magnitude was slightly less than a 100-year recurrence interval at Fargo and slightly greater from Grand Forks downstream (USACE and FEMA 2003). Some tributary floods were even rarer events.

In Manitoba, the flood was the largest flood in the twentieth century. The flow at Winnipeg was 4,580 cubic metres per second (m^3/s). The next largest, in 1950, peaked at 3,050 m^3/s. The largest flood of record in

Canada was in 1826, with an estimated flow of 6,370 m³/s. It wiped out the Selkirk colony in Manitoba. A flood in 1852 was also slightly greater than the one of 1997. The 1997 Red River flood was considered a one-hundred-year flood.

Overland flooding is a major concern. In the United States, Ada, Wahpeton/Breckenridge, and Grand Forks/East Grand Forks suffered, as did Ste. Agathe in Manitoba, which was flooded from the west, not from the river to the east of town.

About 103,000 people were evacuated: 75,000 in the United States and 28,000 in Canada. Flood damages exceeded $3.5 billion in the US and $500 million in Canada. While there was no loss of life, the social costs to the people affected were devastating and are still being felt.

3. THE WINDOW OF OPPORTUNITY

In the follow-up to a flood, the people affected and the officials, experts, and planners responsible for safeguarding against repeated disasters will bring their own experiences, values, openness, and expertise to their understanding of the events and what should be done. People with shared understanding and interests will act and react in ways consistent with their beliefs. The many groups and interests reflecting different views and agendas will shape the planning and political "decision space" for deciding what can be accomplished in preparing for and mitigating future events.

Ideally, the decision space to some degree should encompass the watershed because of the need to integrate and coordinate many aspects of floodplain management. However, river basins are often institutionally complex and lack strong organizational structures. In case of the Red River there are unresolved tensions and issues that have frustrated strong basin wide institutional arrangements.

Kelman and Koukis (2000) address a phenomenon that they term "disaster diplomacy," the potential of disasters to facilitate better relations amongst states in conflict. The cooperative spirit generated from common efforts to deal with disasters may be able to override pre-existing prejudices. Spillover from technical or scientific collaboration or humanitarian assistance to successful diplomatic rapprochement may occur. They recognize, however, that disasters may never eradicate deeply entrenched interstate enmity. The concept raises the question for disasters between friendly nations like Canada and the United States: could the response to the Red River flood be a catalyst to improve ongoing basin-wide cooperation?

A common perception following major disasters is that of a "window of opportunity" for furthering new projects and initiatives. Political leaders are compelled to respond to the damages and suffering, and thus are open to ideas that planners and other interests have long been advocating, often with little success. But the period is short, since new political imperatives will capture public attention and the seeming political imperative to act fades.

The alignment of interests in the Red basin tend to conflict. In international rivers, upstream and downstream states have limited reciprocal interests, since the upstream state has little incentive to care what happens downstream. Laws and other sanctions and incentives generally militate against raw acts of self-interest (LeMarquand 1977). This inherent tension has underlain a number of disputes within the Red basin. For example, the Garrison River Diversion and Devils Lake projects in North Dakota have threatened and antagonized downstream jurisdictions because of the anticipated introduction of alien biota into the Red and wider Saskatchewan-Nelson drainage basin (Kellow and Williamson 2001).

Apart from national divisions, the Red River basin has many local, interstate, bi-national and international arrangements, as well as committees and working groups responsible for aspects of the basin, as well as environmental and other groups interested in basin management issues.

Despite all the institutions with mandates related to river basin management, there was no immediate consensus or basin-wide vision to offer political leaders during the "window of opportunity" following the 1997 flood. The challenge for the IJC was to create the decision space for political discussion by establishing a common understanding of the scientific and technical facts and the case for various floodplain management options.

4.   THE IJC STUDY

The IJC was asked to examine and report on the causes and effects of damaging floods, and to make recommendations on means to reduce, mitigate, and prevent harm from future flooding in the basin (IJC 2000).

The IJC, established under the Canada-USA Boundary Waters Treaty of 1909, has a reputation for impartiality, which serves it well in its fact finding tasks, allowing it to serve as an arbiter of fact on issues at dispute or under consideration by both countries (LeMarquand 1993). The IJC uses technical boards, usually appointing officials from both sides of the border. For the Red study, it formed the International Red River Basin Task Force (Task Force). The ten-person board had an equal mix of American and Canadians, with members from state, provincial and federal governments, as well as two academics. Based on the Task Force's report, the IJC prepared its own report and recommendation to the two national governments.

The IJC and its Task Force had a basin-wide mandate from President Clinton and Prime Minister Chrétien. However, that mandate did not guarantee full support of the jurisdictions involved. While no jurisdiction opposed the study, the states and their Congressional representatives had their own priorities; for example, funding for recovery and new flood defence projects, and for North Dakota, support for its plans to divert waters from Devils Lake into the Red basin. As well, North Dakota officials had a lingering suspicion of the IJC because of its past role in opposing parts of the Garrison Diversion project. In the end, the American direct contribution to the study was $887,000. The Canadian portion was $2,640,000 (IJC, 2000). Other indirect funding came from partners, such as the Global Disaster Information Network (GDIN) in the United States, the city of Winnipeg, and the province of Manitoba.

Within the US administration, the US Army Corps of Engineers (USACE) remained committed to the study, but it was constrained in what it could do by limited funding. The Federal Emergency Management Agency (FEMA) showed only sporadic interest in supporting the international effort.

In Canada, political support was more solid, in part because the then Minister of Foreign Affairs, Lloyd Axworthy, was from Winnipeg and it was on his initiative that the Prime Minister and President asked the IJC to conduct the study. The province of Manitoba and the city of Winnipeg remained supportive of the work, despite the fact the Province had commissioned its own reviews of the flood.

The Task Force prepared two reports for the IJC (Task Force 1997, 2000) and conducted a research program using the expertise of academics, government agency technical staff, and consultants to develop the information and recommendations for the reports. In conjunction with the IJC, it held briefings and fact-finding meetings with officials, as well as open sessions seeking public input. The Task Force also held workshops and other information and analytic activities over the three years of the study.

While the IJC was taking a bilateral basin-wide approach to the flood issues, other jurisdictions continued with their own studies and mitigation projects. For example, the USACE worked on projects estimated to cost $402 million to construct permanent flood defences for Grand Forks, North Dakota and East Grand Forks, Minnesota.

5.   ISSUES

Much of the immediate public reaction that the IJC and the Task Force heard following the 1997 flood concerned peoples' experiences during the flood and in the post-flood recovery. Residents had much to say about flood relief operations, such as operation of flood plans, evacuation procedures, reliability of flood forecasting information, and financial assistance for flood victims (Grant 1997).

The public, officials, academics, interest groups, and other organizations raised with the IJC a range of issues relevant to mitigation initiatives to safeguard the population against large floods. A widely held goal was for a permanent solution to flooding in the valley. Many groups argued for more dams and dikes to protect residents. Many residents and experts felt that the 1997 flood peak could have been reduced if much of the runoff could have been stored or delayed (Task Force 2000). Other groups were concerned about frequent flooding of tributaries and felt that the cumulative effect of constructing new impoundments on smaller streams would decrease the peak flows of major floods.

Environmentalist saw changes in land use in the valley as responsible for the severity of the flood. Most wetlands had been drained for agriculture. Their restoration could restore inherent storage capacity, which would slow the release of the spring runoff and thus avoid the extreme peak flows experienced in 1997. There was also criticism of the amount of drainage and the size of drainage ditches used to hasten the drying of the fields in the spring or after summer rainstorms.

In the United States, plans to construct higher permanent dikes or levees, particularly for Grand Forks/East Grand Forks, raised concern that such protective measures would reduce floodplain storage and increase flood peaks.

The major issue in Canada was enhancing the flood defences for Winnipeg, the largest city in the basin (population 650,000). Following the flooding of Winnipeg in 1950, the city, province, and federal government constructed new flood prevention works, including dikes and a diversion channel (the Red River Floodway) around the city. In 1997, only quick action in raising the major dikes south of the city and operation of the Floodway saved Winnipeg from a major disaster. Less benign weather during the flood would likely have resulted in the defences being overwhelmed, leading to the evacuation of 300,000 people and several billion dollars in flood damages. A related issue was operation of the flood defences, which neighbouring communities upstream complained increased their flooding.

The flood highlighted a local transboundary issue, flooding in Pembina tributary, which flows from Manitoba into North Dakota and enters the Red River near the international boundary. For many years, farmers and communities on both sides of the border unilaterally built dikes and roads to control flooding. These works often aggravated flooding across the border. A number of initiatives by local, state, provincial, and federal governments failed to remedy this long-standing bilateral issue.

Other issues included biota transfer from watersheds outside the basin into the Red and Saskatchewan-Nelson watershed that drains much of western Canada. On occasion (estimated at a 10 percent chance or greater during the spring runoff) water spills over from the Little Minnesota River in the Mississippi basin and flows across the continental divide into Lake Traverse. Consequently, biota may flow from the Mississippi-Missouri watershed into the Hudson Bay drainage system, raising the possibility of alien organisms damaging aquatic life, including commercial fishery stocks in Lake Winnipeg (Spalding 2000).

Another water quality issue concerned water quality in the river, in Lake Winnipeg and in groundwater because of flooding of chemical storage sites and waste sites within the floodplain. Flooding also led to increased movement of nutrients from the floodplain into Lake Winnipeg.

The flood accelerated consideration of the appropriate institutional arrangements for flood management and other river basin development concerns. The IJC would seem to be well placed to take on a much more active role. However, the IJC and its two permanent Red River boards were largely limited to reporting functions. A grass roots level organization, the Red River Basin Board, a coalition of local authorities with representation from state, provincial organizations, sought a much more prominent role in developing comprehensive flood management plans for the basin. Also, the International Flood Mitigation Initiative, created following the flood with funding from FEMA and the Province of Manitoba, promoted the idea of an international arrangement for basin wide cooperation, coordination and citizen participation (IFMI 2001).

The many views and proposals were put forward in a political environment of competing views, concerns, tensions, and grievances. Rural areas were suspicious of plans for urban flood protection, particularly with regard to Winnipeg. Environmental interests were critical of current agricultural practices on drainage. Long standing disputes over the Garrison Diversion project and Devils Lake flooding coloured bilateral relations. Some of the differences were basin-wide; others were of local concern.

6.   PLANNING OUTCOMES

In addressing the wide range of issues that people and governments raised, the Task Force sought to establish facts of the issues and to develop a legacy of tools and data that could be used in future basin planning.

As the basis to revise flood protection measures, the Task Force established the frequency of 1997 magnitude floods. Studies were conducted looking at the evidence of paleo-floods (Kroker 1999), the historical floods (Rannie 1998), and statistically simulated floods (Warkentin 1999). The studies showed that the 1997 flood was consistent with a history of periodic large floods. At Winnipeg, for example, the 1997 flood was a 90-year event; the flood of 1826 a 300-year event. The message to the public was that although the 1997 flood was a rare event, floods of the same size or greater can be expected again and people and property remain at risk (IJC 2000).

To understand the hydrologic and hydraulic conditions in the basin, the Task Force used "unsteady flow" hydrodynamic models to simulate floodwater flows. The Danish Hydraulics Institute's MIKE 11 model was used for the main stem from Grand Forks to Selkirk, Manitoba. Similarly, the USACE used their own Unsteady NETwork (UNET) model for the reach from the headwaters to Letellier, Manitoba. The overlapping sections allowed a comparison of the results from both models (Task Force 2000).

The Task Force conducted studies on the feasibility of reducing the magnitude and timing of flood peaks. The IJC and the Task Force concluded that it would be difficult if not impossible to develop enough reservoir, wetland, or micro-storage to reduce the flood peaks as experienced in 1997. No magic bullet could eliminate the flood threat in the basin. People would have to be resilient and use a combination of structural protection and non-structural measures to mitigate the flood risk. The conclusion was controversial because it conflicted with many local views that saw salvation in wetland or tributary storage. The conclusion also ruled out new flood control dams in which both Canada and the United States would have had a common interest and incentive to work for implementation.

A major effort was put into identifying the data available and implementing the first steps towards a Red River Basin Disaster Information Network and a Decision Support System. The effect of these initiatives was to make flood-related information for the basin more readily available. In part because of IJC recommendations, the Canadian and Manitoba governments changed restrictive data distributions policies making government data freely available by means of the Internet.

To address the Pembina River flooding issues along the international boundary, a major effort was devoted to leading-edge laser (LIDAR) and radar digital topographic mapping (IFSAR) of the areas (USACE 2004). The Lower Pembina basin served as a test area in preparation for the eventual seamless Digital Elevation Model (DEM) for the basin. Similarly, a LIDAR-based DEM was prepared for the flood-prone area south of Winnipeg. These DEMs allow very accurate simulations of flooding using hydrodynamic models. RADARSAT images and a differential global positioning system were also used to help develop the topography of the basin. One outcome of this work is that residents of the rural Manitoba floodplain can now obtain personalized flood forecasts via the Internet for their own property.

To deal with enhanced protection for Winnipeg, the Task Force conducted engineering studies and suggested two options. The IJC recommended the option of expanding the Floodway around the city.

The IJC also recommended that it be given a mandate to report on the progress of implementation of its wide range of recommendations. The Commission also looked to the governments for a binational comprehensive flood damage reduction plan and for cooperation among all jurisdictions to coordinate and implement measures for flood mitigation and preparedness activities. The IJC wanted its legacy projects–the hydraulic models, Red River Basin and Disaster Information Network, in particular–to carry on and support basin wide flood reduction efforts.

A major effort was to identify non-structural measures for floodplain management and to help communities and businesses develop the resiliency to adapt to floods and bounce back to normal life quickly following a major event.

In total, the IJC came to seven conclusions and made 28 recommendations to government. It also endorsed without change another 30 recommendations of its Task Force.

Three years later the IJC's International Red River Board reported on government responses to the IJC's recommendations. Halliday (2003), the consultant, concluded: "it would be fair to say that the expenditures since 1997 relating to the IJC recommendations are in the order of hundreds of millions of dollars and that similar amounts will be spent in the next five years. No recommendations have been formally rejected, although a few are unlikely to be implemented." Considerable progress had been made in increasing preparedness and mitigating potential harm from future floods.

Many structural measures aimed at protecting both rural and urban floodplain residents were built, were under construction, or are in advanced planning stages. Rural protection efforts were nearing completion, major levees, such as those for Grand Forks and East Grand Forks, were well in hand, and expansion of the Red River Floodway at Winnipeg was moving ahead.

There were some indications of attempts to initiate floodplain developments in the United States ahead of new definitions of the area at risk (Leitch 2003). Also, more work was needed on tributary flooding, ice jam flooding, and summer floods.

In conclusion, Halliday found the recommendations having achieved the most success were those involving construction of the structural measures identified in the IJC report, even when those measures require collaboration at the federal, state or province, and local levels. A second group of successful recommendations were those aimed at specific agencies, for example those responsible for improvements to flood forecasting.

The less successful recommendations were ones involving multiple agencies and multiple objectives. The report speculated that public priority favours implementation of structural measures prior to the start of "softer" projects. In summary, he found considerable success in projects and programs to keep water away from people, but less success in programs to keep people away from water.

The response to the flood has led to changes in the institutional landscape. The IJC consolidated its two Red River boards into the International Red River Board. The grass roots Red River Basin Board reconstituted itself into the Red River Basin Commission. The Red River Basin Institute was created to conduct flood-related research. These organizations and governments have been collaborating on advancing the IJC legacy projects to the point where there will soon be a single one-dimensional hydrodynamic model of the Red River mainstem extending from Lake Traverse in the headwaters to Breezy Point near Lake Winnipeg. Increased cooperation has lead to periodic meetings of basin governors and premiers. As well, the Red River Basin Commission is drafting a comprehensive flood mitigation plan. While there has been institutional progress, the desire among some people for a strong basin institution pursuing a common vision for the basin remains elusive.

7.   CONCLUSIONS

The 1997 flood provided an impetus to greater cooperation in the Red River basin. The flood highlighted to everyone in the basin their common interest in the river and how it behaves. The flood caught the attention of the two federal governments, who gave the IJC the mandate to conduct a study to understand better the causes of the flooding and what might be done to prevent future damaging consequences. The study found no basis for advocating flood control projects that might have provided the basis for a common undertaking. It did point to a myriad of structural and non-structural measures that need to be implemented and coordinated to make the basin more resilient to future floods.

There was no quick fix. Greater resiliency is a long-term commitment requiring the sustained effort of all jurisdictions within the basin. Progress has been made in establishing basin-wide institutions that will work for continuing change. The new institutional arrangements have been aided by the legacy projects left by the IJC study, which can give technical foundation to advocacy initiatives. Continuing public interest and discussion seven years later has kept flooding on the political agenda and the "window of opportunity" is still open–not an insignificant accomplishment. However, the reciprocal interests throughout the basin remain weak and much of the future progress will depend on the commitment of state, provincial, and local governments.

What lessons are there in how to sustain a basin-wide approach in an international basin following a major flooding disaster? Flood mitigation requires a mix of non-structural and structural policies, projects, and activities. To obtain the best results, all of the impacts of specific initiatives need to be considered in the context of how they might affect other parts of the basin and other initiatives. There is a strong need for coordination and integration of all these activities. In an international river, there is a need for a common framework for planning and management of the floodplain.

From the Red River experience, the lessons relating to developing a basin-wide international approach following a flood disaster include:
1.   Technical understanding is an essential but not sufficient condition for long-term cooperation.
2.   There needs to be sustaining interest in cooperation, that is, there needs to be common and reciprocal interests.
3.   In an international basin like the Red River characterized by upstream-downstream states, flood control projects offer the possibility of sustaining reciprocal interest. Where such projects are not feasible, the incentives for cooperation must be found elsewhere.
4.   Long-term cooperation needs some form of basin-wide continuing institutional arrangements to maintain the technical basis for common understanding and to highlight problems and opportunities.

5. The scope for cooperation may be limited. Much of floodplain management is local and consequently many of the priorities for action are to meet local needs.

6. International endorsement and monitoring of local projects, such as the Red River Floodway around Winnipeg, can be of use in gaining support from national governments.

References

Grant K.R. (1997) Report on a Strategic Research Workshop on the Social Dimensions of the Flood of the Century, IJC, International Red River Basin Task Force: 17.*

Halliday R.H. (2003) Flood Preparedness and Mitigation in the Red River Basin. IJC, International Red River Board 63.*

IFMI – International Flood Mitigation Initiative for the Red River (2001) Vision, Mission and Goals. Bismarck, ND, IFMI..

IJC – International Joint Commission (2000) Living with the Red. Report to the Governments of Canada and the United States on Reducing Flood Impacts in the Red River Basin, Ottawa / Washington, IJC 82.*

Kellow R.L. and Williamson D.A. (2001) Transboundary Considerations in Evaluating Interbasin Water Transfers. Conference on Transbasin Water Transfers, Denver, US Committee on Irrigation and Drainage, 165-192.

Kelman I. and Koukis T. (2000) Disaster Diplomacy. *Cambridge Review of International Affairs* 14 (1), 214-294.

Kroker S. (1999) Flood Sediments and Archaeological Strata. Report prepared for the Geological Survey of Canada, IJC, International Red River Basin Task Force 19.*

Leitch J.A. (2003) Floodplains and the Tyranny of Small Decisions. ASFPM News & Views, Association of State Floodplain Managers 15.

LeMarquand D. (1977) International Rivers: The Politics of Cooperation. University of British Columbia, Vancouver, Westwater.

LeMarquand D. (1993) The International Joint Commission and Changing Canada-United States Boundary Relations. *Natural Resources Journal* 33 (Winter), 59-91.

Rannie W.F. (1998) A Survey of Hydroclimate, Flooding, and Runoff in the Red River Basin Prior to 1870. IJC, International Red River Basin Task Force 189.*

Slmcleod Consulting, Brian Wilkes and Associates, et al. (1999) Review of Red River Basin Floodplain Management Policies and Programs. IJC, International Red River Basin Task Force 73.*

Spading K. (2000) Browns Valley Dike. Browns Valley, Minnesota, Lake Traverse Project, History and Potential for Interbasin Flow, St. Paul, MN, USACE 29.*

Task Force-International Red River Basin Task Force (1997) Red River Flooding: Short-Term Measures. Ottawa / Washington, IJC 65.*

Task Force-International Red River Basin Task Force (2000) The Next Flood: Getting Prepared. Ottawa / Washington, IJC 164.*

USACE – US Army Corps of Engineers (2004) Basin Level Digital Elevation Models, Availability and Applications: The Red River of the North Basin Case Study. (IWR Report 04-R-1), Washington, DC, USACE, Institute for Water Resources 81.

USACE – US Army Corps of Engineers and FEMA – Federal Emergency Management Agency (2003) Regional Red River Flood Assessment Report. Wahpeton, North Dakota/Breckenridge, Minnesota to Emerson, Man, St. Paul, USACE, St Paul District, FEMA Regions V & VIII.

Warkentin A.A. (1999) Red River at Winnipeg: Hydrometeorological Parameter Generated Floods for Design Purposes. IJC, International Red River Basin Task Forces 30.*

* Reports available at the IJC website: http://www.ijc.org

Chapter 18

OVERVIEW OF US NATIONAL FLOOD INSURANCE PROGRAM REGIONAL

BOHUMIL JUZA PH.D.
Baker Engineering, NewYork, USA.

Keywords: Flood damage, floodplain management, insurance programme.

1. INTRODUCTION

Flood protection is a complexity of technical and economical measures which communities, towns or regions implement in order to defend valuable properties from inundation. For many historic reasons and partially as a reaction to society's demand, the protection was established by building dams, levees, pumping stations, dredging, flood plain filling, river training etc. The rapid flood plain development and in some areas dramatic changes of land use produce an ever increasing potential for flooding.

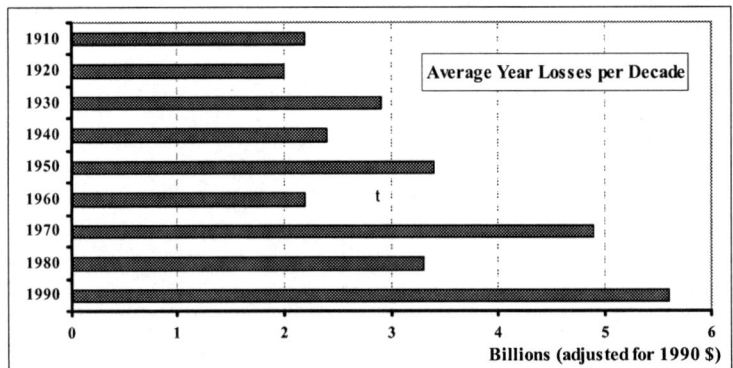

Figure 1. Flood related damages in US

From an engineering perspective, it is possible to build any type of project to fulfill society's requirement for safe and flood-proof floodplain.

J. Schanze et al. (eds.), Flood Risk Management:
Hazards, Vulnerability and Mitigation Measures, 219–228.
© 2006 *Springer.*

But how safe and how successful are those measures? Isn't this just another way of pushing the problem somewhere else or postponing it for another time. As more money gets invested, damage is happening. In the last decade, the flood damage on infrastructure and additional business losses in the U.S. has increased to 6 billion dollars a year.

2. FLOOD INSURANCE COVERAGE

There is a well known fact that flood insurance is not profitable for any type of privately owned company. The statistical probability that a single family house located in a 100 year floodplain will be flooded during a 30 year mortgage is approximately 26%. In comparison, the chance that the same house will be damaged by fire is only 4%.

Extreme flood disasters in the 1920's and 1930's have impacted on the flood related policies in the U.S. As a result of numerous claims the insurance companies stopped providing flood insurance to the public. During that time, only federal flood disaster assistance was available to flood victims. Federal involvement in protecting life and property considerably increased after the passage of the Flood Control Act in 1936. The main focus of this legislation was on structural flood control projects such as dams and levees. Despite the extensive investment in those projects, the losses to life and property and amount of assistance to flood related victims continued to rise.

In recognition of this situation, the U.S. Congress established the National Flood Insurance Program (NFIP) with the passage of the National Flood Insurance Act of 1968. The NFIP is a Federal program enabling property owners in participating communities to purchase insurance as a protection against flood losses in exchange for State and community flood plain management regulations that reduce future flood damages. Participation in the NFIP is based on an agreement between communities and the Federal Government. Flood insurance is available to a community, which adopts and enforces a floodplain management ordinance to reduce future flood risk.

Funding for the NFIP is through the National Flood Insurance Fund, which was established in the Treasury by the 1968 Act. Premiums collected are deposited into the fund, and loses, and operating and administrative costs are paid out of the fund. In addition, the Program has the authority to borrow up to $1.5 billion from Treasury which must be repaid along with interest. Until 1986, Federal salaries and program expenses, as wells the costs associated with flood hazard mapping and floodplain management were paid by an annual appropriation from Congress. From 1987 to 1990, Congress required the Program to pay these expenses out of premium dollars. ($485 million) Since 1991, policy fee ($30) is applied to generate funds for salaries, expenses and mitigation costs. In 1973 the mandatory flood insurance purchase was implemented.

The Federal Emergency Management Agency (FEMA) is the primary Federal agency responsible for assisting state and local governments, private entities, and individuals in preparing for mitigating, responding to and recovering from natural disasters, including floods. The NFIP is the key component of FEMA's efforts to minimize or mitigate the damage and financial impact of floods on the public and to limit Federal expenditures needed after floods occur.

The U.S. territory is divided into 10 regions in order to provide better customer care, localized technical expertise and knowledge (Figure 2).

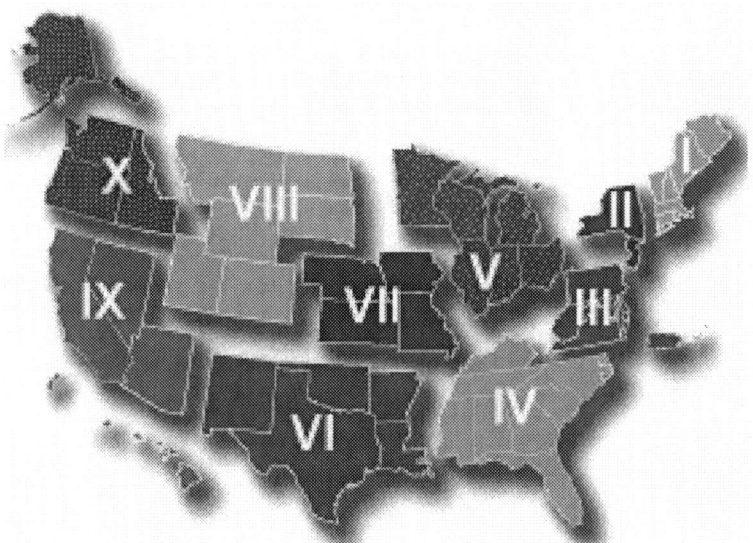

Figure 2. FEMA regions

The NFIP has three basic components: Flood Hazard Identification and Risk Assessment, Flood Management and Flood Insurance.

2.1. Flood hazard identification and risk assessment

In the beginning of the program, the national standard was needed in order to manage and assess flood risks. The "100-year" standard was selected as reasonable on the basis that it provides a higher level of protection to the communities while not imposing overly stringent requirements or the burden of excessive costs on property owners.

Flood Insurance Study (FIS) is conducted in order to provide flood mapping to the community. FIS report contains information regarding flooding in a community and is developed in conjunction with the Flood Insurance Rate Maps (FIRM). The FIS contains a narrative of the flood history of a community and discusses the engineering methods used to delineate Special Flood Hazard Area (SFHA) and produce the FIRMs. The study also contains flood profiles for studied flooding sources and can be used to determine Base Flood Elevations (BFE) for some areas.

Table 1. Flood zone naming and description

Zone A	The 100-year or base floodplain. There are six types of A Zones: **A** The base floodplain mapped by approximate methods, *i.e.,* BFEs are not determined. This is often called an unnumbered A Zone or an approximate A Zone. **A1-30** These are known as numbered A Zones (*e.g.,* A7 or A14). This is the base floodplain where the FIRM shows a BFE (old format). **AE** The base floodplain where base flood elevations are provided. AE Zones are now used with new format FIRMs instead of A1-A30 Zones. **AO** The base floodplain with sheet flow, ponding, or shallow flooding. Base flood depths (feet above ground) are provided. **AH** Shallow flooding base floodplain. BFEs are provided. **A99** Area to be protected from base flood by levees or Federal Flood Protection Systems under construction. BFEs are not determined. **AR** The base floodplain that results from the decertification of a previously accredited flood protection system that is in the process of being restored to provide a 100-year or greater level of flood protection.
Zone V and VE	**V** The coastal area subject to a velocity hazard (wave action) where BFE are not determined on the FIRM. **VE** The coastal area subject to a velocity hazard (wave action) where BFE are provided on the FIRM.
Zone B and Zone X (shaded)	Area of moderate flood hazard, usually the area between the limits of the 100-year and 500-year floods. B Zones are also used to designate base floodplains of lesser hazards, such as areas protected by levees from the 100-year flood, or shallow flooding areas with average depths of less than one foot or drainage areas less than 1 square mile.
Zone C and Zone X (unshaded)	Area of minimal flood hazard, usually depicted on FIRM as above the 500- year flood level. Zone C may have ponding and local drainage problems that don't warrant a detailed study or designation as base floodplain. Zone X is the area determined to be outside the 500-year flood and protected by levee from 100-year flood.
Zone D	Area of undetermined but possible flood hazards.

FEMA currently supports approximately 20,000 communities participating in NFIP and issued roughly 100,000 Flood Insurance Rate Maps. An example of FIRM is in Figure 3.

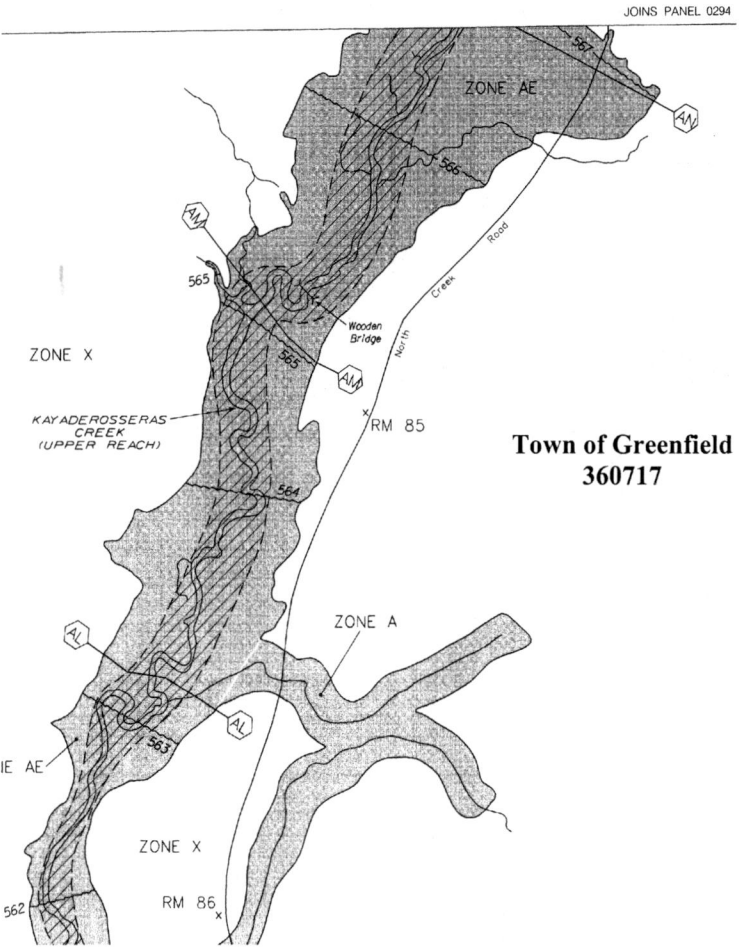

Figure 3. FIRM for Town of Greenfield Saratoga County, NY with Floodway Delineation

FIS is produced by a combination of approximate and detailed study approaches. The detailed study approach uses engineering models to calculate BFE and specify water elevation surface for 10-year, 50-year, 100-year and 500-year floods, additionally regulatory 100-year floodway zone on the map. Regulatory floodway is defended as the channel of the stream plus any adjacent floodplain areas that must be kept free of encroachment so that the entire Base Flood discharge can be conveyed with no greater than a 1.0-foo increase in the BFE. The detailed study is considerably expensive: the cost is approximately $10,000 per river mile. The approximate study provides only BFE extension information without any elevation and depth information.

The zone designation on FIRM indicates magnitude of the flood hazard in the community. Depends on the zone specification, flood insurance may or may not be required. Mandatory flood insurance is required for Zone A and Zone V.

2.2. Floodplain management

Depending on the flood risk condition provided by FEMA for the community, the minimum management requirements must be adopted. A community participating in NFIP is required to regulate and manage all development in the Special Flood Hazard Area. There are structural, non-structural and no adverse (zero impact) management techniques.

- Structural techniques, such as dams, levees, and channel alternation were extensively used in past decades and improved community protection.
- Non-structural measures are focused on land use and subdivision ordinances and zoning regulations, building elevations or retrofitting, land preservation, public education and emergency preparedness.
- Zero impact means that development in the floodplain will not have any impact on flood peak, maximum water level, flow velocity and erosion condition in the river.

2.3. Flood insurance

Flood insurance under NFIP is sold to owners of properties by state licensed property and casualty insurance agents and brokers who deal directly with FEMA and through private insurance companies. Coverage under this program is limited, for example residential 1-4 family units are eligible for up to $250,000 in building and $ 100,000 in personal coverage. Average coverage is approximately $ 140,000.

Since 1978, the NFIP Property Coverage increased 6.2 times but premiums increased only 2 times, when adjusted for inflation. On average there is less than 34,000 insurance claims paid per year for a total of 220 million dollars. Actual numbers obviously vary over the years as can be seen in Figures 4-7.

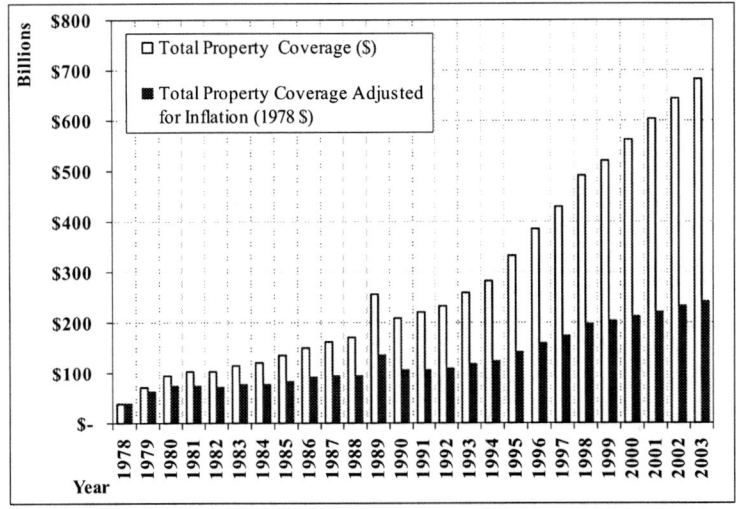

Figure 4. Total property coverage by NFIP

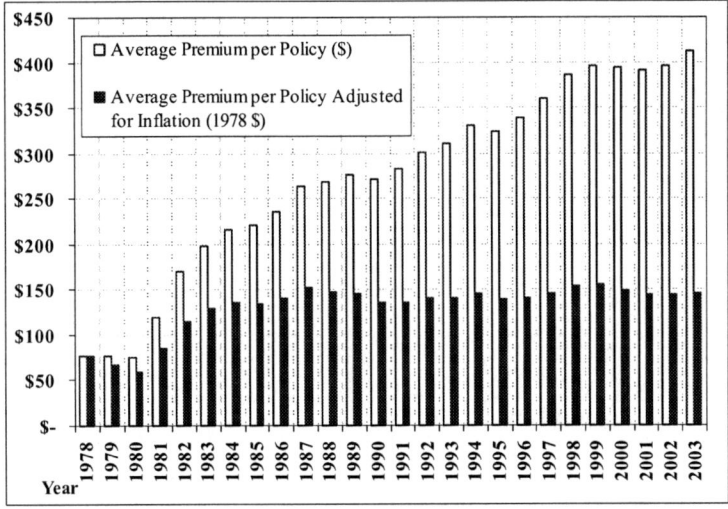

Figure 5. Average Premium per Policy

226

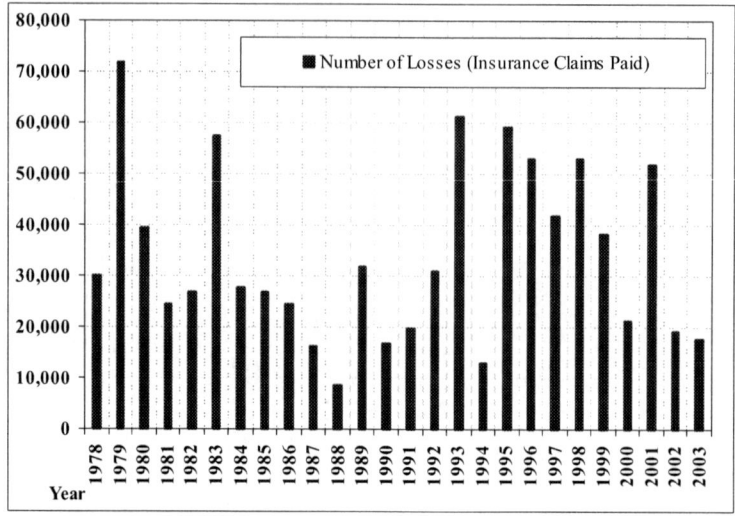

Figure 6. Number of Insurance Claim Paid by NFIP

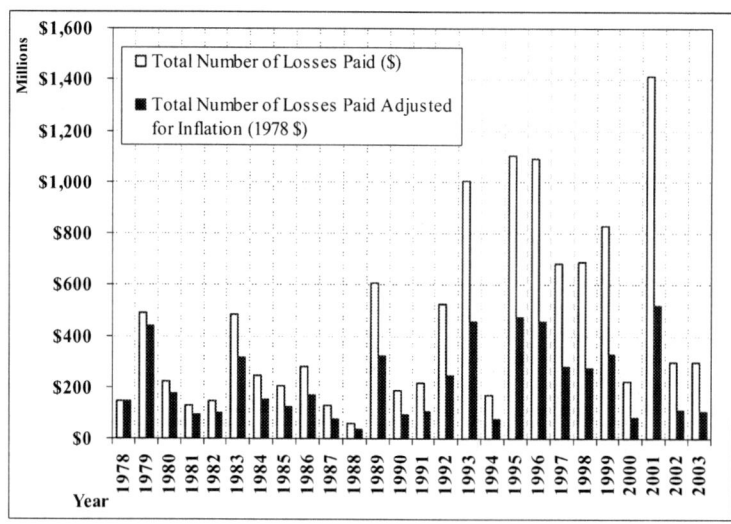

Figure 7. Total Number of Looses Paid by NFIP

3. FLOOD MAP STATUS AND PROCESS OF MODERNISATION

Recently, FEMA initiated a systematic process for modernizing flood insurance rate maps for the entire U.S. territory, taking a more active approach to minimizing flood property loss and more importantly loss of the life. One of the reasons for the modernization is the age of the FIRMs. Approximately 70% of the maps are 10 years old (Figure 8).

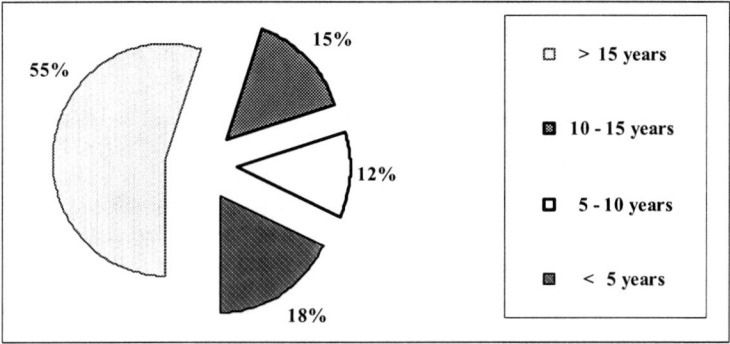

Figure 8. Average age of the FIRM

The main goal of this modernization initiative is to shift from a paper map system to a fully digital system which will be fully integrated into nation-wide multi-layer hazard information system based on GIS.

4. REPETITIVE LOSS PROPERTIES

Property for which the policyholder claimed two or more $ 1,000 flood claims in a 10 year period is called repetitive loss property.

Repetitive loss properties cost the NFIP about $200 million annually. These properties while comprising approximately one percent of the currently insured properties, are expected to account for 25% to 30% of claims paid. There is one example of costly abuse of the insurance coverage. One home, valued at $114,480, has received $806,591 in flood insurance payments over 18 years. Furthermore, twenty five percent of all current NFIP policies do not pay actuarial rate for their coverage and thus are subsidized by the 75% of other policy-holders. Today, the vast majority of repetitive-loss properties are eligible for subsidized flood insurance far below the actuarial risk rate they should be paying.

There are few options in terms of preventing future financial losses from the NFIP program. The first option is to allow residents to move out of the floodplain through a buy out program. The second is to provide grants to owners which will help them improve the structure (for example raising the structure) to decrease flood risk.

5. CONCLUSIONS

Floods are a natural hazard which cannot be avoided or totally mitigated. Flood insurance programs such as NFIP is one of the possible approaches to manage and increase awareness and provide communication between the general public and government agencies. In addition NFIP is helping communities to be more prepared for flood disasters and their consequences. NFIP assist communities to reduce costs of flooding by adopting and enforcing floodplain management ordinances.

References

Bereuter D. (2004) Statement Hearing on the National Flood Insurance Repetitive Losses Senate Committee on Banking, Housing and Urban Affairs, Subcommittee on Economics Policy, March 25, 2004.

FEMA Flood Insurance Statistics (2004) http://www.fema.gov/nfip/sitemap2.shtm.

Howard, M. (2004) FEMA Multi Hazard Flood Map. The National Academies Disasters Roundtable Workshop - Reducing Future Flood Losses: The Role of Human Actions, http://dels.nas.edu/dr/f10.html.

Jenkins, W.O. Jr, (2004) National Flood Insurance Program Action to address Repetitive Loss Properties. United States General Accounting Office, Testimony Before the Subcommittee on Economic Policy, Committee on Banking, Housing, and Urban Affairs, U.S. Senate, GAO-04-401T, March 25, 2004.

Larson L.A., Klitzke M.J. and Brown D.A. (eds.) (2003) No Adverse Impact: A Toolkit For Common Sense Floodplain management. Association of State Floodplain Managers, Synergy Ink, Ltd. www.flood.org.

National Flood Insurance Program: Program Description. August 2, 2002.

Chapter 19

STRATEGIES FOR FLOOD RISK MANAGEMENT
– A PROCESS PERSPECTIVE

GÉRARD HUTTER

Leibniz Institute of Ecological and Regional Development (IOER), Member of the Dresden Flood Research Center (D-FRC), Dresden, Germany

Keywords: Strategy, Flood Risk Management, Process, Adaptive Management, Strategic Planning.

> *"The comment, 'Great strategy; lousy implementation', gives unjustified credit to the strategist. If the strategy has been designed without taking account of the organizations capacity for implementation, it's a lousy strategy."*
> *(Grant 2005, p. 187)*

1. INTRODUCTION

To manage extreme flood events like the Weisseritz flash flood within the Elbe river basin in August 2002 and their adverse impacts on people and properties, practitioners and scientists argue for a shift from the traditional paradigm of flood protection to flood risk management (Schanze 2002, DKKV 2003, Hall et al. 2003). However, developing a risk-based strategy is a difficult task. Directly after a severe flood event the need for an effective strategic approach to flood risk management is usually widely acknowledged. Nevertheless, after some time memories of the event and its causes fade and it is not easy to maintain political support for flood risk management in all relevant policy fields. As a consequence, flood risk issues are often outweighed in political and administrative decision processes (Fleming 2002) such as spatial planning (DKKV 2003). Furthermore, a strategic approach requires continuous co-operation of water authorities, local planning authorities, and regional bodies. But co-operation is costly (e.g., direct costs in terms of time, financial and human resources). Therefore, the request for a strategic approach to flood risk management does not suffice. It should be demonstrated how such an approach can be implemented under real-world conditions (Penning-Rowsell and Peerbolte 1994, Hutter 2005). Questions of *how* to formulate and implement strategies call for a process perspective (Poole 2004) on flood risk management.

229

J. Schanze et al. (eds.), Flood Risk Management:
Hazards, Vulnerability and Mitigation Measures, 229–246.
© 2006 *Springer.*

The purpose of this paper is to describe important aspects of a process perspective on flood risk management and to illustrate this perspective by referring to specific management issues. To realise this purpose, *section 2* outlines a multidimensional framework encompassing the content, process, and context dimensions of strategies for flood risk management. *Section 3* briefly describes the content dimension of strategies. *Section 4* argues that linear and adaptive management processes, strategic planning modes, and learning types can be linked to specific flood risk management issues. *Section 5* refers to the context dimension to illustrate some "givens" in a strategy process. *Section 6* states that strategies can be formulated at different spatial levels (local, regional, and so forth) with important implications for adopting a process perspective. *Section 7* draws con-clusions and provides an outlook on findings from empirical research currently under way to analyse the prospects of spatial planning as policy instrument of pre-flood risk management.

2. STRATEGIES FOR FLOOD RISK MANAGEMENT – A MULTIDIMENSIONAL FRAMEWORK

The term strategy is not new to flood risk management researchers (e.g., Penning-Rowsell and Peerbote 1994, p. 5, Kundzewicz 2002, Hooijer et al. 2004). Seldom is the term explicitly defined (an exception is Hooijer et al. 2004, p. 346). Even when explicit definitions are given, concepts and empirical findings of strategy research (e.g., Burgelman 2002, Pettigrew et al. 2002) are not taken into account. Strategy research fosters a comprehensive, multidimensional understanding of strategy. Such an understanding is necessary to solve non-routine management problems that imply – by definition – different understandings of cause-effect relation-ships and diverging interests.

For example, various researchers argue for deploying spatial planning to reduce vulnerability in flood-prone areas through discouraging development on floodplains and by channelling demands for housing and infrastructure to areas unlikely to be flooded (e.g., Burby et al. 2000, Penning-Rowsell 2001, Hutter 2005). Especially to reduce the negative consequences of extreme flood events, a more systematic use of planning instruments is recommended (Hooijer et al. 2004). However, research shows that local authorities make very limited use of spatial planning to reduce flood risk (Olshansky and Kartez 1998, Hutter 2005). There is an enduring discrepancy between *recommendations* to use planning as effective risk-reduction instrument and management *practice*. Up to now, concepts and findings from strategy research have not been used to shed light on this discrepancy and to find ways of using planning more effectively. To show how strategy research could improve analysis of flood risk management practice, we start with what could be seen as common ground for using the term strategy.

The term strategy derives from the Greek word strategia, meaning "general-ship", itself formed from stratos, meaning "army", and –ag, "to lead" (Grant 2005). Military strategy and strategies in other fields of societal development (e.g., business economics, public affairs) share a number of common concepts and principles, the most basic being the distinction between strategy and tactics. *Strategy* is the overall solution for deploying resources to establish a favourable position; a *tactic* is a scheme for a specific action. Whereas tactics are concerned with winning battles, strategy is about winning the war. Thus, strategic decisions share three common characteristics (Grant 2005, p. 14):

- They are important.
- They involve a significant commitment of resources.
- They are not easily reversible.

Especially the third characteristic is important to understand how flood risk management can benefit from strategy research. If external pressures are dominant, there is only very little decision space for an actor to choose his strategy (Burgelman 2002). Therefore, strategic decisions imply a significant degree of freedom for choice. They are fundamental decisions that influence the overall welfare of an actor. They refer to the interface of internal and external context conditions (Grant 2005).

For example, making strategic decisions a business organisation chooses in which industries it will compete against other enterprises. The decision to compete in a specific industry heavily influences the deployment of resources and the building of distinctive competencies over time. Once a strategic decision has been made fundamental decision possibilities decrease and one can concentrate on delivering the net benefits the strategy is promising. But context can change. Therefore, contexts conditions have to be monitored to assess if they still support the chosen strategy.

The tension between a) defining strategy to determine resource deployments and to build distinctive competencies on the one hand and b) to preserve flexibility and future fundamental decision possibilities on the other cannot be avoided. It has to be handled within the process of strategy making (Volberda 1998).

We conclude that strategy is a multidimensional phenomenon that encompasses the dimensions of content (*"Deciding what to do"*), process (*"Deciding how to do it"*), and context (*"Aligning strategic decisions with internal and external conditions"*). Consequently, strategic change requires time, resources, changing power structures, and capabilities to learn under enabling and inhibiting conditions for learning. In line with this multidimensional understanding of strategy the following definition is proposed:

> *A strategy for flood risk management is defined as a consistent combination of long-term goals, aims, and measures, as well as process patterns that is continuously aligned with the societal context.*

The rationale for this definition is as follows: Changing from the paradigm of flood protection to flood risk management raises challenging questions of formulating and implementing strategies within society. In particular, reducing vulnerability and increasing preparedness require a comprehensive understanding of flood risk management.

Strategy as multidimensional phenomenon is distinct from other possibilities to define strategy. Firstly, the definition of strategy is distinct from the classic definition from business economics. The classic approach defines strategy as combination of measures and resources for actions to implement the basic long-term goals of a business organisation (see Whipp 2001, p. 15151). This definition is closely linked to strategic planning as long-term planning under conditions which are largely predictable (Volberda 1998, p. 37, p. 206). The classic definition does not explicitly comprise process patterns of strategy making. Yet under conditions of increasing uncertainty process becomes more important to consider different views on a complex phenomenon, to exploit different possibilities of formulating and implementing strategies, and to adapt swiftly to unforeseen conditions. More and more, uncertainty is becoming a central topic for modelling and managing flood risk (Sayers et al. 2002, Hall et al. 2003). Hence, we should take process patterns into account in a systematic way and not ad hoc.

Secondly, Hooijer et al. see strategy as "a consistent set of measures, aiming to influence developments in a specific way" (2004, p. 346). This definition is restricted to the content dimension of strategies (e.g., aims, measures, scenarios, and so forth). Process and societal context issues are not included. In this report the term *strategic alternative* is used for combinations of structural measures and policy instruments aiming to influence developments in a specific way. Strategic alternatives are tactics within an overall strategy.

Thirdly, in the flood risk management literature you can find a simple daily live definition. In this case, a strategy is defined as a statement indicating the direction of using structural measures and policy instruments (e.g., "Do-nothing strategy", "Do-minimum strategy", "Use new measures to enhance flood risk management standard"). This daily live definition should not be confused with the scientific understanding of strategy as multidimensional phenomenon combining content, process, and context.

To specify this multidimensional definition of strategy, Figure 1 shows the three dimensions of content, process, and context. Important categories within the dimensions are listed.

It is crucial to recognise that planning and plans are elements of a strategy. *They are not themselves the strategy.* Usually, especially within the public sector, planning is necessary for "winning the game", but it is not sufficient (Mintzberg et al. 1999). Planning has to be complemented by implementation and learning processes at various levels of societal development. Process is about learning how to deal with diverse political interests, resource scarcity, existing responsibility of actors, cultural "world views" (Hooijer et al. 2004, p. 353), and limited capabilities to act and interpret the often complex processes of strategy and especially strategic change.

In what follows, the three dimensions of strategies for flood risk management are described in some detail. Thereby, the paper focuses on the process dimension. To consider the close relationship between process and context, the context dimension is described *after* elucidating various process categories.

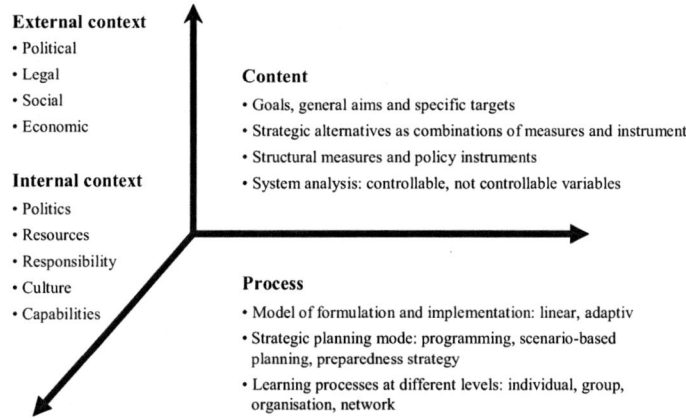

Figure 1: Dimensions of Strategies for Flood Risk Management (Hutter and Schanze, i.p.)

3. THE CONTENT OF STRATEGIES

Flood risk management aims at a continuous and holistic societal analysis, assessment and reduction of flood risk (for a basic framework depicting the main features of flood risk management, see Schanze, this issue, cf. Sayers et al. 2002, Hall et al. 2003). In developed ("formal") societies, flood risk is primarily managed by politicians and officials representing public institutions. Scientific institutions and non-profit organisations may play an important role.

From a process perspective on strategies, the nexus between goals, aims, and targets, as well as measures and instruments is of special importance.

Goals are defined as long-term goals of an actor or a set of actors (group, organisation, network of organisations). They are grounded in external conditions of strategies. They refer to the identity of the actor(s). For instance, in democratic political systems politicians pursue the goal of winning elections for different reasons (e.g., political belief that a specific strategy for flood risk management should be adopted, acquiring reputation and political power, increasing income, and so forth). This example illustrates that goals can be adapted to different interests of actors. Compared to aims and targets, they are a more abstract entity. They serve as a stable basis for evaluating different tactics despite changing societal context conditions.

Whereas goals refer to conditions with long-term stability, *aims* reflect conditions that can be changed in the medium term. More than goals, aims refer to the desire of actors to change the outside world. For instance, just to mention a hypothetical case, reacting to a recently experienced local flood event, politicians can be urged by citizens and organised stakeholders to pursue the aim of controlling development on floodplains through spatial planning. Politicians continue trying to win elections; because of changed external context conditions to win elections, the aim of avoiding a further increase of development in flood-prone areas becomes more important for reaching this political goal.

Targets specify aims. Targets can be defined temporally, spatially, and with regard to the possibility of quantifying the desired effects of flood risk management. For instance, the Environment Agency (EA) of the UK states that key targets of its strategy for flood risk management in the first three years (beginning in year 2003) are to

- "have no loss of life through flooding,
- reduce risk of flooding to life, major infrastructure, environmental assets and some 80,000 homes" (EA 2003, p. 6)

Targets are preconditions for programming specific actions, for answering questions if overall aims were realised, and for allocating responsibility with regard to intended and unintended consequences. Despite these benefits of targets for precise and transparent strategy making, they are not always to be found in practice. Formulating targets increases the risk to identify unsuccessful policies. Therefore (and for other reasons), flood risk managers sometimes prefer more abstract intentions in the form of aims.

Strategic alternatives are combinations ("bundles") of structural measures and policy instruments for flood risk management. They are tactics for using measures. Specific management problems of flood risk mitigation are due to various mutually reinforcing causes (e.g., the tendency to develop built-up areas on floodplains is facilitated, for instance, by development plans that do *not* address flood risk as a strategic problem, limited preparedness of citizens, no participatory planning process at local level, and so forth). These problems have to be addressed by consistent bundles of various measures and instruments.

Activities for reducing flood risk encompass *physical measures* and *policy instruments* (Schanze, this issue, further elucidates the distinction between measures and instruments). Generally, structural measures involve some engineering work (Kundzewicz 2002). They are capital-intense investments that require maintenance over a long life time. Policy instruments do not involve engineering work. To mention only a few examples, they are used for controlling land use, insuring flood losses, and informing people about the risk of flooding (further examples are given by Penning-Rowsell and Peerbolte 1994). The distinction between structural measures and policy instruments corresponds to the notion that flood risk management encompasses measures to control the flood hazard (the water that produces the flood) and instruments to reduce vulnerability (e.g., discouraging new development on floodplains through spatial planning).

Traditionally, flood management focused on water control through structural measures like dykes, dams, and reservoirs. Currently, we can see a shift from flood control to more holistic approaches for managing flood risk (Schanze, this issue, Hall et al. 2003) within a European perspective on integrated governance and water basin management (Bressers and Kuks 2004). The risk-based approaches underline the importance of considering policy instruments for flood risk management. For instance, spatial planning for controlling development in flood-prone areas is a key issue for reducing vulnerability with regard to extreme flood events (Hooijer et al. 2004). This shift implies a *strategic* and *processual* (Pettigrew 1997) approach because risk management requires developing *consistent* combinations of measures and instruments *over time*. The balanced combination of structural measures and policy instruments is an essential aspect of "integrated management" (Hooijer et al. 2004, p. 355).

4. ANALYSING THE PROCESS DIMENSION

Up to now, flood risk management research has a quite narrow view on the process of strategy making. Often, a linear model of strategy is assumed (explicitly or, more often, implicitly). However, there are alternatives that can contribute to improving the strategy process. Especially to tackle non-routine management problems, more attention should be allocated to issues of *how* strategy is made in practice. To show the link between different process categories and specific flood risk management issues, the following focuses on the linear and adaptive strategy model and strategic planning modes. Furthermore, the challenge of learning for flood risk management is mentioned.

4.1. The linear and the adaptive process model of strategy

In text books of strategy analysis (e.g., Weihrich and Koontz 1992) the strategy process is usually described as a *step-by-step* process of (1) *analysing* the internal and external forces of an organisation, (2) *formulating* aims and targets on the basis of the fundamental values, beliefs, and goals of decision makers, (3) *defining* and *evaluating* strategic alternatives as well as *deciding* on a preferred alternative, and (4) *implementing* the chosen alternative. (5) The fifth step comprises the subtasks of *controlling* and *learning* to gather information on consequences for answering the question what features of the process should be changed for future strategy making.

There is no problem in using such a linear approach as a heuristic device to understand what different factors and aspects are important for strategy analysis. But nowadays it has become all to clear from extensive theoretical discussion (e.g., Mintzberg et al. 1999, Volberda 1998) and empirical findings (Burgelman 2002, Grant 2005) that strategy processes do not always follow a simple step-by-step logic to solve complex and dynamic problems. In line with this, in the field of flood risk management the notion is propagated that decision making (e.g., of water authorities) should be understood as an iterative (e.g., Hall el. 2003, Hooijer et al. 2004) and adaptive process (see Pahl-Wostl et al., i. p.).

To look at the real-world conditions of strategies for flood risk management, it is a start to consider *two models* that make salient the process dimension of strategies. Here, a process model is a category that refers to organisational decisions and actions (Poole 2004). A process model explicitly describes *how* an organisation or a subunit decide and act to formulate and implement their aims and targets. Two models are of special importance (Chaffee 1985, Healey 1997, p. 248, Volberda 1998): the *linear* and the *adaptive* model (see Table 1).

Table 1. Two Process Models of Strategy

	Linear model of strategy	Adaptive model of strategy
Process	Sequential process of planning, programming, and implementation	Continuous alignment of content and process with context
	Top-down strategy making	Combination of bottom-up initiatives and top-down strategic decisions
Content	System of aims, targets, and strategic alternatives	System of strategic alternatives, aims, and targets
	Integrated set of strategic, operative and resource plans	Flexible configuration of resources
Context	Stable	Unstable
	Predictable	Limited predictability

According to the *linear model* the process of strategy making consists of a well-structured sequence. Using a linear model one would describe the making of flood risk management strategies as follows:

1. *Analysing the flood hazard and its actual and/or possible damages:* Analysis in the linear model is understood as an inquiry into objective relationships. Analysing facts and cause-effect relationships is clearly separated from discussing societal values.
2. *Formulating aims and targets:* On the basis of values and key principles of society, aims and targets for flood risk mitigation are formulated. It is crucial that aims and targets are defined before action is taken. Hence, in principle, the evaluative basis of mitigating flood risk has to be known.
3. *Defining, assessing, and deciding on strategic alternatives:* A decision to implement a certain alternative is only made after comparing and assessing the effects of possible bundles of measures and instruments. No action is taken until a preferable solution has been found ("First decide, then act").
4. *Implementing the chosen strategic alternative:* In a narrow sense, implementation can be defined as realising a formally defined programme consisting of more or less specified aims, measures and instruments to reach the desired effects of the programme. The programme comprises decisions to organise, staff, and direct for effective implementation.
5. *Controlling and learning:* Of course, the linear model of strategy considers that context conditions can change and expected effects will not occur, but unexpected will. Characteristically, activities to

control, evaluate, and learn are mainly undertaken *after* the implementation of the strategic alternative is finished.

The linear process model of strategy making has its merits. Aims are formulated on a thorough understanding of the flood risk system. The evaluation of a broad range of strategic alternatives is undertaken to avoid restricting strategy making to current practices. Decision makers are encouraged to look for new solutions. The model can be seen as a disciplined effort to produce fundamental decisions about how flood risk mitigation should be conducted.

The linear model corresponds with a top-down view on strategy making. Strategic decisions are mainly made by the decision makers formally responsible for the overall welfare of an organisation or a set of organisations. The model is based on the assumption that decision makers are willing to undertake a comprehensive analysis and assessment of the flood risk system and mitigation alternatives despite the possibility that context conditions can change (e.g., political support to deploy spatial planning for flood risk management can fade). To identify the critical elements for improving the current state of management, the linear model assumes a stable societal context.

The hallmark of the *adaptive process model* is that strategy does not move forward in a direct way through easily identifiable sequential phases (Pettigrew 1997). The process pattern is much more appropriately seen as continuous, iterative and uncertain. An adaptive process is characterised by *parallel* processes of formulating and implementing strategic alternatives. Hence, formulation and implementation are more difficult to differentiate. These difficulties can be observed in strategic management (Chakravarthy and White 2002, p. 201) and policy science studies (Jann and Wegrich 2003, p. 81). Decisions for formulating aims and targets, for analysing the internal and external context, and for combining measures and instruments are continuously aligned with the changing societal context (Chaffee 1985). Especially political context conditions facilitate or inhibit effective flood risk management. They do not only affect the implementation of strategic alternatives. They influence analysis and assessment too because political conditions restrict what actors notice and, therefore, what elements of the flood risk system they ignore.

The linear model works on the assumption that a single decision maker or an elite of decision makers can design an explicit "grand" strategy based on a highly top-down, deliberate, analytical process (Volberda 1998, p. 38). Top-down forces are strong. The adaptive model assumes that effective strategy making requires both *strong bottom-up and strong top-down forces* (Burgelman 2002). Empirical work has revealed very different roles of managers within such a strategy process: For instance, (1) *managers close to stakeholders* develop bottom-up initiatives how to improve operational decisions; (2) *middle managers* connect these initiatives with the highly political process at the top level of decision making; (3) *decision makers at the top* are responsible for challenging the status quo, setting broad performance aims, as well as co-ordinating, evaluating, and legitimising bottom-up initiatives. Hence, origins of strategic contents often lie in lower levels of decision making. Decision makers at the top heavily influence enduring resource commitments. Adaptive strategy is the joint outcome of content-based bottom-up initiatives and more formal top-down influences (Burgelman 2002).

Similar processes of maintaining strategic focus on the one hand and of expanding the range of strategy makers and considering the interests of stakeholders on the other are to be observed in participatory water management (e.g., House 1999), flood management at local and regional level (Fordham 2000, Hutter and Schanze 2004) and strategic spatial planning (Healey 1997, 2003).

This paper argues for a continuous adjustment of societal decision making and context. Hence, under specific conditions it could be appropriate to use the linear process model of strategy. It follows that the linear and adaptive models of strategy are complements. We already pointed in this direction. The linear model assumes a high degree of stability of the context conditions. Stability can be achieved, for instance, through a dominant coalition of political and administrative decision makers at regional level in favour of comprehensively analysing the flood risk system and implementing strategic alternatives that effectively mitigate flood risk. Thus, the question whether the linear or the adaptive model is appropriate for effective strategy making should be answered on the basis of empirical findings that consider the context of flood risk management in sufficient detail.

4.2. Strategic planning modes

The linear and the adaptive model comprise strategy formulation and implementation. Strategic planning is solely about formulation. It can be defined as a "disciplined effort to produce fundamental decisions and actions that shape and guide what an organisation is, what it does, and why it does it." (Bryson 1998, p. 2160) Fundamental decisions can be made in informal processes too. Hence, it is crucial to define strategic planning *not* with regard to its substance ("fundamental decisions"), but as a formal procedure (Mintzberg et al. 1999) that usually leads to some sort of strategic plan. Especially if the predictability of the context is very limited strategic planning is more about coordinating and codifying basic decisions than about producing these decisions (Grant 2005).

Strategic planning is *not* an unitary phenomenon. It is a *set* of organisational regulations, procedures, and process patterns that are partly applicable to the non-profit and public sector of developed, "formal" societies (Bryson 1998). Strategic planning can be fashioned in different ways. With regard to the turbulence of context it is useful to distinguish between three strategic planning modes (Volberda 1998, p. 36):

- *Programming* is appropriate under conditions which are highly predictable.
- *Scenario-based planning* considers different plausible futures; it is appropriate in case of complex and dynamic context conditions that can be predicted with a sophisticated set of routines and information systems.

- *Preparedness strategies* increase organisational activities for coping with strategic surprise; a minimum of planning devices gives the strategy process a broad, overall direction in which actions and decisions can adaptively emerge.

These three strategic planning modes can be related to specific problems of using structural measures and policy instruments for flood risk management. Planning as *programming* is based on an extrapolation of trends. No fundamental changes of context conditions are accounted for. Organisational activities are directed towards implementing a predefined strategic alternative. The programming mode corresponds with the linear model of strategy making.

For instance, after the severe flash flood event in the Weisseritz catchment, a medium-sized tributary of the Elbe river within the Region of Dresden, the water authority is planning to build additional water retention basins (Hutter and Schanze 2004). To make an informed decision about the location and capacity of the basins, a decision making process that resembles the programming mode is now taking place. Thereby, the water authority is communicating with local and regional actors (e.g., municipalities, the regional planning office) about the conflict potential of different locations (e.g., conflict between nature conservation aims and flood protection). The water authority is combining the structural measure of building additional water retention basins with the policy instrument of communicating with stakeholders to reach a decision with high probability of implementation.

Scenario-based planning considers systematically that various developments of external context conditions are possible. Scenarios are neither predictions nor forecasts. A scenario is defined as a *plausible description* of how the future may develop, based on a coherent and internally consistent set of assumptions about key relationships and driving forces (e.g., rate of technology changes, prices). Scenarios serve different purposes. Firstly, scenarios can be formulated to identify *robust* analytical findings with regard to future flood risk and strategic decisions despite very different possible future context conditions. Secondly, they can be defined for matching strategic alternatives to different possible external developments. In this case scenarios increase the *flexibility* of flood risk management (Hooijer et al. 2004). Scenario-based planning can be deployed on the basis of a linear or an adaptive strategy model.

Within the context of flood risk management, a *preparedness strategy* aims at ensuring effective responses to the impact of hazards, including timely and effective early warnings and the evacuation of people and property from threatened locations. Preparedness strategies are developed to cope with strategic surprise. *Actors expect the unexpected* (Weick and Sutcliffe 2001). Therefore, planning activities are decreased, whereas organisational activities are increased. Actors focus less on the question "What should we precisely do?", but on the questions "What should we be capable of doing?" and "With whom should we be able to communicate swiftly?". Preparedness strategies connect pre-flood risk management and event management (see Schanze, this issue). They are essential for coping with flash floods (Gruntfest and Ripps 2000). They enable decision makers and citizens to build in advance of the event the capabilities of effective communication under the condition of a fast developing flood hazard. But swift communication has its limits. Flash floods can be too fast to have no loss of life through flooding. Therefore, preparedness strategies should be compared with pure pre-flood strategic alternatives, for instance, instruments to discourage further development in flood-prone areas (Hall et al. 2003, p. 133) such as spatial planning (Hutter 2005). In case of comparing pre-flood and event strategic alternatives a broad range of actors has to be taken into account (e.g., insurance companies, developers, and municipalities with regard to pre-flood risk management, municipalities and emergency institutions for event management).

4.3. Learning for flood risk management

Strategic planning and learning are sometimes seen as antipodes (e.g., Mintzberg et al. 1999). In fact, planning can be more a way of preserving old views "how things are down and ought to be down around here" than exploring new strategic alternatives. It can enable organisations to gain focused action on a limited set of strategic issues, related aims, and targets. Especially planning procedures within large-sized organisations are prone to become rigid rituals that inhibit strategic thinking (Volberda 1998). However, planning and learning are not necessarily opposites. Planning as linear programming is appropriate if the relevant context conditions are stable and largely predictable. Strategic planning can be fashioned as scenario-based learning which fosters thinking in multiple possible futures and which enables decision makers to think about their "mental models" (Grant 2005). Especially spatial planning can be understood as process of learning with plans (not despite of plans) for taking into account the diversity of actors relevant for urban and regional development as well as the limited resources of spatial planners to implement spatial aims and targets (Mastop and Faludi 1997).

"Learning to live with rivers" has become a well-known expression within the policy field of flood risk management (e.g., Fleming 2002). Thereby, the term learning points to the challenge of acquiring something new (new knowledge, new behavioural potentials). Hence, the outcome of learning is in the foreground.

In the scientific literature on organisational learning and the "learning organisation" (see the contributions in Easterby-Smith and Lyles 2003) more emphasis is put on understanding learning as a complex process involving different learning types. Learning is defined as history-dependent, aim-oriented process drawing inferences from experiences of self or others (March and Levitt 1999). Learning encodes inferences into routines that control subsequent behaviour. Very different types of learning can be distinguished (e.g., simple learning, complex learning, and learning to learn; cognitive, behavioural, semantic learning). Thus, usually, actors learn (Volberda 1998). The question is *who* learns and *what*, *how*, *when*, and *why* learners learn and if the current learning processes are appropriate to foster an effective flood risk management.

Learning is context-sensitive because it a) refines *existing* knowledge and affirms existing beliefs or b) leads to fundamentally *new* knowledge and beliefs about causal and evaluative relationships. Whatever the case, affirmation or fundamental change, learning refers to the context of its starting conditions.

5. THE CONTEXT OF STRATEGIES

Some elements of societal context relevant for flood risk management are listed in Figure 1. The list is not exhaustive and it is doubtful if such a list could ever be defined. The context of flood risk management is not a fixed entity, but follows from which management problem is in the foreground of inquiry.

The context comprises variables that cannot be altered within the chosen time perspective. This is not to say that there is no dynamic relationship between context, content, and process. Context reduces the complexity of possible decisions and actions to a manageable amount. Hence, context enables actors to make informed decisions because many relevant variables can be treated as given (e.g., responsibility of water authorities in Germany for defining floodplains and responsibility of spatial planning authorities to formulate development plans for areas behind structural measures as dykes etc.).

Context enables and restricts human agency. Human agency and societal processes can change context (Bressers and Kuks 2004). However, it is very demanding to study the dynamic relationship between context and process. In many cases of strategy research, a pragmatic approach to consider context seems appropriate:

External context factors are understood as static boundary conditions for analysing the content and processes of flood risk management at local level or regional level. From the viewpoint of strategy making within small- to medium sized catchments, external context comprises political, legal, economic and social conditions (e.g., policy guidelines of federal and state government, national spatial planning system, mean real estate market prices).

Internal conditions influence urban and regional development from within. *Political* factors determine the extent and duration of support for flood risk management in various policy fields and the way of dealing with (potential) conflicts between public and private interests. For instance, discouraging further development of settlements on floodplains through spatial planning requires a stable dominant political coalition at local level in favour of this policy initiative as well as facilitating external context conditions such as legal requirements. In this case, growing demands for housing are channelled to areas unlikely to be flooded; the economic interest of landowners in using their property on floodplains as built-up area is outweighed in local political decision making.

Internal *financial resources* have an important impact on how pre-flood risk management is developed. A high extent of resources currently not needed to accomplish predefined tasks ("high extent of organisational slack") facilitates experimentation with *new* structural measures and policy instruments because projects can be implemented without significantly changing existing priorities.

Local and regional actors can decide to take the *responsibility* for pre-flood risk management without referring primarily to the external political and administrative context (Mileti 1999). They believe that adverse impacts of flooding are partly consequences of prior decisions to invest in flood-prone areas.

These are only simple examples how the societal context of strategies for flood risk management can be considered. A more detailed account of context and the interplay of context, process, and contents requires a closer look at specific cases (see Hutter 2005 for a comparative case study design that analyses local spatial planning and flood risk management in three European countries: Germany, UK, and Italy).

6. INTEGRATING SPATIAL SCALES, POLICY ISSUES, AND PLANNING HORIZONS

So far, the paper proposes a flexible three-dimensional framework for strategy analysis. The framework can be applied to different spatial scales, actor constellations and situations. This leaves open the question at what spatial scale strategy making takes place and who the strategy makers are. Scales, actors and policy issues, as well as planning horizons are crucial for defining the level of integration.

Strategies of single actors (for example, local planning authorities, water authorities, and regional bodies) are very diverse with regard to spatial scale and policy issues. For instance, to put it simple, the responsibility of water authorities ends where the responsibility of local planning authorities begins ("in front of and behind the dykes"). For pre-flood risk reduction through reducing vulnerability, integrating the strategies of these actors becomes a crucial challenge. We consider two possibilities: developing a strategy for *a part* of the catchment and crafting a strategy for *the* whole catchment.

A *strategy for a part of the catchment* is defined as a strategy to manage risk within a geographical part of the catchment. Notwithstanding the heterogeneity of single-actor strategies, a consistent pattern of decisions and actions of multiple organisations can be observed. The scale can vary between very limited ("project level"), limited ("area"), and covering more than half of the catchment. The geographical part of the catchment serves as a common reference for integrating the relevant decisions and actions of various actors.

For instance, in some catchments an enduring overall consensus can be observed that a specific area is of high political, cultural, and economic value for regional development (e.g., historical centre in Dresden). For this reason, structural measures and policy instruments are combined to guarantee a safety level above average compared with other areas of the catchment. This example illustrates the assumption that strategies for a part are more likely to emerge than strategies for the whole catchment.

Relevant organisations can be identified with regard to the geographical boundaries of the area. It is easier to include citizens and private enterprises into the strategy process.

A *strategy to manage flood risk on catchment scale* is a complex endeavour. This is so because, more often than not, flood risk management practice in whole catchments is fragmented (Hall et al. 2003). To overcome a state of fragmentation, various challenges have to be met. Within the *content* dimension, different strategic aims and measures of all relevant actors have to be integrated (e.g., interdependencies between strategies to reach "good water quality" according to the European Water Framework Directive and "protecting society against flooding"). Strategy making takes place at different interrelated spatial levels.

Process integration aims at combining different time horizons and process patterns (linear / adaptive model, planning modes, and so forth). Even in one municipality time perspectives can differ between a department responsible for fostering economic development (dominant short-term planning horizon) and a department responsible for restructuring open spaces (dominant long-term planning horizon). *Context* conditions are needed that address the double challenge of (1) fostering autonomous actions of single organisations to reap the benefits of specialisation and (2) establishing formal and informal institutions for common orientation and constant flood risk communication (Grabowski and Roberts 1997).

In sum, a catchment strategy that comprises multiple actors, spatial scales, policy issues and planning horizons ought to be established empirically rather than rhetorically.

7.   CONCLUSIONS AND OUTLOOK

Flood risk management research is moving in the direction of a continuous and holistic approach that stresses the need of analysing, assessing, and mitigating flood risk on catchment scale (Schanze, this issue). From a process perspective, three issues stand out to be considered in future research work:

- Deploying a linear and top-down *or* an adaptive and participatory management process should be analysed with regard to different societal contexts and spatial scales.
- To design effective strategies, the relationships between strategic planning modes (e.g., scenario-based planning, preparedness strategy) and flood types (plain floods, flash floods) need further investigation.
- Changing from flood protection to flood risk management is a complex and long-term learning process in which the unique management histories of European countries should be considered (Handmer et al. 1999). Therefore, process research takes history-dependent contexts into account and looks for management principles with high versatility (Poole 2004).

244

This process perspective on strategies for flood risk management needs elaboration with regard to specific management issues. In this respect, local spatial planning for flood risk management is a promising issue (Penning-Rowsell 2001). Currently, there is evidence within the research community and related policy fields that leeway to exploit strategic choices inherent in planning is being recognised and used more intensely than before for pre-flood risk management. This evidence has to be enhanced through case studies showing the prospects of planning-based strategies in the real world. Thereby, it is inappropriate to expect major improvements from deploying only one measure or one dominant combination of structural measures and policy instruments. It is more likely that robust improvements will come from changing various elements over time (Pettigrew 1997).

In line with these expectations, Burby and associates argue that local governments "must take care in carrying out hazard reduction planning, minding both the political and technical details. Some of the lessons from local experience show that communities must be both visionary and pragmatic. They need to be far-sighted in gathering credible data, preparing maps, building consensus through planning, and paying attention to development management well before pressures built to use hazard areas more intensively. They also must be practical in using site-specific approaches, integrating hazard mitigation into their normal development review procedures, taking advantage of post-disaster windows of opportunity, and being prepared to purchase property if necessary." (Burby et al. 2000, p. 103f.) Case studies within the Integrated Project (IP) FLOODsite under the 6th European Framework (see Samuels, this issue) will show how these recommendations can be specified with regard to context conditions of cities as different as Dresden and London (Hutter 2005).

ACKNOWLEDGEMENTS

I acknowledge the support of the NATO-ARW in funding my presentation of the paper, and the FLOODsite project (contract GOCE-CT2004-505420) for its preparation.

References

Bressers H. and Kuks St. (2004) *Integrated Governance and Water Basin Management. Conditions for Regime Change and Sustainability.* Kluwer, Dordrecht.

Bryson J. M. (1998) Strategic Planning. Shafritz J. (ed.) *International Encyclopedia of Public Policy and Administration*, Holt and Co., New York, 2160-2169.

Burby R. J., Deyle R. E., Godschalk D. R. and Olshansky R. B. (2000) Creating Hazard Resilient Communities through Land-Use Planning. *Natural Hazards Review*, 1(2), 99-106.

Burgelman R. (2002) *Strategy is Destiny. How Strategy-Making Shapes a Company's Future.* The Free Press, New York.

Chaffee E. E. (1985) Three Models of Strategy. *Academy of Management Review* 10(1), 89-98.

Chakravarthy B. S. and White R. E. (2002) Strategy Process: Forming, Implementing and Changing Strategies, Pettigrew A., Thomas H. and Whittington R. (eds.) *Handbook of Strategy and Management*. Sage, London, 182-205.

DKKV – Deutsches Komitee Für Katastrophenvorsorge (2003) *Hochwasservorsorge in Deutschland. Lernen aus der Katastrophe 2002 im Elbegebiet* [Preventive Flood Management in Germany. Lessons Learned from the Catastrophe 2002 within the Elbe River Basin]. DKKV, Bonn.

EA – Environment Agency (2003) *Strategy for Flood Risk Management (2003/4 – 2007/8). Version 1.2*. EA, London.

Easterby-Smith M. and Lyles M. (eds.) (2003) *Handbook of Organizational Learning and Knowledge Management*. Basil Blackwell, Malden/USA.

Fleming G. (2002) Learning to Live with Rivers – the ICE's Report to Government. *Proceedings of ICE*, 15-21.

Fordham M. (2000) Participatory Planning for Flood Mitigation: Models and Approaches. Parker D.J. (ed.) *Floods. Volume II*, Routledge, London, 66-79.

Grabowski M. and Roberts K. (1997) Risk Mitigation in Large-Scale Systems: Lessons from High Reliability Organizations. *California Management Review* 39(4), 152-162.

Grant R. M. (2005) *Contemporary Strategy Analysis*. Basil Blackwell, Malden/USA.

Gruntfest E. and Ripps A. (2000) Flash Floods: Warning and Mitigation Efforts and Prospects. Parker D. J. (ed.) *Floods, Volume I*, Routledge, London, 377-390.

Hall J. W., Meadowcroft I. C., Sayers P. B. and Bramley M E (2003) Integrated Flood Risk Management in England and Wales. *Natural Hazards Review* 4(3), 126-135.

Handmer J., Penning-Rowsell E. and Tapsell S. (1999) Flooding in A Warmer World: The View from Europe. Downing T. E., Olsthoorn A. A. and Tol R. S. J. (eds.) *Climate, Change and Risk*, Routledge, London, 125-161.

Healey P. (1997) Collaborative Planning. Shaping Places in Fragmented Societies. Macmillan Press, Houndmills.

Healey P. (2003) Collaborative Planning in Perspective. *Planning Theory* 2(2), 101-123.

Hooijer A., Klijn F., Pedroli B. and Van Os A. (2004) Towards Sustainable Flood Risk Management in the Rhine and Meuse River Basins: Synopsis of the Findings of IRMA-SPONGE. *River Research and Applications* 20, 343-357.

House M. A. (1999) Citizen Participation in Water Management. *Wat. Sci. Tech.* 40(10), 125-130.

Hutter G. (2005) *Strategies for Pre-Flood Risk Management*. FLOODsite Report T13-05-01, Dresden.

Hutter G. and Schanze J. (i.p) *A Framework for Understanding and Improving Strategies for Flood Risk Management*. Dresden.

Hutter G. and Schanze J. (2004) Potenziale kooperativen Lernens für das Hochwasserrisikomanagement – am Beispiel der Vorsorge gegenüber Sturzfluten im Flussgebiet der Weißeritz [Collaborative Learning for Pre-Flood Risk Management – The Case of the Weisseritz]. Felgentreff C. and Glade Th. (eds.) *Von der Analyse natürlicher Prozesse zur gesellschaftlichen Praxis*, Universitätsverlag Potsdam, Potsdam, 63-87.

Jann W. and Wegrich K. (2003) Phasenmodelle und Politikprozesse: Der Policy Cycle [Phase Models and Politics: The Policy Cycle]. Schubert K. and Bandelow N. C. (eds.) *Lehrbuch der Politikfeldanalyse*, Oldenbourg, München, 71-104.

Kundzewicz Z. W. (2002) Non-structural Flood Protection and Sustainability. *Water International* 27(1), 3-13.

March J. G. and Levitt B. (1999) Organizational Learning. March J. G. *The Pursuit of Organizational Intelligence*, Blackwell, Malden, 75-99.

Mastop H. and Faludi A. (1997) Evaluation of Strategic Plans: the Performance Principle. *Environment and Planning B: Planning and Design* 24, 815-832.

Mileti D. S. (1999) *Disasters by Design. A Reassessment of Natural Hazards in the United States*. Joseph Henry Press, Washington D.C..

Mintzberg H., Ahlstrand B. and Lampel J. (1999) *Strategy Safari: Eine Reise durch die Wildnis des Strategischen Managements* [Strategy Safary: A Guided Tour Through the Wilds of Strategic Management]. Ueberreuter, Wien.

Olshansky R. B. and Kartez J. D. (1998) Managing Land Use to Build Resilience. Burby R. J. (ed.) *Cooperating with Nature. Confronting Natural Hazards with Land-Use Planning for Sustainable Communities*, Joseph Henry Press, Washington, D. C., 167-201.

Pahl-Wostl C., Downing T., Kabat P., Magnuszewski P., Meigh J., Schlueter M., Sendzimir J. and Werners S. (i.p.) Transitions to Adaptive Water Management: The NeWater Project. *Water Policy*.

Penning-Rowsell E. (2001) Flooding and Planning: Conflict and Confusion. *Town and Country Planning*, April, 108-110.

Penning-Rowsell E. and Peerbolte B. (1994) Concepts, Policies and Research. Penning-Rowsell E. and Fordham M. (eds.) *Floods Across Europe. Flood Hazard Assessment, Modelling and Management*, Middlesex University Press, London, 1-17.

Pettigrew A. (1997) What is a Processual Analysis? *Scandinavian Journal of Management* 13(4), 227-248.

Pettigrew A., Thomas H. and Whittington R. (eds.) (2002) *Handbook of Strategy and Management*. Sage, London.

Poole M. S. (2004) Central Issues in the Study of Change and Innovation. Poole M. S. and Van de Ven A. H. (eds.) *Handbook of Organizational Change and Innovation*, Oxford University Press, Oxford, 3-31.

Sayers P. B., Hall J. W. and Meadowcroft I. C. (2002) Towards Risk-based Flood Hazard Management in the UK. *Proceedings of ICE,* 36-42.

Schanze J. (2002) Nach der Elbeflut – die gesellschaftliche Risikovorsorge bedarf einer transdisziplinären Hochwasserforschung [After the Elbe flood – Societal Risk Reduction Requires Transdisciplinary Flood Risk Management Research]. *GAIA* 11 (4), 247-254.

Volberda H (1998) *Building the Flexible Firm. How to Remain Competitive*. Oxford University Press, Oxford.

Weick K. E. and Sutcliffe K. (2001) *Managing the Unexpected. Assuring High Performance in an Age of Complexity*. Jossey-Bass, San Francisco.

Weihrich R. and Koontz H. (1992) *Management*. McGraw-Hill, New York.

Whipp R. (2001) Strategy: Organizational. Smelser N. J. and Baltes P. B. (eds.) *International Encyclopedia of the Social and Behavioral Sciences. Volume 22*, Amsterdam, 15151-15154.

PART 6

HISTORICAL FLOODS AND TRANSBOUNDARY ISSUES

Chapter 20

HISTORICAL AND RECENT FLOODS IN THE CZECH REPUBLIC: CAUSES, SEASONALITY, TRENDS, IMPACTS

RUDOLF BRÁSDIL[1], PETR DOBROVOLNY[1],
VILIBALA KAKOS[2] AND OLDRICK KOTYZA[3]
[1]*Institute of Geography, Masaryk University, Brno, Czech Republic,*
[2]*Institute of Atmospheric Physics, Czech Academy of Sciences, Prague,*
Czech Republic, [3]*Regional Museum, Litoměřice, Czech Republic*

Keywords: Historical floods, recent floods, principal component analysis, Elbe River, Vltava River, Czech Republic.

1. INTRODUCTION

In the Czech Republic the most destructive natural disasters result from floods. Evidence for this was the flood of July 1997 in Moravia and Silesia which took 52 lives and caused damage in excess of 63 billions Kč (Czech Crowns). Similarly the flood in Bohemia in August 2002, took 19 lives and damage ran over 70 billions Kč. The flooding in eastern Bohemia in July 1998 (6 victims, damage 2 billions Kč) was lesser in scope. The occurrence of destructive floods following a long period of relative calm raises the question as to what extent these floods are the results of natural climatic variability or the consequence of anthropogenic conditioned global warming that could result in future increase in frequency and intensity of floods (Houghton et al. 2001, Beniston 2002, Milly et al. 2002, Christensen and Christensen 2003). In order to implement protective measure that would tend to minimise human and material losses it is necessary to establish a knowledge base that would be sufficiently comprehensive, and when necessary to combine the information from systematic (instrumental) hydrological observations with those documentary sources of the pre-instrumental period.

This contribution discusses briefly some aspects of major floods in the Czech Republic over the past millennium which resulted from the research project of the Grant Agency of the Czech Republic "Historical and recent floods in the Czech Republic: causes, seasonality, trends, impacts", investigated during 2003.

J. Schanze et al. (eds.), Flood Risk Management:
Hazards, Vulnerability and Mitigation Measures, 247–259.
© 2006 *Springer.*

2.   DATA ABOUT FLOODS

The beginnings of hydrological measurements in the Czech Lands date to the year 1825, when the city administration of Prague built the first water-gauge station in the profile of the Old Town weir on the river Vltava (Novotný 1963). Systematic hydrological observations began on the so-called imperial rivers (the Vltava from České Budějovice to Mělník and the Elbe from Mělník to the country's frontier) in 1851 at water-gauge stations at Mělník, Litoměřice, Ústí nad Labem and Děčín. Measurements were carried out almost exclusively by workers of the building division of the royal and imperial governor's office. After a disastrous flood on the Berounka and the Ohře on 25–27 May 1872 and after an unusually dry year 1874, a Hydrographical commission for the Czech Kingdom was established at the instigation of the land parliament. From 1875 its hydrometrical department was responsible for measurements of water levels and discharge rates. For 20 years thereafter it published year-books relating to water-gauge observations in Bohemia. When the Hydrographical commission was disbanded in January of 1888, 15 water-gauge stations were in operation on the imperial rivers and further 32 stations on other streams. Subsequently the hydrographical division of the department of the technical office of the agricultural council for the Czech Kingdom was entrusted with the organisation of water-gauge observations.

In 1893 a centralised hydrographical service was established in Vienna (Hydrographischer Dienst). The hydrographical office began publishing a year-book relating to the status of all rivers of the empire (Jahrbuch des K. k. hydrographischen Central-Bureaus), entries began with the year 1895 (with data for 1893).

After the rise of independent Czechoslovakia the State Hydrological Institute was established in Prague along with an executive hydrographical division in each Land (in Prague for Bohemia, in Brno for Moravia, in Opava for Silesia). In 1954, following various reorganisations, the hydrological service was merged with the State Hydrological Institute resulting in the creation of a new Hydrometeorological Institute. Renamed in 1969 to the Czech Hydrometeorological Institute, this institute remains responsible for hydrological observations in the Czech Republic.

Analysis of historical floods on the major Czech Republic's rivers was made possible by data obtained from water-gauge stations often with more than a hundred-year long observation series: Vltava-Prague (1825–2003), Labe-Děčín (1851–2003), Ohře-Louny (1884–2003), Odra-Bohumín (1896–2003) and Morava-Kroměříž (1916–2003) (Figure 1). It is not unreasonable to assume that with the passage of time the river's observational sites change and along with the watershed area. One needs to be mindful of factors relating to the re-location of the water-gauge stations, changes in land-use and the riverbed as well as the construction of water reservoirs. Thus naturally occurring changes in cross-section of rivers and the man-made interventions make the precise quantifying difficult, although above facts were doubtlessly projected into the homogeneity of the flood series.

The intensity of a flood can be characterised by the value of the so-called peak discharge rate. From those values during individual floods the N-annual peak discharge rate Q_N is determined, which is reached or exceeded on the average once in N-years. Since the Q_N values are derived from fitting a theoretical curve onto measured values, Q_N values are in effect subject to the researcher's choices of periods and lengths of observation, homogeneity, and the selection of theoretical distributions. Further, the Q_N values allow for modelling of the extent of an inundated area during the N-annual peak discharge rate.

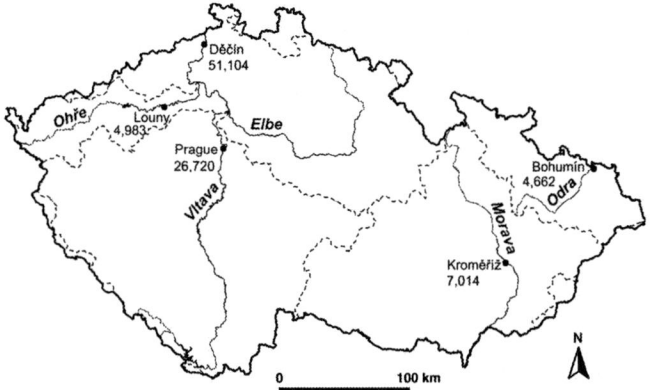

Figure 1. A schematic map of studied river watersheds with hydrological stations in the Czech Republic. The number at a station indicates the size of the corresponding part of the catchment area in km²

3. SYNOPTIC-CLIMATOLOGICAL ANALYSIS OF FLOODS

Typical floods on the Czech Republic's rivers can result from short intense precipitation (flash floods), from substantial continuous precipitation or from snow melting. In case of sudden melting after heavy colds, flooding can occur during formation of ice barriers on river due to movement of ice. These types of floods can exhibit different transitory forms, mainly occurring in the spring months and resulting from the combination of above causative and other physical-geographical factors (e.g. the freezing of the soil). Since the absolute majority of mixed floods (snow-melting or ice movement combined with precipitation) in the Czech Republic occur from December to March, these floods are further denoted as floods of the winter synoptic type. On the other hand, the absolute majority of rain floods occur from May to October, hence these floods ascribed as floods of the summer synoptic type (Kakos 1978).

Using the method of Principal Component Analysis (PCA) it was possible to characterise typical features of the sea level pressure (SLP) field on days D-5 to D (D indicates the day with the occurrence of the culmination discharge rate at the given station) for the floods of the winter synoptic type for the Vltava in Prague (Figure 2) and on days D-3 to D for floods of the summer synoptic type for the Odra at Bohumín (Figure 3). The first component describes in the two cases about 1/3 variability of the SLP field. In the first case there is an advection of the warm oceanic air into Central Europe is situated in the frontal zone between the belt of low air

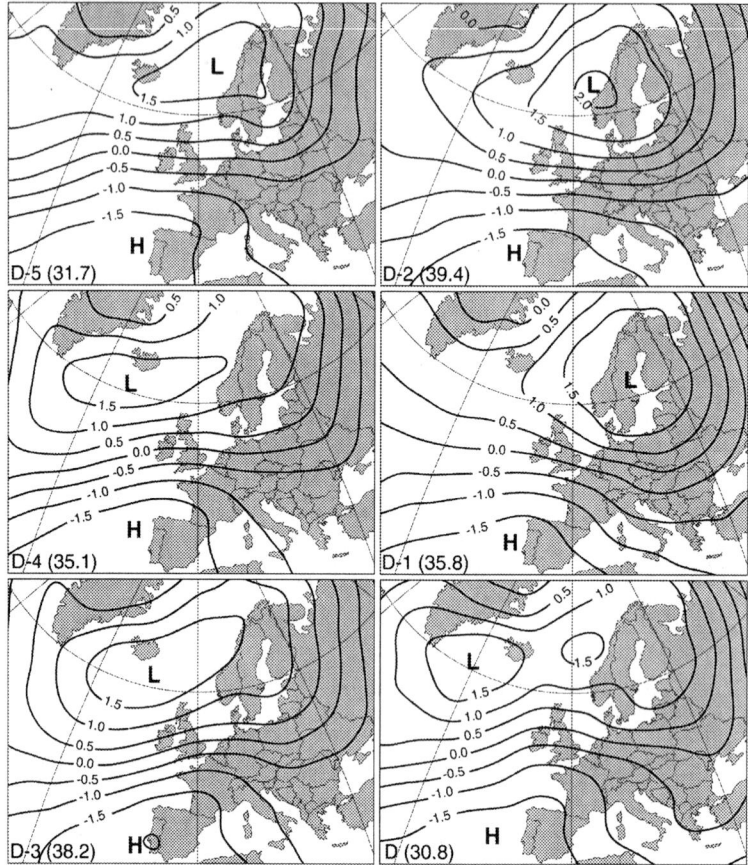

Figure 2. Component scores of the first component of the SLP field on days D-5 to D calculated by the PCA method for 37 floods of the winter synoptic type on the river Vltava in Prague in the period 1881–2000. In brackets the percentage of the explained variance of the SLP field is given (H – high, L – low)

pressure stretching from southern Greenland to Iceland and Scandinavia and a high air pressure above south-western Europe and the adjacent part of the Atlantic Ocean. The snow-melting due to the warm airflow is frequently intensified by the falling liquid precipitation. On the other hand, in the case of floods of the summer synoptic type, an important role is played by the progress of cyclones from the region of the Mediterranean to Central Europe. Their centre at the time of maximum precipitation is located easterly and north-easterly of the territory of the Czech Republic.

Extraordinarily high precipitation is recorded in the regions of the Moravian-Silesian Beskids and the Hrubý Jeseník Mts., and likewise in the Jizerské hory Mts. and the Giant Mts. (Štekl et al. 2001). Exceptionally high precipitation can also affect the Ore Mountains, as was the case prior the flood in August 2002 (e.g. the extreme precipitation total of 312.0 mm measured on 12 August 2002 at the station Zinnwald, 882 m a.s.l. – Meteorologické zprávy 2002).

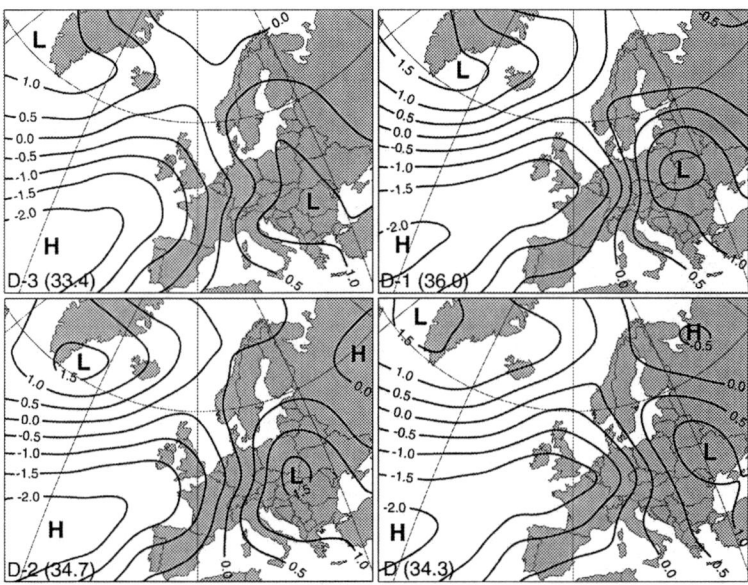

Figure 3. Component scores of the first component of the SLP field on days D-3 to D calculated by the PCA method for 40 floods of the summer synoptic type on the river Odra in Bohumín in the period 1896–2000. In brackets percentage of the explained variance of the SLP field is given (H – high, L – low)

4.   TRENDS IN THE OCCURRENCE OF FLOODS IN THE INSTRUMENTAL PERIOD

Figure 4 shows the fluctuation of the culmination discharge rates higher than Q_2 for 5 selected rivers. The drops in frequency as well as N-year peak discharge rates can be easy seen, particularly on the well documented two longest series, i.e. on the Vltava at Prague and the Elbe at Děčín. At both measuring stations the number of recorded floods has steadily decreased (Table 1): 1851–1900 (47 in Prague, 36 at Děčín); 1901–1950 (26 and 21, respectively); 1951–2000 (14 and 16, respectively). At Prague the occurrences of flooding decreased 3.4 times, at Děčín 2.3 times. With respect to the two stations, homogenisation has been in effect since 1954 due to construction of water reservoirs (the so-called "Vltava cascade"). This system contributes to reduction of peak discharges by different-size values (Kašpárek and Bušek 1990). The drop in the number of floods and peak discharges in the last 50 years was actually greater than stated. The overall rise in air temperature (Brázdil et al. 2004) bears directly on the marked decrease in the number of floods on these rivers. This trend was caused by a reduction of the winter type flooding, mainly due to warmer winters and smaller water content in the snow cover, and/or to a later onset of constant frosts with snowfall. For the stated reasons and due to increased river pollution, freezing-over of rivers and the formation of ice barriers thus became less frequent. For example, the formerly frequently occurring February and April floods were not noted in the latter half of the 20th century. Further, the number of March floods dropped conspicuously. Moreover, a decrease of January floods was noted in Prague (see Table 1).

Table 1. *Monthly frequencies of floods with peak discharge rates higher than Q_2 (corrected data according to Kakos 2001)*

Vltava - Prague ($Q_2 = 1090$ m^3.s^{-1}), period 1825–2000

Period	J	F	M	A	M	J	J	A	S	O	N	D	Year
1825-1850	3	4	3	0	1	2	1	0	0	0	0	1	15
1851-1900	5	11	10	4	3	2	1	4	3	1	1	2	47
1901-1950	1	8	6	3	2	1	0	1	1	1	0	2	26
1951-2000	1	0	4	0	0	2	3	1	0	0	0	3	14
1825-2000	10	23	23	7	6	7	5	6	4	2	1	8	102

Elbe - Děčín ($Q_2 = 1830$ m^3.s^{-1}), period 1851–2000

Period	J	F	M	A	M	J	J	A	S	O	N	D	Year
1851-1900	3	8	10	5	2	1	0	2	2	0	2	1	36
1901-1950	3	5	5	3	0	2	0	0	0	1	0	2	21
1951-2000	3	0	5	0	0	1	4	1	0	0	0	2	16
1851-2000	9	13	20	8	2	4	4	3	2	1	2	5	73

During the latter half of 1851–2000 the Vltava and the Elbe floods were characterised by lower peak discharge rates (Figure 4). In this context, the flood of August 2002 is an exception, with waters reaching their highest levels on 14 August on the Vltava in Prague and on 16 August on the Elbe at Děčín (during the second flood wave) (Meteorologické zprávy 2002). On the Elbe at Děčín, however, the August flood fell short of the peak discharges of floods of 30 March 1845 and of 3 February 1862.

5. MILLENNIAL CHRONOLOGIES OF FLOODS

In order to ascertain the frequency and extent of floods prior to instrumental period, different documentary sources (chronicles, memoirs, diaries, economic records, etc.) can be utilised. The first credible report about a flood in the Czech Lands is given by the chronicler Kosmas for September 1118 in Prague (Bretholz 1923): *"In the year of our Lord 1118 in the month of September there was such a flood as, I think, has not been on the Earth since the Deluge. This river of ours, the Vltava, suddenly breaking out of its bed, how many villages, how many houses in the suburbia, huts and churches did it take away! At other times, although it happens rarely, the water reaches only the floor of the bridges, but this flood raised to the height of ten ells* [i.e. approximately 6 m] *over the bridge."*

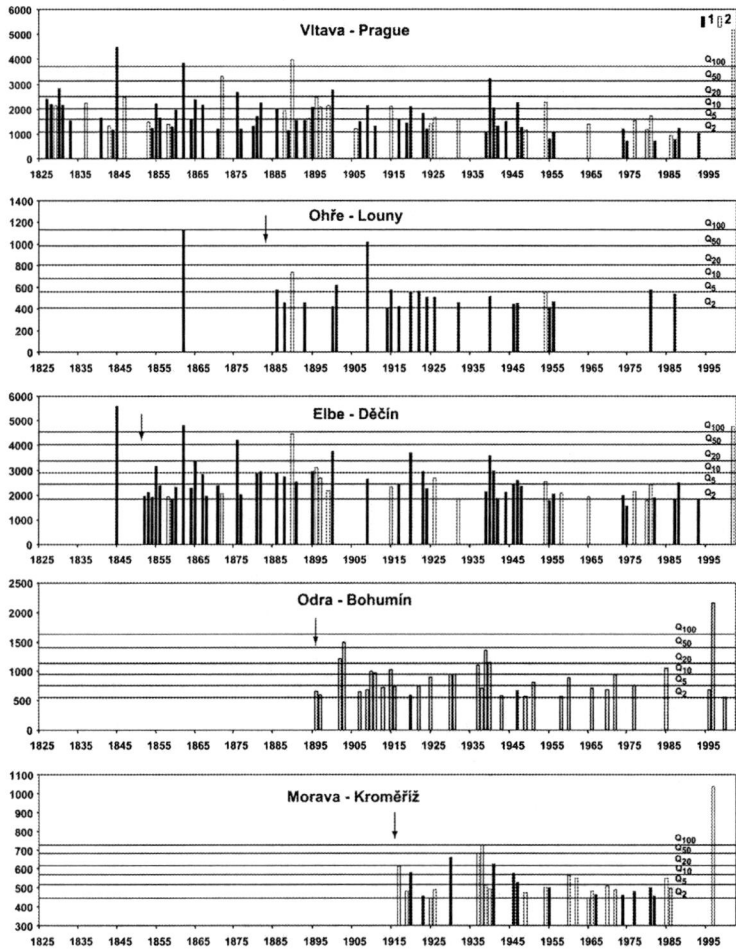

Figure 4. The variation of floods with at least a two-year peak discharge rates (Q_2; $m^3.s^{-1}$) on the studied rivers in the Czech Republic in the period 1825–2003. Arrow marks the beginning of systematic measurements. In the case of several floods during one year only the largest flood is indicated.
Explanation: 1 – flood of the winter synoptic type, 2 – flood of the summer synoptic type

Such limited evidence, depending on the accuracy and completeness of the records, nevertheless flags most of the historic floods with significant impacts. More often than not, these reports contain information about water levels reached during a flood (watermarks). If no continuous series of watermarks are preserved for individual floods, then their evaluation is possible only be reference to qualitative indices (see e.g. Sturm et al. 2001).

Although historical reports can extend the chronology of floods (e.g. Kotyza et al. 1995, Brázdil 1998), information prior to year 1500 is rather sporadic. As for the Vltava, 36 floods are documented; on the Elbe from Roudnice nad Labem to Děčín 25 are. With respect to Louny, only 8 events were recorded. In contrast, substantially richer documentation of historical floods also exists, such as data from the 16th century presented as decadal frequencies for selected streams (Figure 5). Whereas on the Vltava in Prague the highest frequency of floods was found in 1851–1900, with a dominant maximum of winter floods, the maximum number of the summer floods took place in the latter half of the 16th century. Ranking individual centuries, the 19th century (68 floods) stands on top, followed by the 16th century with 54 floods. With respect to the Elbe, the 19th century was dominant with 74 floods (in two-thirds of the winter synoptic type), followed by the 16th century with 40 floods, and the 20th century with 37 cases. Due to particularly detailed records keeping at Litoměřice (Katzerowsky 1886), the inclusion of lesser floods might tend to overestimate the total number of floods in the first half of the 19th century. For the river Ohře, the highest number of floods was found between 1551 and 1650 (out of 52 cases 32 of them occurred in 1551–1600) with a predominant occurrence of the summer type floods. In comparison with 40 floods of the 19th century (with the prevalence of winter floods) it seems that many floods of the summer synoptic type on the Ohře were rather of local character, being connected with local spate rains.

From the above analysis it follows that since 1501 the maximum flood activity in the Czech Republic has so far concentrated on the 19th century with an evident prevalence of floods of the winter synoptic type and to the latter half of the 16th century with a prevailing share of floods of the summer synoptic type (see also Brázdil et al. 1999). The number of floods of the 20th century is comparable to their frequency in the 17th century, even though it is necessary to keep in mind the evidently incomplete database of data for that century. It is noteworthy that documentary evidence about lesser floods as well as about the Morava and the Odra flooding are far from complete.

The conspicuously variable number of floods in the individual time intervals indicates the non-stationarity of the flood series, which, however, Thorndycraft et al. (2003) do not consider surprising. Besides, the non-stationarity is forecasted in the scenarios for future climatic changes.

The task at hand is to compare the historical floods with the present ones. Studying preserved watermarks may aid in that process. For example, culmination water table marks of the Elbe floods are chiselled into the castle rock at Děčín (Značky velkých vod 1966, Brázdil 1998, Brázdil et al. 2004). In determining a problematic water level of the 1118 flood, the following order of watermarks was established with the highest flood on March 1845, followed by August 2002, July 1432, February 1805, February 1862, February 1784, February 1655, September 1890, etc.

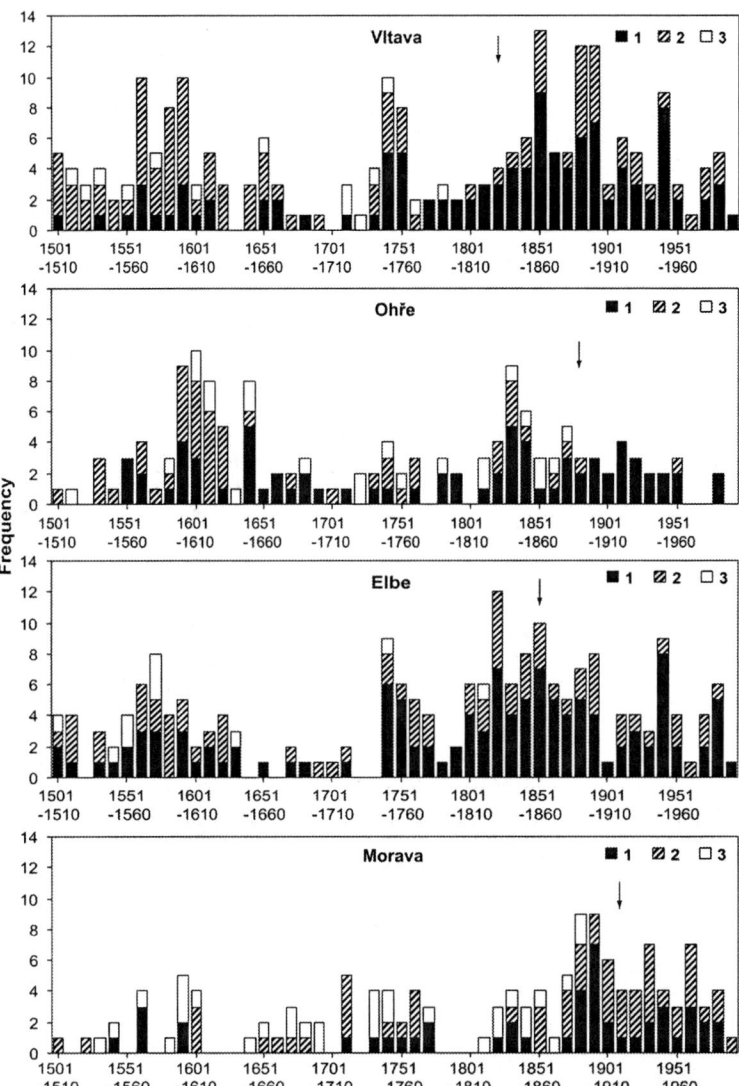

Figure 5. Decadal frequencies of floods on the Vltava (from České Budějovice to its mouth into the Elbe near Mělník), on the Ohře (from Kadaň to its mouth into the Elbe at Litoměřice), on the Elbe (from Brandýs nad Labem to the state frontier) and on the Morava in the 16th–20th centuries with differentiation according to the synoptic type of the flood (1 – winter synoptic type, 2 – summer synoptic type, 3 – without specification). Arrow marks the beginning of systematic measurements

In Prague flood levels were recorded at a stone statue of a bearded man (so-called "Bradáč"), originally located on the pillar of the former Judith Bridge, and later transferred on the right-side embankment wall of the Charles Bridge. First reference to a flood level measurement at

256

"Bradáč" appears in June 1481, more precisely in an ex post reference of February 1342, when the water was said to have reached the level of "Bradáč's nose" (Palacký ed. 1941, Šimek and Kaňák eds. 1959). Figure 6 shows several Prague floods in which the water levels exceeded those measured at "Bradáč".

The greatest event of August 2002 can be compared to the flood of 30–31 July 1432 (based on the Gregorian calendar) which to date has been considered to be one of the greatest floods of the last millennium in the Czech Lands. According to authentic sources, the Vltava, after 4 days and 3 nights of rain, flooded Prague's Old Town and the lower parts of the

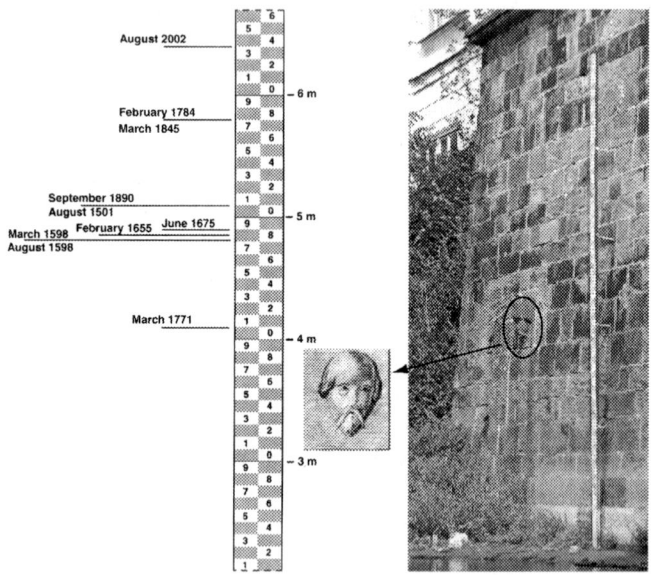

Figure 6. The recorded and estimated levels of the greatest floods on the Vltava at Prague with the projection to the position of "Bradáč". Marks for floods of July 1432 and February 1862 are not indicated (Brázdil et al. 2004)

Little Town were allowing for transport by boat only. Mills on the Vltava were washed away, many houses demolished, people and cattle drowned, fields flooded, etc. The stone Charles Bridge was destroyed in 5 places. Historical sources also make reference to a flood on the Berounka at Beroun, to the Otava at Písek (where waters washed away the railing from a stone bridge) and to the whole drainage area of the Elbe, as well as to floods along rivers in Moravia, Austria and Hungary (Brázdil and Kotyza 1995). It remains an open question whether the two mentioned floods are also comparable to the 1118 flood (Bretholz 1923), when the waters reached about 6 m above the level of the wooden bridge across the Vltava (predecessor of the Judith Bridge and then the Charles Bridge).

On the other hand, comparison of individual floods according to corresponding watermarks seems to be difficult due to unknown riverbed conditions, overall morphology and municipal riverside flooding. Despite some difficulties with the watermarks, the data helps to estimate the culmination discharge rates during historical floods (Benito et al. 2003) or in the determination of the probable extent of the flooded territory.

6. CONCLUSIONS

The fluctuation of the climate and changes in the anthropogenic activity in the cultural landscape influenced the variability of floods in the Czech Lands during the past millennium. The existing chronology of floods compiled from documentary data and from systematic hydrological measurements presents a picture about long-term fluctuations from the view of the frequency of occurrence, seasonality, synoptic causes, intensity and impacts. The main points from the documented research can be summarised into the following points:

a) In the period of systematic water-gauge observations (i.e. since about the mid-19th century) the total number of floods in the Czech Republic has been dropping, as has the extremity expressed by the N-annual culmination discharge rates. This drop can be ascribed to less frequent occurrence of floods of the winter synoptic type due to global warming, when, as a consequence of a later onset of winters and a lower accumulation of snow cover, the total number of floods from February to April is decreased.

b) Documentary data about floods in the Czech Lands enhances understanding about flooding to centuries prior to the onset of systematic water-gauge observations. This knowledge can be useful in the study of the frequency of floods, their seasonality, synoptic typology and impacts. As a result, it is possible to document with relative reliability the occurrence of disastrous floods in the Czech Republic for the last eight or nine centuries. Analysis shows that with respect to the occurrence of destructive floods, the 20th century can be seen somewhat positively (with the exception of the July 1997 flood in the eastern part of the Czech Republic).

c) The most disastrous, well documented floods affected the Czech Lands in September 1118, March 1272, February 1342, July 1432, August 1501, March and August 1598, February 1655, June 1675, February 1784, March 1845, February 1862, May 1872, September 1890, July 1897, July 1903, July 1997 and August 2002. From the May 1872 flood up to the present, that is to say during the period of global warming, floods have been only of the summer synoptic type (i.e. rain floods). The disastrous floods of the winter synoptic type of February 1784, March 1845 and February 1862 fall into the latter part of the so-called Little Ice Age, whose onset was in the early 14th century.

d) Available data do not yet permit one to render an opinion as to what extent the disastrous floods of 1997, 1998 and 2002, which came after a relatively very quiet period, are a possible consequence of anthropogenic conditioned process of global warming and whether similar disasters will be repeated in the coming years with greater intensity and frequency.

e) The flood events of the equivalent intensity have nowadays more catastrophic consequences than in the past in connection with the more and more complex infrastructure of human society and with the growing degree of anthropogenic re-shaping of the cultural landscape. Floods, as in the past, will also arise in the future after reaching a favourable constellation of meteorological, physical-geographical and anthropogenic patterns.

f) Problems with anti-flood protection measures can become even more pressing in the future if the assumption of the continuing anthropogenic conditioned climatic change is projected into a more conspicuous fluctuation of the runoff processes and general flood activity. This trend can enhance the importance of the development of the anti-flood protection measures in the form of warning and forecasting systems, including the hydrosynoptic analyses of historical floods (Hladný 1995).

ACKNOWLEDGEMENTS

We acknowledge the financial support of the Grant Agency of the Czech Republic for Grant No. 205/03/Z016.

References

Beniston M. ed. (2002) *Climatic Change: Implications for the Hydrological Cycle and for Water Management.* Kluwer Academic Publishers, Dordrecht, Boston, London.

Benito G., Diéz-Herrero A. and Fernández de Villalta M. (2003) Magnitude and frequency of flooding in the Tagus basin (Central Spain) over the last millennium. *Climatic Change* 58, 171-192.

Brázdil R. (1998) The history of floods on the rivers Elbe and Vltava in Bohemia. *Erfurter Geographische Studien* 7, 93-108.

Brázdil R., Dobrovolný P. and Kotyza O. (2004) Floods in the Czech Republic during the past millennium. *Colloque SHF "Etiages et crues extrêmes régionaux en Europe perspectives historiques".* Societe Hydrotechnique de France, Paris, 81-88.

Brázdil R., Glaser R., Pfister C., Dobrovolný P., Antoine J.-M., Barriendos M., Camuffo D., Deutsch M., Enzi S., Guidoboni E., Kotyza O. and Rodrigo F.S. (1999) Flood events of selected European rivers in the sixteenth century. *Climatic Change* 43, 239-285.

Brázdil R. and Kotyza O. (1995) *History of Weather and Climate in the Czech Lands I. Period 1000–1500.* Zürcher Geographische Schriften 62, Zürich.

Bretholz B., ed. (1923) *Die Chronik der Böhmen des Cosmas von Prag*, Weidmann, MGH SRG NS II, Berlin.

Christensen J.H. and Christensen O.B. (2003) Severe summertime flooding in Europe. *Nature* 421, 805-806.

Hladný J. (1995) Odhad vývoje povodňových situací analýzou historických případů, in *Povodňová ochrana na Labi.* Český hydrometeorologický ústav, Ministerstvo životního prostředí, Povodí Labe, Ústí nad Labem, 161-183.

259

Houghton J.T., Ding Y., Griggs D.J., Noguer M., van der Linden P.J. and Xiaosu D. (eds.) (2001) *Climate Change 2001: The Scientific Basis*, Cambridge University Press, Cambridge.

Kakos V. (1978) Výskyt povodní na Vltavě v Praze ve vztahu k pražským meteorologickým pozorováním v Klementinu. *Meteorologické zprávy* 31, 119-126.

Kakos V. (2001) Přehled povodní na Vltavě v Praze a na Labi v Děčíně v období přístrojových měření. Manuscript.

Kašpárek L. and Bušek M. (1990) Vliv vltavské kaskády na povodňový režim Vltavy v Praze. *Vodní hospodářství* 40, 280-286.

Katzerowsky W. (1886) Periodicität der Überschwemmungen. *Mitteilungen des Vereines für Geschichte der Deutschen in Böhmen* 25, 156-171.

Kotyza O., Cvrk F. and Pažourek, V. (1995) *Historické povodně na dolním Labi a na Vltavě*. Okresní muzeum, Děčín.

Meteorologické zprávy [Special volume devoted to the 2002 flood], 55, Prague 2002, 161-203.

Milly P.C.D., Wetherald R.T., Dunne K.A. and Delworth T.L. (2002) Increasing risk of great floods in a changing climate. *Nature* 415, 514-517.

Novotný J. (1963) *Dvě stoleté hydrologické řady průtokové na českých řekách*. Sborník prací Hydrometeorologického ústavu ČSSR 2, Praha.

Palacký F., ed. (1941) Staří letopisové čeští od roku 1378 do roku 1527 čili pokračování v kronikách Přibíka Pulkavy a Beneše z Hořovic z rukopisů starých vydané. J. Charvát, *Dílo Františka Palackého*, Díl 2. L. Mazáč, Praha.

Sturm K., Glaser R., Jacobeit J., Deutsch M., Brázdil R., Pfister C., Luterbacher J. and Wanner H. (2001) Hochwasser in Mitteleuropa seit 1500 und ihre Beziehung zur atmosphärischen Zirkulation. *Petermanns Geographische Mitteilungen* 145, 14-23.

Šimek F. and Kaňák M. (eds.) (1959) *Staré letopisy české z rukopisu křížovnického*, Státní nakladatelství krásné literatury a umění, Praha.

Štekl J., Brázdil R., Kakos V. Jež J., Tolasz R. and Sokol Z. (2001) *Extrémní denní srážkové úhrny na území ČR v období 1879–2000 a jejich synoptické příčiny*. Národní klimatický program České republiky 31, Praha.

Thorndycraft V.R., Benito G., Llasat M.C. and Barriendos M. (2003) Palaeofloods, historical data & climatic variability: applications in flood risk assessment. Thorndycraft V.R., Benito G., Barriendos M. and Llasat M.C. (eds.) *Palaeofloods, Historical Data and Climatic Variability: Applications in Flood Risk Assessment*. CSIC, Madrid, 3-9.

Značky velkých vod na Labi v úseku od státní hranice u Hřenska po ústí Vltavy. Ředitelství vodních toků v Praze. Správa vodohospodářského rozvoje, Praha 1966.

Chapter 21

HISTORICAL 2002 FLOOD IN CENTRAL EUROPE AND FLOOD DEFENCE
Situation in the Czech Republic, illustrated with the Elbe River Basin Experience

VACLAV JIRASEK
Povodi Labe, statni podnik, Hradec, Kralove, Czech Republic

Keywords: Floods, climate, flood protection, prevention.

1.   INTRODUCTION

The floods that the Czech Republic has experienced in recent years were mostly of the 50 to 100 year occurrence rate. Especially between 1997 and 2002, the Czech Republic suffered five major floods and two of them, the floods of 1997 and 2002, were of catastrophic magnitude and brought devastating effects to large land area.

Analyses of meteorological situations prevailing during the floods are now a common fact that has been discussed by the professional community many times. Speaking on these terms, let me please point out Mr. Milankovitch's hypothesis in which he attributes the radical changes in the global climate to the planetary system (Harvey 1992), namely to periodic changes in tilt of the Earth's axis and in orbit excentricity. If, while taking into account this hypothesis, we project the anticipated development of the graph depicting the statistical evaluation of the runoff in the major European watercourses (Figure 1) (Russ 1992) into present time, the pluvial period should extend from the 1990s to about 2030. It seems, however, that the increasing amount of greenhouse gases in the atmosphere might correct this assumption in the sense of word that we can expect shifts in the European continent climate character. We will experience noticeable regional precipitation fluctuations that, in extreme situations, will induce floods and prolong the periods of draught. After all, the character of 2003 weather already confirms this thesis. In the event of such climatic change in Central Europe, the society will have to re-evaluate the existing water management systems and put in place all the necessary measures in time.

J. Schanze et al. (eds.), Flood Risk Management:
Hazards, Vulnerability and Mitigation Measures, 261–265.
© 2006 *Springer.*

262

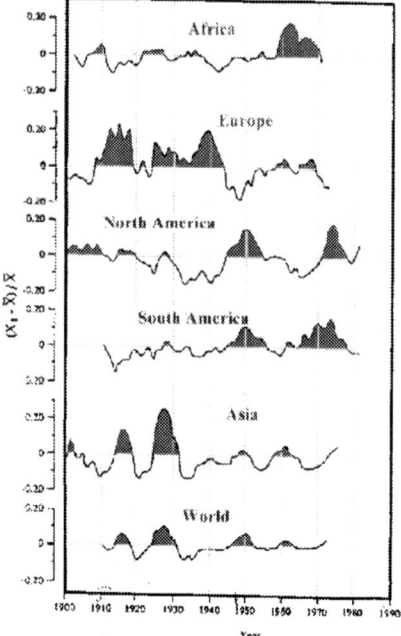

Figure 1. Variation of runoff of major global watercourses in the 20th Century

2.   2002 FLOOD IN THE CZECH REPUBLIC

Let's take a brief look at the 2002 flood. Like in 1997, similar meteorological situation occurred twice in a row. Heavy rainfall belt slowly advanced from south (Italy, Austria) across central Bohemia and continued north. The highest daily precipitation totals exceeded the 400 mm mark. Those two precipitation waves were matched by two flood waves (Ministry of the Environment 2004). The first flood wave's peak discharges on small watercourses exceeded the 500-year return period. The second wave rainfall landed on already saturated basin area and caused catastrophic flood situations in lower-lying areas along the basin's major rivers (the Moldau River, the Elbe River).

The Moldau River peak discharge in Prague of the rate of 5,160 m^3/s corresponded approximately to the flow of 500-year water. When leaving the Czech Republic in the town of Hrensko on the Elbe River, after having flooded large areas at the Elbe-Moldau confluence near the town of Melnik (66.8 km^2) and downstream near the town of Litomerice (51.2 km^2), the peak discharge exceeded the 100-year water flow with 4,780 m^3/s rate (Ministry of the Environment 2004).

This meant that the river water surface in the town of Melnik was 8 m higher, compared to the standard conditions, and the year 1845 recorded mark was passed; the situation downstream was similar – the water surface in the city of Usti nad Labem was measured at the level of 1,126 cm.

Thus, for the first time in the last century, the Czech Republic experienced a major flood whose rate significantly exceeded the passage of a 100-year water. Employing the Elbe River mathematic model of the area between the Elbe – Moldau confluence and state border, we and DHI Hydroinform, a.s. Prague Company were able to recalculate the previously delineated Q100 (4,400 m³/s) flood plain to the predicted peak discharge (5,500 m³/s) and, on the following day, we handed over the newly delineated maps to the flood committees involved to base their emergency rescue operations on.

The expected danger of uncontrollable escapes of harmful and polluting substances made us establish emergency monitoring of surface water and the results were made available in MKOL international measuring program indicators. There were 166 water quality parameters monitored and only in isolated cases did the recorded values exceed the legally-binding immission limits.

If we compare this flood with the 1997 flood, we can clearly see the areas that came out a lot better off in 2002.

Characteristics	1997	2002
Affected land area (in thou. km³)	11	17
Number of affected inhabitants (thou.)	2,855	3,200
Total damage (in bill. CZK)	62.6	70
Human lives lost	60	16

Unfortunately, the 2002 flood could not be averted. But the lessons learnt from the 1997 flood led to technological improvements in weather forecasting service (performed by the Czech Hydrometeorological Institute), to improvements in the effectiveness of early flood warning service (performed by the basin's managers in cooperation with the flood committees), and also to the formation of the Integrated Rescue System (operated by the Ministry of the Interior). Taking such steps enabled timely evacuation of citizens, thus significantly reducing the number of casualties of this catastrophic event. The 1997 flood also led to systematic delineation of flood-prone areas in the Czech Republic (to these days, the delineation has been carried out on 50 % of the length of major Czech watercourses – i.e. with watershed over 3 km²) (Ministry of Agriculture 2003) and to the initiation of state program to develop adequate flood protection measures.

3. FLOOD PROTECTION

After the floods of 1997 and 1998, the Czech government initiated the preparation of flood prevention programs within individual departments. In addition to the already mentioned changes, let me add information on the Ministry of Agriculture's program focused at the implementation of technical elements. The Phase 1 of this program (reference No. 229060) is co-financed by the European Investment Bank and the program's goal consists of five basic elements (Ministry of Agriculture 2003) that are aimed, above of all, at the creation of water storage capacity and the protection of dwellings:

- Construction of polders, reservoirs, and levees
- Increasing watercourses capacity in developed areas
- Delineation of flood-prone areas
- Run-off condition studies
- Delineation of areas threatened by extraordinary floods

When selecting localities for Phase 1 Schemes, the amount of risk was evaluated in each given locality and the selection process was based on the assessment of the dwelling priority (number of affected inhabitants, the extent of probable damage, historical relations), on the existing capacity of the riverbed and on the locality's proneness to floods (bank erosion, the number of days during which a flood emergency is declared). The proposed Schemes were evaluated by an independent environmental expert for possible environmental impact (the Ramsar Directive and Natura 2000) and by an independent strategic expert who assessed the effects of the proposed flood protection measures. Individual Schemes of the program had to be based on a study or a mathematic model that evaluates formerly mentioned effects. The Phase 1 began in 2002 and is scheduled to end in 2005. At present time, we are preparing materials for the eventual launch of Phase 2 that will deal with the areas and localities affected by the 2002 flood.

4. LESSONS LEARNT

The 2002 flood situation underwent a comprehensive evaluation in a multidisciplinary project conducted by Water Management Research Institute of TGM in Prague. The output report was, after having been approved by the Cabinet, published in January 2004. We deem it necessary to point out the following particular conclusions of the report (Ministry of the Environment 2004):

- Consistently enforce the flood prevention principles; in particular, regulate the land utilization in the delineated flood-prone areas and in their active zones
- Safeguard the waste water treatment plants (WWTPs) against flood damage since nonfunctional WWTPs were among the leading sources of the surface water quality impairment
- The comprehensive evaluation indicates that the rainfall in the sub-basins affected during the 2002 flood situation reached 68% of the probable maximum precipitation (PMP); in future, heavier precipitations can be expected thus it is necessary to supply major urban areas authorities and industrial plants managements with flood plans in case there is a flood of the occurrence rate exceeding the 100-year return period

- Build up storage capacity by suitable land modifications and by revitalization of the regulated sections of small watercourses
- Continue in delineation of flood-prone areas and in the implementation of flood-prevention and protection measures
- Improve the information system by means of automation of the precipitation and hydrometric stations, and utilize the close linkage onto the European forecasting system

References

Harvey J. (1992) ENVR 133 – Determinants of Solar Insolation. University of North Carolina, USA, 7-11.

Ministry of the Environment (2004) Assessment of Catastrophic Flood of August 2002 and Flood Prevention System Improvement Proposal. 3-7, 25-28, 77-78.

Ministry of Agriculture (2003) Report on the State of Water Management in the Czech Republic. 38-40.

Russ C. (1992) ENVR 133 – Hydrologic Cycle and Weathering. University of North Carolina, USA, 5.

Chapter 22

THE FLOOD IN AUGUST 2002 – CONSEQUENCES ON FLOOD PROTECTION FOR THE CITY OF DRESDEN

CHRISTIAN KORNDÖRFER

Head of Environmental Office, City of Dresden, Germany

Keywords: Plain flood, flash flood, groundwater flood, comprehensive flood prevention.

1. INTRODUCTION

Flooding in Dresden in August 2002 occurred in four stages caused by a sequence of two strong Adria depths within two weeks (Korndörfer 2003a). The first depth with intensive precipitation in South Bohemia in August 4th to 6th was followed by the depth whirl "ILSE" hitting the Vltava basin again and causing extreme rain fall in the eastern Ore Mountains close to Dresden. August 12th, we registered 140 to 180 mm rain within 16 hours in Dresden. At this time, the water level of the Elbe River started to rise significantly because of the run off from Moldavia basin due to the precipitation there at the beginning of the month.

In Dresden, about 1 km^2 of residential area and 8.11 km^2 of business and industrial area were inundated. Four persons came to death. Residential building damages of 304 mill. €, business building damages of 467 mill. €, and damages of public infrastructure to the extent of 468 mill. € had to be accounted for. The flood in August 2002 had its greatest impacts within the region of Dresden. Immediately after the flood event the City of Dresden started analysing the event, its causes, and consequences. Based on these findings, a comprehensive flood management concept was developed.

2. THE FLOOD IN DRESDEN AUGUST 2002

2.1. The first stage: The municipal brooks

In August 12th at noon the municipal brooks with a total length of 440 km as well as the sewage system have been overloaded. Streets and housing

J. Schanze et al. (eds.), Flood Risk Management:
Hazards, Vulnerability and Mitigation Measures, 267–273.
© 2006 *Springer.*

areas have been inundated, causing damages in public infrastructure of about 7 Mio. €.

2.2. The second stage: Ore Mountains rivers (Weisseritz, Lockwitz)

In the afternoon of August 12th the run off from the Ore Mountains reached Dresden. At first the Lockwitz river with a length of 30 km and an basin area of about 50 km^2 flooded the eastern part of the city and filled the yet existing old branch of the Elbe, an important retention area in case of an Elbe flood.

In the evening, the Weißeritz River, springing in the same region as Lockwitz, filled up the 3 existing storage reservoirs and flew over the dams. Without any possibility left to control the discharge, a drainage of about 500 m^3/s passed the border of the City. The tremendous violence of the streaming water as well as of the drift-wood and bed load caused heavy damages of about 400 Mio. €. Two people died. Within 20 hours, 9 from a total amount of 11 bridges were destroyed. Furthermore, the power plant, the Central Railway Station, streets, hours, enterprises, and so forth were flooded. Energy supply and telecommunication broke down. Town traffic and the railway relations to Dresden were interrupted or seriously disturbed.

2.3. The third stage: Elbe flood August 13th to August 21st

The disastrous flash flood of the Weißeritz and Lockwitz was followed by the rapid rise of the Elbe water level. Because of the destroyed communication system, energy supply and traffic routes the struggle against the threatening Elbe flood was rendered extremely difficult. Since 1890 the water level in Dresden did not exceed the 8-m level (Kirchbach 2003). In August 2002 the official flood warning system failed at a level of about 8 m. The municipal environmental office dared to calculate an own prognosis, using the values of Czech stations. The predicted values actually occurred. These results have been used by the Lord Mayor of Dresden for fundamental decisions like the evacuation of hospitals with 2400 patients and of parts of the City. So he could avoid damages even much higher. The Elbe flood culminated with 9.40 m as we predicted 20 hours before. Once again the Old Town and the historical Friedrichstadt, already damaged a few days before by the Weißeritz has been flooded, but a lot of other housing and business areas, too. At 9.30 m, the Elbe reached the sewage plant. In consequence, the sewage system partially broke down, thus, accelerating the inundation of wide areas. The Dresden flood protection system properly worked up to a water level of about 8.70 m, this is the mark of 1845, the highest level ever observed before 2002. Though the discharge amount had not been significantly higher than in 1890, the water level in Dresden became 85 cm higher due to modified building structures on the floodplains.

2.4. The fourth stage: Rise of groundwater

The outstanding rain fall from August 12th and the wide spread inundation lasting several days caused a rapid groundwater rise even in parts of the City which are some kilometres away from the Elbe river. The groundwater reached levels never observed before, in some cases 4 m above the highest value before. Yet before the Elbe flood culminated some houses had to be flooded on purpose to avoid heavy static damages. In spite of the drought year 2003, it took more than 12 months to return to normal groundwater levels.

3. SOME LESSONS LEARNED

From a didactic point of view, the August flood was a masterly lesson for all people in Dresden (Korndörfer 2003b). Here are some lessons we had to learn:

Weather events with extreme precipitation are possible in our region too and can cause so-called century floods now and in the future. Discharge and water level of so called 100 year floods are statistic values, which will change if the fundamentals of the calculation are drifting. Within the time span of August 12th to August 17th, the water flood culminated in the municipal creeks all over the town as well as in the rivers Weißeritz and Lockwitz running down from the Ore Mountains and at least in the Elbe River itself. This coincidence and these extreme values of precipitation and water discharges in the Dresden region and in the eastern Ore Mountains have never been reported before.

To improve flood prevention, the whole system of waters in town has to be taken into account, including the sewerage as well as the groundwater. The available retention volume in the basins cannot keep the discharges of extreme floods away from Dresden. The agricultural fields and meadows as well as damaged or devastated pine forests do not reduce surface drainage as healthy and well mixed forests on the slopes of the mountains proved to be able. Regarding the tributaries to the Weißeritz storage basins the 2002 event had a probability of less than 1 in 10,000 years (LfUG 2004). This may be a hint on a certain drift of climate parameters. The ongoing climate change (Enke et al. 2001) will increase flood risk of the Ore Mountain Rivers as well as of the municipal waters.

The observed curves of discharge and water levels at Dresden gauge was stronger than reported from 1890 or 1940. There must be serious obstacles in the Elbe river bed hindering the drain and so elevating the water level in a dangerous manner. The neglect of a regular maintenance of the creeks and rivers caused inundations that could have been avoided. The citizens and companies have to pay more attention to private precautionary measures. The must be enabled to act in the right way by good and easily available information about their individual flood risk. The official flood warning and prognosis system in Saxony, which failed at the Weißeritz as well as at the Elbe river, must be improved and complemented by a own warning system of the municipalities situated along the course of the rivers.

4. COMPREHENSIVE PLAN ON FLOOD PREVENTION FOR THE CITY OF DRESDEN (PLAN HOCHWASSERVORSORGE DRESDEN PHD)

4.1. Principles of the plan on flood prevention

By order of the Lord Mayor the Environmental Office of Dresden worked out a concept for a comprehensive plan for improving the flood protection of Dresden. The principles of this plan have been worked out and were accepted by the City Council in April 2004. The total system of waters in town including groundwater and sewage system is being investigated to develop a comprehensive and optimum flood protection system.

The Plan comprises the following:

- Improvement of flood forecasting and warning for the citizens of Dresden and support of individual precautionary measures
- Reconstruction of water courses and embankments, preferably as a near-natural river development
- Careful and regular maintenance of rivers, dykes, and sewage channels
- Determination of flood areas for a 100year inundation for all relevant waters, and strict control of land utilisation to prevent further reductions of the river cross section and in retention volume
- Development of additional flood protection measurements for a protection level adapted to the damage potential on the one hand and to the hazards on the other hand
- Flood prevention control in the city by on-site water management incl. infiltration to the groundwater, operation and maintenance of retention basins and limitation or reduction of sealing (Korndörfer 2003c)
- Control of groundwater rise and groundwater pumping by a network of gauges
- Develop of the sewage system to ensure the drainage of waste water in case of a 100year flood and simultaneously of storm water in case of a 1year rainfall during a flood
- The priority of flood control measures has to be derived by cost-benefit investigations. Measures with superior effects will be realized immediately ("Urgent Measures")

This comprehensive plan has to be worked out in detail for the whole town area within the next two years. Then, it has to be presented to the City Council for amendment. In parallel, the Flood Prevention Plan has to be developed and actualised. Total costs for an effective flood prevention system are estimated to be about 90 mill. €.

4.2. Actual considerations and urgent measures

4.2.1. Flood areas

With the aim of keeping the needed space for the water flow in case of a high flood, the river bed as well as the retention areas must be kept free from obstacles and further building areas. Thus, the legally fixed flood areas are of outstanding importance.

In 2003 the determined flood area of the Elbe has been newly fixed acc. to the significant higher water level. In 2004 we will complete this by legally determined flood areas for all other relevant waters in town. These official flood areas are to be marked in the real estate register. Restrictions will be gathered, especially in land use.

All parts of the town, that could be inundated by failure of the flood protection walls or dams or by a 200 year flood are marked in published map as flood endangered areas. Inhabitants and land owners have to pay attention for risk reduction measures.

4.2.2. Flood prevention at municipal waters

The risks by inundation from the local creeks are higher than estimated before. Inundation from local creeks can efficiently be reduced by a set of measures, which is already current business of the local authorities:
- The local planning authority will channel urban development to brown fields in towns.
- On-site storm water managements all over the city.
- Strict separation of storm water drainage from waste water.
- Retention of precipitation water at least for a 10 year rain and providence of a risk - free discharge for the overflow.

4.2.3. Flood protection at the ore mountains rivers

The rivers from the Ore Mountains need a set of measures to provide better retention in the upper course as well as more room in the City. A rise in the probability for the occurrence of extreme precipitation events is predicted being caused by regional climate change. Thus, investments in preventive measurements for the mountainous rivers as well as for the municipal waters are of highest priority.

Actual the river bed is widened and hazardous windings are newly shaped. A protection wall for the downtown area is already pre-planned. A risk-benefit analysis is on the way.

4.2.4. Protection against Elbe floods

First of all, the reasons for the unexpected high water level of 2002 must be discovered and, if possible, reduced. The sediment banks in the bed of the river and the flood channels which were slowly grown over many years are figured out and shall be removed in 2005.

The historical flood protection system of the Elbe consisting of wide riverside meadows restraining incoming waters and flood channels for accelerated drain has to be kept and re-constructed and completed by protection walls in the historical down town area. The pre-planning stage is finished now and the funds for realization are raised.

Currently, the numeric calculation of several states is on the way.

4.2.5. Groundwater monitoring system

The analysis of groundwater rise and flow during flooding as well as of the chemical reactions which occurred is completed and will be published soon. Now we are going to plan the network of groundwater gauges in the inner city.

4.2.6. Sewage system

The sewage system plays an important role during flood periods. A well outlined and maintained system can reduce the hazards by inundation and by groundwater rise. To ensure a free level drainage of waste water and additional rain water in case of a flood, two pumping stations and a lot of additional measures are necessary.

5. FINAL REMARKS

The devastating damages of the August 2002 flood are removed to a high extent. Despite budgetary difficulties of the City of Dresden as well as of the Free State of Saxony, the flood prevention is reclaimed to be of extraordinary importance. Within the next years, Dresden is going to develop a comprehensive flood protection system. The overall costs are estimated to be about 90 mill. €. The achievable reduction of the damage potential will depend on the available funds as well as on the strong will of the Dresden authorities to accomplish the task of controlling urban development on flood plains and land use.

References

Enke W., Küchler W. and Sommer T. (2001) *Klimaprognose für Sachsen.* Dresden, Landesamt für Umwelt und Geologie.

Kirchbach-Kommission (2003) *Bericht der Unabhängigen Kommission der Sächsischen Staatsregierung,* Dresden.

Korndörfer C. (2003a) Hochwasserschutz für Dresden. Deutsches Institut für Urbanistik (ed.).

Korndörfer C. (2003b) Möglichkeiten und Grenzen des vorsorgenden Hochwasserschutzes am Beispiel des Augusthochwassers 2002. Institut für Städtebau Berlin, www.umweltbundesamt.de\rup\hochwasser-workshop.

Körndorfer C. (2003c) Festgesetztes Überschwemmungsgebiet für ein 100-jähliches Elbehochwasser und Siedlungsentwicklung in der Stadt Dresden. Roch I. (ed.) *Flusslandschaften an Elbe und Rhein.* Verlag für Wissenschaft und Forschung Berlin, 2003, 103-114.

LfUG – Landesamt für Umwelt und Geologie (2004) *Ereignisanalyse Hochwasser August 2002.* Dresden.

Chapter 23

FLOOD RISK IN CITIES OF BELARUS: SPECIFIC CASES AND PROBLEMS OF MANAGEMENT

TAMARA I. KUKHARCHYK
Institute for Problems of Use of Natural Resources & Ecology of National Academy of Sciences of Belarus, Minsk, Belarus

Keywords: Urban floods, recurrence intervals, maximum rainfall, storm water sewage system.

1. INTRODUCTION

Floods in Belarus are caused by spring snowmelt or heavy rains in summer and autumn. Approximately 6.8 % of the total area of the country (14.1 thousand km^2) is periodically flooded with 1-10 % repetition frequency (Taratunin 2001). The duration of floods vary between 30-120 days and depends on hydro-meteorological factors, as well as the river size, morphological peculiarities of the valley and the form of the riverbed.

According to data of the monitoring (Ministry of natural resources and nature protection of Republic of Belarus 2002), the following types of floods occur on the territory of Belarus:

- raging floods (with the repetition once in 10-50 years and maximum water level provision 2-10 %);
- prominent floods (with the repetition once in 55-100 years and maximum water level provision 1-2 %);
- catastrophic floods (with the repetition once in 100-200 years and maximum water level provision less than 1 %).

Floods, which caused considerable damages for the last 50-70 years, took place at least 10-12 times. The most prominent among them were the floods in the years of 1956, 1958, 1974, 1979, 1993 and 1999 (Cherepanski 1998, Rutkovski 2001). Large urban areas were also flooded in 2004. Numerous built-up areas, households, some sections of highways, power lines and lines of communication are subjected to partial or complete floods from time to time. For example, floods happen in each of the 34 settlements where hydrological stations are situated.

J. Schanze et al. (eds.), Flood Risk Management:
Hazards, Vulnerability and Mitigation Measures, 275–282.
© 2006 *Springer.*

The information indicated above concerns riverine floods only. But there are also a lot of flood situations that are not so well-understood: they are of rather small scale and as a rule are short-term but occur regularly from year to year. In the paper different types of floods in cities are described; the main attention is devoted to analysis of reasons, consequences and management possibilities of floods in urban regions, built on hollows and peatlands.

2. TYPES OF FLOODS IN CITIES AND THEIR FACTORS

2.1. Riverine floods

Floods that happen as a result of water overflowing rivers (streams) and reservoir banks are probably the most common type of floods and in general represent a natural phenomenon, which is caused by a number of regional and local factors (Chebotarev 1978, Nezihovski 1988, Watt 2003). Floods become natural disaster when they damage people's interests. In this case the more developed the civilisation is, the greater the losses are (Taratunin 2001, Kupriyanov 1977). According to (Integrated flood management 2003), around 70% of all global disasters are connected with hydro-meteorological events.

Riverine floods in Belarus are caused by disproportionate surface runoff distribution (about 55 % of annual runoff occurs during the spring), by poor carrying capacity of rivers and by uneven atmospheric precipitation. The area of floods depends, first of all, on geomorphologic peculiarities of the region. Thus, vast territories are flooded in Polesie region, which represents a flat alluvial plain with large wetlands and hollows. The average width of floods here is to 1.5-2.0 km, but sometimes reaches up to 15 km.

According to the data obtained (Ministry of natural resources and nature protection of Republic of Belarus 2002), the following cities are subjected to riverine floods: Gomel (the Sozh), Rechitsa (the Goryn), village Chernichi (the Pripiat) – once every 2 years; Petrikov and Pinsk, village Koroby (the Pripiat), Loyev (the Dnieper), Verkhnedvinsk (the Zapadnaya Dvina) – once every 2-3 years; Stolbtsy (the Neman), Mogilev (the Dnieper), Borisov (the Berezina), Mozyr (the Pripiat) – once every 4-5 years. During riverine floods, vast urban areas can be covered with water. Thus, about 5-10% of the territory of Orsha was flooded in 1956 and 1958, up to 20-30% of the territory of Borisov – in 1931, Surazh – in 1929, 1931, 1956 and Sharkovschina – in 1956, up to 45% of the territory of Verhnedvinsk – in 1956, 2004, and up to 75% of the territory of Ulla – in 1931, 1941, 1951 and 1956.

2.2. Floods caused by urbanisation

Floods of urban areas caused by large volumes of storm waters, formed mainly during heavy rains, represent another type of floods. This is a result of changes of land use and rapid expansion of urban and industrial areas (Kupriyanov 1977, Woloszyn 2003). This category of floods is typical of cities of Belarus, though it should be mentioned that their statistics and monitoring are not done, and in most cases there is no estimation of the damages they cause. Meanwhile, different functional urban areas are subjected to floods, including transport and build-up areas, and this leads to severe financial difficulties.

Floods differ in their duration, area, water-level, which is closely connected both with meteorological conditions and with local conditions of runoff formation: the character of a building, geomorphologic peculiarities, the capacity of storm water sewage system. Taking into account the statistics on heavy rains (when the amount of the rainfall can reach up to 50 mm during 12 hours), it is obvious that the possibility of floods appearance in urban territories is rather high. Thus, according to the data obtained (Ministry of natural resources and nature protection of Republic of Belarus 2002), the repetition of heavy rains in several cities (including Minsk) for the last 35 years of close research has reached 20%. In Vitebsk district, for example, the repetition of heavy rains at least in one of the cities reaches 74%. Daily maximum rainfall values in the cities of Belarus, obtained during the research period (45-103 years), reach up to 102-145 mm. Uneven precipitation during the year, the increase of extreme meteorological events, including heavy rains, are characteristic of global warming process (Institute for problems of natural resources use and ecology 2003).

Figure 1. Derzhinski avenue after the heavy rain on July 27ᵗʰ, 2004 in Minsk (the pictures were kindly presented by Belarusian television)

Vast flooded areas can be formed during heavy rains even in 1-2.5 hours. For example, the depth of heavy rains in Minsk on July 25th and 27th, 2004 reached approximately 40 mm (the norm of rainfall amount in July is 90 mm). Several streets and squares, basements of dwelling houses and production areas, warehouses were completely or partially flooded.

Some streets (for instance, Derzhinski avenue, Bogdanovicha street, Nemiga, Kuibysheva, R. Luksemburg, etc.) basically turned into rivers (Figure 1). The level of water above the surface in some of them achieved up to 1.5–2 m.

Even though it was a short-term heavy rain, the damages caused by it were estimated to the amount of more than 1 million US dollars.

The analysis of the recent flood in Minsk has proved that mainly those streets or their sections were flooded, which had once been constructed on one of two rivers – the Perespa or the Nemiga (today they have basically turned into storm waters sewer), or built on the territory of the former Komarovski peatland. We have analyzed the peculiarities of land use within the basin of the Perespa in 1944 when the riverbed remained open and in 1988 when it was destroyed.

It was established that in 1988 the area, used for building up, increased by 9 times, while the area of unused and agricultural land decreased by 3 times. As a result the amount of rain-off has been increased by 1.5 times. On whole the increase of rain-off coefficient in urbanized areas is a natural process. The annual rain-off coefficients taken for a big industrial city can reach up to 200% (Kupriyanov 1977).

Another factor, which caused the recent flood in Minsk, is connected with the system of storm water collectors. It was created in 1960-1970s and certainly is greatly worn out and become out of date since than.

2.3. Floods caused by building up wetlands in cities

Wetlands and bogged areas are typical elements of natural landscape of Belarus. Quite often they can be found within urban areas as well (Kukharchyk et al. 2003). The results of a many years study in various Belarusian cities have showed that one of the reasons of floods occurrence in cities has to do with the fact that these cities had been built on the wetlands.

Figure 2. Microdistrict Zalineiny in Borisov (on the left) and Gromy in Polotsk (on the right) during spring snowmelt; April 1st, 2004. The flood lasted for more than a week and the initial water level was significantly higher than shown on the picture (pictures by T.Kukharchyk)

It has been established that floods can happen several times a year on such areas: during snowmelt and heavy rains periods. It is no doubt that this type of floods can be studied as one of the cases of land urbanisation and transformation, though to our minds it requires greater attention because of the peculiarities of its results. This category of floods does not have any statistics; the damages caused by them are not evaluated either. Local people indicate that floods can last up to 2-3 weeks; the roads and the land attached to houses get flooded, and sometimes water level gets even higher than the basement of houses (Figure 2).

Below we would like to illustrate several peculiarities of 2 microdistricts, which are flooded especially often.

2.3.1. Microdistrict gromy (Polotsk city)

In 1950s the dwellers of Polotsk city got permission to building up in the outskirts of the city, which represented a wetland at that time. The wetland had formed in a hollow and was characterised by high watering; peat depth in some parts reached up to 3 meters. Wetland drainage was done with the help of open canals. The construction of dwelling houses was started almost at the same time as the drainage procedure of this territory. Peat was not extracted, the ground was actively added up. At present the whole territory of the former wetland is built up, the peat is covered by mineral ground and it is difficult to identify it on the basis of outward appearance (Figure 3).

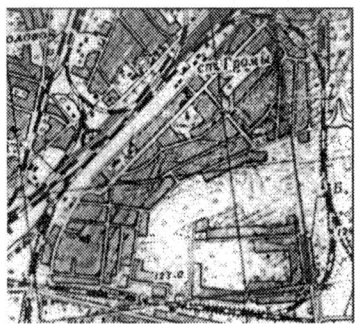 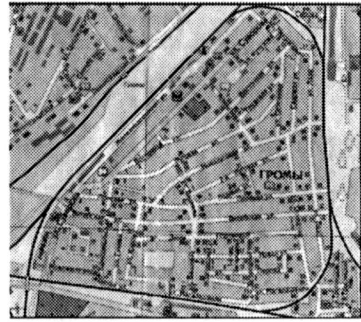

Figure 3. Overall view of the microdistrict Gromy in Polotsk city before its intensive building up in 1950 (on the left) and today (on the right)

Meanwhile today peat can be found beneath the ground, which was once added, at a depth of only 20-50 cm. It differs in botanical composition (wood, reed and mixed), various decomposition degree (10-45%) and ash content (12-33%). The level of ground water even at low season gets close to the surface and amounts to 0.5-0.7 meters.

280

2.3.2. Microdistrict zalineiny (Borisov city)

The process of building up the wetland began at the end of 1950s – the beginning of 1960s. This is a floodplain wetland with peat depth up to 1.5 meters. Here the peat was not extracted either; the ground that was brought here was often poured right into the water. At present only small areas of the shrubby wetland have remained untouched (Figure 4).

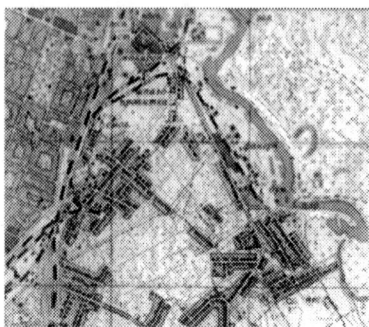

Figure 4. The overall view of microdistrict Zalineiny in Borisov city before its intensive building up in 1950 (on the left) and today (on the right)

Today a lot of dwelling houses have basically "sank" into the ground in both microdistricts as a result of peat slump and its mineralization. Some of the houses are inappropriate for human habitation due to frequent floods and constant moisture. The matter is that generally in 20-30 years after building, the processes of rewetting and restoration of wetlands occur and dwelling houses become non-suitable for living. Wetland flora (reed, sedge, rush) quickly restores on neglected territories (Figure 5).

Figure 5. The restoration of wetland flora on a neglected area in Polotsk city (pictures by T.Kukharchyk)

3.   WATER POLLUTION AS A RESULT OF FLOODS

Another problem typical of residential areas, which are built on wetlands, is water pollution. It is mainly determined by the fact that pollutants enter surface and ground water with domestic wastes and wastewater discharges. The majority of old residential areas in the cities are not equipped with sewage system. Therefore if the level of ground waters is high and when the household area is flooded, the pollutants enter surface reservoir and streams together with runoff. In such cases ground water is almost not protected. Thus, typical pollutants of water in residential areas, which indicate the source of the pollution quite precisely, are the following: organic pollutants, compounds of nitrogen and phosphorus, chlorides, sulphates, potassium (Khomich ct al. 2004).

What is worse is that in some residential areas, built on wetlands, water taken from shaft wells is used for drinking purposes. According to the data obtained by the Centre of sanitation and epidemiology in Polotsk city, the water in the majority of controlled shafts does not correspond to sanitation standards on microbiological and organoleptic indices. Thus, almost in 100 % of cases the content of enteric bacteria exceeds the maximum permissible concentration, the color exceeds the standard excepted in 68-97%, and hardness of water – in 60-70% of cases. Some shafts are also polluted with nitrates: the occurrence of samples with exceeding MPC varies from 7% (in 1999) up to 23% (in 2001).

4.   CONCLUSION

Flood prevention is not possible without knowing such factors as scales, probability, and periodicity and their reasons. In Belarus regular studies, register of water levels and scientific-technologic projects on protection of territories from floods are done only for riverine floods. Besides they mainly concern the region of Polesie, where certain measures such as dyking and polder construction are taken.

Other types of floods remain almost not investigated. A detailed analysis of reasons of flood occurrence is required, considering water collection systems of small rivers and streams (even former), land use structure, and capacity of storm water sewers.

For dwelling areas built on wetlands, it is advisable to work out certain preventive measures: conservation or restoration of natural regime of stream, limitation of agricultural activities on certain areas, etc. Normative documents regulating land use of flooded areas are needed, because even though floods usually have local character, they might have significant consequences for the city and/or the region on the whole. One of the aims of flood prevention is to prevent the spread of pollutants with runoff and their transfer over large distances.

References

Chebotarev A.I. (1978) *Hydrological glossary.* Third edition. Leningrad (in Russian).

Cherepanski M.M. (1998) Hydroecological problems of Belorussian Polesie. *Natural Resources* 2, 90-97 (in Russian).

Golberg M.A. (2002) Natural hydro-meteorological events on the territory of Belarus. Reference-book of the Ministry of natural resources and nature protection of Republic of Belarus (in Russian).

Khomich V., Kakareka S. and Kukharchyk, T. (2004) *Ecogeochemistry of urban landscapes of Belarus.* Minsk, Minsktipproekt (in Russian).

Kukharchyk T., Khomich V. and Kakareka S. (2003) Urban Wetlands in Belarus: State, Threats and Perspectives. Proc. 'Enhancing Urban Environment by Environmental Upgrading and Restoration', November 5-8, 2003, Rome, Italy, 343-352.

Kupriyanov V.V. (1977) *Hydrological aspects of urbanisation - hydrology of cities and urbanised territories.* Leningrad (in Russian).

Loginov V.F. (ed.) (2003) *Changes of climate of Belarus and their consequence.* Institute for problems of natural resources use and ecology, Minsk (in Russian).

Nezihovski R.A. (1988) *Floods on rivers and lakes.* Leningrad (in Russian).

Rutkovski P.P. (2001) Problems of floods in the Republic of Belarus and ways of its solution. *Natural Resources* 2, 59-63 (in Russian).

Taratunin A.A. (2001) Contemporary strategy of protection from floods in the Republic of Belarus and reduction of losses. *Natural Resources* 2, 64-69 (in Russian).

The Associated Programme on Flood Management, Technical support UNIT (ed.) (2003) Integrated flood management. Concept paper, Switzerland.

Watt E. (2003) Reduction of flood damages in urban areas of Canada. Arsov R., Marasalek J., Watt E. and Zeman E. (eds.) Urban Water Management: Science Technology and Service Delivery. *NATO Science Series. IV: Earth and Environmental Sciences* 25, Kluwer Academic Publishers. The Netherlands, 105-114.

Woloszyn E. (2003) The catastrophic flood in Gdansk on July 2001. Arsov R., Marasalek J., Watt E. and Zeman E. (eds.) Urban Water Management: Science Technology and Service Delivery. *NATO Science Series. IV: Earth and Environmental Sciences* 25, Kluwer Academic Publishers. The Netherlands, 115-124.

Chapter 24

NATO SCIENCE FOR PEACE PROJECT ON
MANAGEMENT OF TRANSBOUNDARY FLOODS IN
THE CRISUL-KÖRÖS RIVER SYSTEM

JIRI MARSALEK[1], GHEORGHE STANCALIE[2], ROBERT
G. BRAKENRIDGE[3], VALENTINA UNGUREANU[4],
J. KERÉNÝI AND JANOS SZEKERÉS
[1]*National Water Research Institute, Burlington, ON, Canada,*
[2]*Romanian Meteorological Administration, Bucharest, Romania,*
[3]*Dartmouth Flood Observatory, Dartmouth College, Hanover, NH, USA,*
[4]*Romanian Waters Administration, Bucharest, Romania,*
[5]*Hungarian Meteorological Service, Budapest, Hungary,*
[6]*Research Center for Water Resources (VITUKI) Plc, Budapest, Hungary*

Keywords: Flood management, transboundary collaboration, earth observation.

1. INTRODUCTION

Experience of recent years indicates frequent occurrence of large and extreme floods in many parts of the world, including Europe (Marsalek 2000). One region, which suffers from flood damages on a regular basis, is the transboundary area of the Crisul Alb and Crisul Negro rivers flowing from Romania into Hungary, where they are known as Kőrös rivers. Floods in this area typically start in the mountainous terrain of the upper parts of the basin in Romania and propagate to the plains in Hungary. Recent floods in this area include the two spring 2000 floods, which caused on the Romanian territory damages of more than $US 20 million (Brakenridge et al. 2001). These losses included damages to houses, roads and railways, bridges, hydraulic structures, loss of domestic animals, and business losses. On the Hungarian territory, a particularly notable flood occurred in the summer of 1980, with total losses of $US 15 million, including destruction of farmhouses and large losses in agriculture (Brakenridge et al. 2001).

Historically, there has been a close co-operation between both countries in flood management in this area. The issues connected with the trans-boundary rivers crossing the Romanian – Hungarian border are covered by

283

J. Schanze et al. (eds.), Flood Risk Management:
Hazards, Vulnerability and Mitigation Measures, 283–296.
© 2006 *Springer.*

the bilateral Agreement for the settlement of hydrotechnical problems, which was issued on Nov. 20, 1986. This document includes eight specific regulations addressing flood defence, water quality protection, hydrological and meteorological data exchange, etc. To facilitate the implementation of this agreement, working groups from the Crisuri Water Authority in Oradea, Romania and Körös Valley District Water Authority (KOVIZIG) in Gyula, Hungary meet regularly to address the issues of mutual interest (Brakenridge et al. 2001).

The flood forecast and defence related information provided by Romania to Hungary (downstream) is presently based entirely upon the ground-observed data, which are mostly collected by non-automatic hydrometeorological stations. Such data are somewhat limited in terms of spatial distribution, temporal detail, and speed of collection and transmission, and these limitations should be remedied.

Recognising the threat of floods and the need for further improvement of flood management in this area, at the initiative of the Romanian National Institute of Meteorology and Hydrology (now known as the Romanian Meteorological Administration), an international team was formed, with representatives of Hungary, Romania and USA, and proposed a project on "Monitoring of Extreme Flood Events in Romania and Hungary Using Earth Observation (EO) Data" to the NATO Science for Peace (SfP) Programme (Brakenridge et al. 2001). After some modifications, this proposal was accepted by NATO and the 3-year project started in November 2002. The paper that follows provides a brief overview of this SfP project, including the progress to date (Marsalek 2004).

2. STUDY AREA: CRISUL/KÖRÖS BASIN

The study area represents the Crisul Alb/Negru/Körös transboundary basin spanning across the Romanian–Hungarian border, with a total area of 26,600 km^2 (14,900 km^2 on the Romanian territory). In Romania, the catchment (basin) comprises mountainous areas (38 %), hilly areas (20 %) and plains (42 %). About 30 % of the catchment is forested. On the Hungarian side, the catchment relief represents plains. Annual precipitation ranges from 600-800 mm/year in the plain and plateau areas to over 1200 mm/year in the mountainous areas of Romania. This precipitation distribution can be explained by the fact that humid air masses brought by fronts from the Icelandic Low frequently enter this area. The orography of the area (Apuseni Mountains) amplifies the precipitation on the western side of the mountain range. Thus, the Crisuri Rivers Basin frequently experiences large precipitation amounts in short time intervals and the frequency of such events seem to be increasing in recent years (Brakenridge et al. 2003).

In terms of hydrography, there is a marked difference between high rates of mountain runoff and low rates of runoff in plains. Thus, runoff flood waves formed quickly in the Romanian part of the basin move rapidly to the plains in the Hungarian part of the basin, which is characterised by relatively slow flows and a potential for inundation. In terms of flood forecasting, the

Romanian part of the basin is of greater interest with respect to flood formation, which is also reflected in this paper. The hydrography of the study area is well established. There are 62 hydrometric stations on the Crisul Alb and Negru (and their tributaries); 7 of these stations have flow records longer than 80 years. The list of significant floods includes the events of June 1974, July-August 1980, March 1981, December 1995-January 1996, March 2000, April 2000 and April 2001. On the Hungarian territory, the hydrometric stations at Gyula and Sarkad are particularly of interest. A review of the significant stage and discharge data for the Körös Rivers at these two hydrometric stations shows some trends in the flood data. In Gyula, the flow was decreasing in time, but the stage was rising (this could be caused by hydrotechnical structures); at Sarkad, both discharge and stage were increasing (reflecting more natural conditions, without much change in the river channel geometry) (Brakenridge et al. 2004).

Thus, the frequency and importance of floods in the study region require further work to reduce flood damages and improve flood monitoring by the agencies in charge of flood protection, such as government agencies, civil protection authorities or municipalities. To mitigate flood impacts in the study area, structural and non-structural measures have been undertaken in the past. The Romanian area is defended by dikes along the Crisul Alb River and Crisul Negru River. These dikes were built in the 19th century for a 20-year design return period and further improved in later years. Currently, the dikes on the right bank of the Crisul Negru River and the Teuz River (43 km) are designed for a 50-year return period, and on the Crisul Alb, 67 km of dikes on the right bank and 59 km on the left bank are designed for a 100-year return period. In spite of these improvements, in April 2000, the right bank dike of the Crisul Negru broke near the village Tipari (a 130 m breach) and caused significant flooding of, and damages in, the adjacent territory. Other structural flood protection measures include permanent retention storage facilities (total volume of $34 \times 10^6 \, m^3$) and temporary storage facilities (a total storage volume of almost $80 \times 10^6 \, m^3$) (Brakenridge et al. 2003, 2004).

On the Hungarian side, in the Körös valley, high flood potential is recognised and exacerbated by low flood plains. Much of the area is, therefore, protected by flood dikes, of which construction started in the 18th century. More than 440 km of dikes are maintained by the KOVIZIG. Following the 1979 flood, construction of detention reservoirs started. Altogether, these reservoirs provide storage capacity of 188 million m^3 and serve to reduce critical flood levels. The reservoirs are activated during floods, by a controlled explosion opening a protected spillway (a side weir) in flood dikes. Detained water inundates areas with lower intensity of agricultural activities and causes limited damages. Nevertheless, the reservoirs are activated only when necessary to avoid higher losses caused otherwise (Brakenridge et al. 2003).

The analysis and management of floods constitute the first indispensable step towards, and a rational basis for, the development of flood protection. Where certain flood risk levels are inevitable, the affected parties must know it and be appropriately warned. To reduce the frequency and magnitude of the damages due to flooding, comprehensive, realistic and integrated strategies must be developed and implemented.

The flood forecasting and monitoring systems existing in the study area do not reflect well the spatial distribution of floods and the related phenomena (pertaining to geographic distances or patterns) in both pre- and post crisis phases. To mitigate these limitations, the SfP project was initiated with emphasis on a satellite-based surveillance system connected to a dedicated GIS database that will offer a much more comprehensive evaluation of the extreme flood effects. Also, so far, the flood potential, including the risk and the vulnerability of flood-prone areas, have not been yet quantitatively assessed. An inventory of the past floods observed by the EO facilities would allow a more cost-effective design of structural and non-structural measures for flood protection and disaster relief. Finally, such data also provide important validation of the hydrological modelling-based flood risk assessment, because they show the actual extent of past flooding (Brakenridge et al. 2001).

3. SFP PROJECT OBJECTIVES

The main goal of the project is to reduce flood damages in the study area by improved flood forecasting and flood defence, and to deliver on other programmatic criteria, including enhancing co-operation among scientific personnel in the participating countries, training young researchers, disseminating results to the international scientific community, and transferring the tools developed in the study area to another river basin.

A flood forecasting system (non-automated hydro-meteorological stations, which transmit data by phone or radiotelephone) already exists in the study area. Using the observed peak discharges in the headwaters of the Crisul Alb and Crisul Negru and their tributaries, a flood forecasting procedure based on time-lagging of the corresponding discharges is applied. The deficiencies of this method include limited lead-times of the forecasts and improper consideration of tributary inputs. Consequently, this forecasting procedure needs to be improved (Brakenridge et al. 2001).

The success of risk management largely depends on the availability, dissemination and effective use of timely information. In flood risk management, orbital sensing technologies, used in conjunction with the traditional means, can greatly improve the management of flood hazards (Brakenridge et al. 1998, Muller et al. 1993).

The SfP project aims to provide an efficient and powerful flood monitoring tool to a broad range of stakeholders, and thereby significantly improve the efficiency and effectiveness of the action plans for flood defence. Apart from ground information on the occurrence and evolution of the flood, locally received NOAA/AVHRR satellite data, microwave data from the U.S. DMSP and Quikscat and follow-on satellites, and the high resolution images supplied by the European and American orbital platforms (SPOT, ERS, LANDSAT–7, EOS-AM "TERRA" and EOS–PM "AQUA"), will substantially contribute to determining the flood-prone areas. Furthermore, the information obtained from optical and radar images will be used in determination of certain parameters required in flood monitoring, such as the hydrographic network, water accumulation, size of the flood-prone area, and land cover/land use features (Brakenridge et al. 2004).

A database containing EO data, as well as hydrological and meteorological parameters related to significant flood events, will be established and used to test the processing and analytical algorithms, in order to establish an operational methodology for the detection, mapping and analysis of flooding. The management of remote sensing data related to river flooding largely relies on the functional facilities provided by the GIS, in combination with satellite data and hydrological models. The SfP project adopted this integrated approach, in order to establish a methodology that would allow the assessment of the flood risk on the basis of representative regional indices (Brakenridge et al. 2001).

The project will provide the decision-makers with updated maps of land cover/land use, hydrological networks and more accurate/comprehensive thematic maps at various spatial scales, indicating the extent of the flooded areas and affected zones. Due to the upcoming release of the NASA SRTM topographic data for this part of Europe, the SfP project may also provide updated and badly needed information concerning the detailed topography of the study area and especially of the low-relief floodplains.

The proposed hydrological forecasting model will increase the lead time and thereby provide additional time for implementing the measures needed to protect the human life and property in the study area. Thus, the use of the recorded rainfall in the model will provide extra time, varying from 6 to 12 hours, depending on the storm duration. Consequently, the time available to the emergency personnel in flood defence will be considerably increased. This will lead to direct and indirect reduction of economic losses and an earlier flood warning for the Hungarian territory, which is affected by the Crisul Alb and Crisul Negru floods (Brakenridge et al. 2001). Also, the newly proposed model allows the assessment of hydrograph shape, which is needed for hydraulic computations, but is not available in the traditional methods applied in the past.

Finally, the project results described here could also help in restoration and rehabilitation of some river courses, which were adversely altered by floods, and also in the future analysis and selection of structural flood protection works (Muller et al. 1993). The SfP project will greatly benefit from an ongoing Phare Project on Flood Prevention in the Crisuri Basin. The main objective of the Phare project is to implement a new system for automatic weather radar and ground surface measurements and data transmission of meteorological and hydrological parameters in the Crisuri basins. The telemetry system will provide continuous measurement and transmission of main hydrometeorological data (precipitation, air temperature and water levels) received from 28 automatic stations. Thus, these stations will provide, in real time, input data to the flood forecasting model, with further increase in lead times. Specific objectives of the SfP project are described in more detail below (Brakenridge et al. 2001).

3.1.   Objective 1

To analyze, adapt and test methods and algorithms for processing and interpretation of the medium and high spectral and spatial satellite resolution data, from the optical and microwave regions (provided by the existing platforms like NOAA-AVHRR, DMSP, SPOT, IRS, ERS, RADARSAT and by the newly available ones such as TERRA-MODIS, TERRA-ASTER and QUIKSCAT). The specific goal is to identify, delineate and map floodwater and excess soil moisture areas. The selected algorithms and methods for satellite data processing and interpretation that are dedicated to the analysis of severe flooding will improve the scientific and technological capabilities of the Romanian and Hungarian remote sensing teams.

3.2.   Objective 2

Development of a dedicated sub-system, based on remote sensing and GIS technology, to improve flood management and mitigate flood effects in the Romanian – Hungarian transboundary study area. This sub-system will allow the storage, management and exchange of raster and vector graphic information, and also of related attribute data for the flood monitoring activities. This dedicated information sub-system will contribute to the regional quantitative risk assessment (using flood hazard and vulnerability characteristics), and will serve for monitoring and hydrological validation of risk simulations. Another important result will be the preventive consideration of flood events in land development and special planning of the flood-prone areas.

3.3.   Objective 3

Development of integrated methods, encompassing hydrological modelling, EO data and GIS facilities, for flood management before and after flood events, and their adaptation to local conditions in the Crisul Alb and Crisul Negru transboundary basins. These methods will include a statistical model of flood wave hydrographs of various frequencies of occurrence, algorithms for flood hazard mapping (using representative parameters obtained from historical images and other geo-referenced data) and objective documents for post-crisis analysis and damage assessment.

3.4.   Objective 4

Adaptation of a semi-distributed rainfall-runoff model to the study area and its improvement with respect to increasing the flood forecast accuracy and lead-times. The model outputs are stage and discharge hydrographs in the sections chosen by the user. The model will be provided with an updating procedure and a routine assessment of the effects of the storage reservoirs (permanent and temporary reservoirs) on the peak flood discharge. The forecasted hydrograph will serve as the input to a hydraulic model for computing transport and transformation of flood waves along the river, using descriptions of the geometry and hydraulic characteristics of the river channel.

3.5. Objective 5

Development of new satellite-based applications and products in the study area, for water management and civil protection authorities, environmental agencies and private users: updated digital maps of the hydrological network and land cover/land use, hazard maps at various spatial scales (1:200,000 to 1:5,000) with the extent of the flooded areas and the affected zones in detailed scales and with the possibility of managing and displaying the geographical database and observing rainfall-runoff model outputs. These products will contribute to preventive consideration of flooding in land development and special planning in the flood-prone areas, and for optimising the distribution of flood-related spatial information to end-users.

3.6. Programmatic objectives

While the first five objectives were of the scientific/technical nature, the programmatic objectives include the general objectives of the NATO SfP Programme, such as fostering international cooperation, training young scientists, disseminating results to the international scientific community, and applying key project results in another river catchment.

4. PROGRESS ACHIEVED AT HALF-TIME OF PROJECT DURATION

4.1. Image processing for flood analysis

EO images have wide applications in flood analysis, in such tasks as producing catchment maps, detecting water surface and soil moisture, detecting inundated areas, and assisting with remote flow measurement (Brakenridge et al. 1998, Muller et al. 1993). Thus, image processing is important for developing such products and using them in flood analysis and management. Towards this end, the project team undertook an inventory and documentation of image processing methods, set up an experimental image database, tested and evaluated processing methods, initiated the selection of the best processing method, and conducted training in image processing (Marsalek 2004).

General methods and algorithms used for EO data processing and interpretation, needed for identifying, delineating and mapping water and excess soil moisture areas, were analysed and tested (Brackenridge et al. 2003, 2004). An image database with high spectral and spatial satellite data was established, and the images were analysed with respect to correct identification of water bodies, accurate estimation of flooded (inundated) areas, and the dynamic monitoring of flood processes. In this work, emphasis was placed on medium spatial resolution (e.g., obtained from NOAA-AVHRR and TERRA-MODIS).

MODIS (Moderate Resolution Imaging Spectroradiometer) is a key instrument aboard the Terra (EOS AM) and Aqua (EOS PM) satellites. Terra MODIS and Aqua MODIS view the entire Earth's surface every 1 to 2 days, acquiring data in 36 spectral bands, or in groups of wavelengths. The data available from MODIS are highly suitable for flood warning and management, because they are available in real time (or nearly real time), can be rapidly processed and disseminated, cover a wide area, and are abundant and inexpensive, which is important when dealing with longer duration floods (Brakenridge et al. 2003).

In processing MODIS data, major technical challenges include proper water classification and the development/refinement of the MODIS-based hydrological monitoring system (detection of changes, and satellite stream gauging stations). MODIS provides very accurate geo-coding, clouds can be removed by composition techniques, and the data are distributed via the EOS Gateway. In water classification, strong water/land differentiation is needed. The Romanian Project team has developed a methodology, which allows co-processing the ASTER and MODIS data, essentially by superposing the degraded ASTER mask over the MODIS water mask and obtaining a MODIS mask with the "percentage of water" pixels.

The algorithm serving to create a water mask using a threshold technique on multispectral MODIS images was adopted from the Dartmouth Flood Observatory (Hanover, USA) and further modified in early 2004 (Brakenridge et al. 2004). The objective of these modifications was to develop an automatic method for creating the water mask using multispectral MODIS images. An algorithm was developed to detect pixels that were either 100% water or partial water, using a threshold technique and investigating the spectral characteristics of the pixels individually, without taking into account the adjacent pixels. Calibrated shortwave images with 250 and 500 m spatial resolution were used for this purpose. The 1.6 μm-channel data were used, because they are characterised by the lowest water reflectance and are independent of turbidity, which is typical for flood waters. However, the 1.6 μm channel allows separating water from low-land areas, but not from snow in mountainous areas. To mask out snowy areas, a threshold on the 0.87 μm channel was used (Brakenridge et al. 2003, 2004).

The other problems encountered were caused by orographic and cloud shadows, and by melting snow. Their spectral characteristics were quite similar to those of water, so only partial separation was achieved. Large, clear water surfaces without vegetation have NDVIs (Normalised Difference Vegetation Index) < -0.2 and can be easily separated from shadows. However, using this threshold would mask out not only the shadows but also many pixels containing water, such as the pixels with high vegetation fractions, turbid waters, and mixed pixels along the coasts. To avoid the loss of such pixels, a NDVI threshold > -0.2 had to be used. After experimenting with different threshold values, the best suited values were adopted (Brakenridge et al. 2004).

4.2. Development of a dedicated subsystem (DSS) based on remote sensing (RS) and GIS

The main functions of this subsystem are acquisition, analysis and interpretation of data; management of data; handling and preparation of data for rapid access; updating of information; data restoring, and preparation of value-added information (Stancalie et al. 2003). The proposed DSS system is based on both remote sensing and GIS. A central platform information DSS server was set up at the NIMH, where satellite data are received and processed, and further distributed, e.g., to the end-users. Several satellite data providers were incorporated into the Project and other EO data can be accessed via FTP from European and American satellite database centres. The dedicated subsystem allows data exchange among the Project partners. Standardized products and project results are also distributed via the central network Internet interface (Brakenridge et al. 2003).

A GIS database needs to be created for the study area (the catchments of the Crisul Alb, the Crisul Negru and the Körös Rivers). In preparation for GIS construction, the data needs were assessed, the existing map data were inventoried, and the needs for their updating from satellite images or field measurements were identified. The structure of the GIS was designed to meet the Project requirements with respect to evaluation and management of the information pertinent to flood occurrence and development, as well as for the assessment of damages caused by floods. In this regard the database contains a spatial geo-referenced information ensemble, with satellite images, thematic maps, time series of meteo-rological and hydrological parameters data, and other data. The GIS database will be connected with the hydrological database, which will facilitate synthesized representations of the hydrological risk using separate or combined parameters (Brakenridge et al. 2004).

While the Hungarian team activities concerning the GIS development are only in the planning stage, the Romanian team proceeded with developing a GIS for the entire Romanian study area of the Crisul Alb and Crisul Negru basins using cartographic maps in the scale of 1:100,000. Recognising that the information in such maps may be outdated, it was updated on the basis of recent satellite images (e.g., the hydrographic network, land cover/land use) and/or by field measurements (e.g., dike and canal network) (Brakenridge et al. 2004).

As being currently developed, the GIS database contains the following info-layers: (a) sub-basin and basin limits; (b) land topography (organised in DEM); (c) hydrographic network, dikes and canals networks; (d)

transportation network (roads, railways); (e) municipalities; (f) meteorological station network, rain-gauge network, hydrometric station network;
and, (g) updated land cover/land use.

Inventories of all meteorological and hydrometric stations, hydrotechnical structures (including dikes, drainage works, agricultural drainage,
irrigation systems, flood retention reservoirs, and polders) in the Crisul Alb
and Crisul Negru basins have been completed. Also, cross-sectional profiles
of main river beds, corresponding to the Crisul Negru, Crisul Alb, White,
Black and Double Koros, have been completed through field measurements.
The GIS info-layers related to the meteorological and hydrometric station
networks in the Crisul Alb and Negru basins are shown in Figure 1.

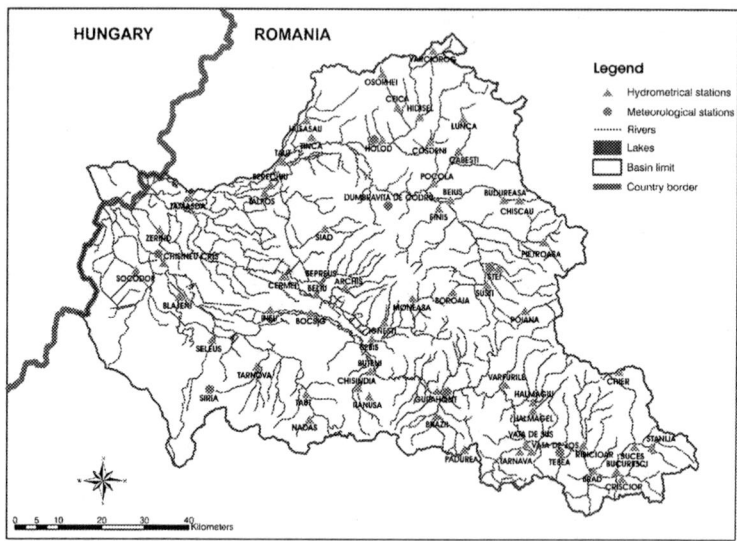

*Figure 1. The GIS info-layers related to the hydrometeorological station network in the
Crisul Alb and Crisul Negru basins* (Brakenridge et al. 2004)

4.3. Integrated methods for flood management

In this task, a flood data base was established for the study area (from the
Romanian and Hungarian national databases) and validated, maximum
discharges of various return periods were calculated, synthetic flood
hydrographs were developed, and land cover/land use maps are under
preparation.

The characteristics of extreme floods, i.e., peak flows, volumes and
durations, and their probabilistic distributions, are needed and were
determined by the Flow (Q)-Duration (d)-Frequency (F) method. The
estimates of low-frequency flood quantiles were produced by the GRADEX
method, in which maximum rainfall distributions are used to extrapolate
hydrometric data (Marsalek 2004, Brackenridge et al. 2003).

For each flood event, characteristic flows were determined, partial-duration series of these variables were fitted by the exponential law, and extrapolated to lower frequencies by gradually replacing the flow distribution slope by the rainfall gradex. Synthesized (design) flood hydrographs are needed as inputs to hydraulic models for establishing flood maps and were derived with hydrograph shapes consisting of two segments; the linear rising limb with a time to peak ≤ D, where D is defined as the median value of flow durations corresponding to the 10-year flood peak, and the falling limb, of which shape is determined from threshold discharges of the same occurrence and various durations. Examples of synthetic flood hydrographs for return periods varying from 10 to 1,000 years are given in Figure 2.

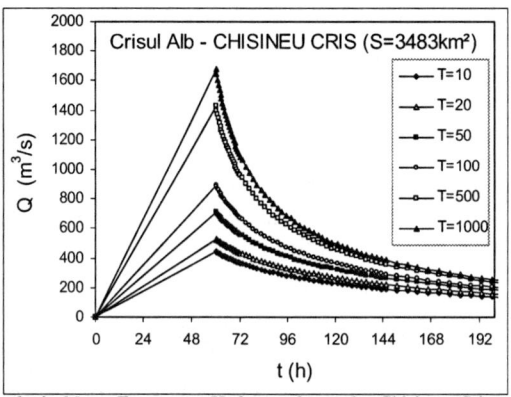

Figure 2. Synthetic Mono-Frequency Hydrographs at the Chisineu Cris gauging station (Crişul Alb River) (Brakenridge et al. 2003)

Finally, for updating the land cover/land use in the study area, TERRA/ASTER data were used and were found suitable for producing detailed maps of land cover/land use, especially when using the visible and near infrared bands (1, 2, 3B) data with a 15-m resolution. The method for the land cover/land use mapping based on the TERRA/ASTER data included three steps: (a) Geo-referencing of the ASTER data, (b) Detection of clouds and water surface, and (c) Data classification, in which both supervised and unsupervised classifications were tested. Better results were obtained with the unsupervised classification, in which different numbers of classes and iterations were tested.

Some progress was reported on starting the Digital Terrain Model (DTM) construction and integration into the GIS database (Stancalie 2004, et al. 2004). DTM is a dataset representing ground surface elevations (bald earth), without portraying any above ground objects. For the Hungarian catchment, a DTM is available with a 50 x 50 or 10 x 10 m frame distance, and the following ranges of the vertical accuracy: hilly area (slope > 6 %) < 5.0 m; medium relief areas < 2.5 m; and, flat areas < 0.8 m. A digital map of the Hungarian study area produced may have resolution which is inadequate for the Project. In the (Romanian) area vulnerable to flooding, which is situated in the floodplain regions of the Crisuri River basin and demarcated at its eastern boundary by the Ineu-Talpos and its northern boundary by the Crisul Repede basin, the most refined DTM will be developed.

There are several techniques for DTM generation: Shuttle Radar Topography Mission (SRTM) products, radar interferometry, lidar altimetry and DTM construction from topographical maps. The final DTM will be constructed in a UTM projection, with reference to the Baltic Sea, and cell resolutions of < 7 m in the horizontal plane and < 1 m in the vertical plane (Brakenridge et al. 2004).

4.4. Methodologies for flood forecasting

Two types of models are used in the project – a flood forecasting model VIDRA on the Romanian territory and the output hydrographs from this model will be routed in the Hungarian part of the catchment by the HEC-RAS model of the U.S. Army Corps of Engineers. The VIDRA model (Brakenridge et al. 1998) simulates the rainfall-runoff processes taking place in a watershed by conducting the following computations:

*   Sub-basin snowmelt estimation, using the degree-day method;
*   Computation of the average rainfall in each sub-basin, by weighing the rainfall and snowmelt data measured in the meteorological network;
*   Calculation of the effective rainfall over each sub-basin by subtraction of infiltration and evapotranspiration abstractions from the average water inflow, using the deterministic reservoir model PNET;
*   Integration of the effective rainfall on hill slopes and in the primary river network, which results in runoff hydrograph formation in each sub-basin, using the instantaneous unit hydrograph as a transfer function of the hydrographical system;
*   Superposition of the flood waves formed in each sub-basin and their routing along the main river channel using a non-linear model based on the analytical solution of the Muskingum method; and,
*   Flood wave attenuation through reservoirs, using a reservoir co-ordinated operation method.

The VIDRA model has a variable computational step (from one to 24 hours) and simulates all the main hydrological processes taking place in a watershed (Brakenridge et al. 2004).

4.5. Programmatic objective achievements

So far, great progress has been achieved towards meeting the programmatic objectives (Marsalek 2004). Project visibility is maintained through a number of presentations at various international meetings and the project website. There are about 15 young scientists working on various project tasks at the five agencies involved. Staff members have been trained at the Dartmouth Flood Observatory (USA), DHI (Denmark), and elsewhere.

5. CONCLUDING OBSERVATIONS

The NATO Science for Peace Programme provides an important mechanism for supporting scientific research in the transition countries in Central and Eastern Europe. The project featured here is helping deliver on that programme and is particularly worthwhile because of its focus on research with high societal value (protection of human life and property), conduct of leading-edge research in terms of Earth Observations and their use in flood management, enhancing collaboration between the neighbouring countries and others, and fostering collaboration among international experts. The progress achieved under this project so far holds promise that all these goals and objectives will be met.

References

Brakenridge R.G, Stancalie G., Ungureanu V., Diamandi A., Streng O., Barbos A., Lucaciu M., Kerenyi J. and Szekeres J. (2001) Monitoring of extreme flood events in Romania and Hungary using EO data. project plan, September, Bucharest, Romania.

Brakenridge R.G, Stancalie G., Ungureanu V., Diamandi A., Streng O., Barbos A., Lucaciu M., Kerenyi J. and Szekeres J. (2003) Monitoring of extreme flood events in Romania and Hungary using EO data. Progress report, May. Hanover, NH, USA.

Brakenridge R.G, Stancalie G., Ungureanu V., Diamandi A., Streng O., Barbos A., Lucaciu M., Kerenyi J. and Szekeres J. (2004) Monitoring of extreme flood events in Romania and Hungary using EO data. Progress report, May. Hanover, NH, USA.

Brakenridge R. G., Tracy B. T. and Knox J. C. (1998) Orbital SAR remote sensing of a river flood wave. *Int. J. Remote Sensing* 19 (7), 1439-1445.

Marsalek J. (2000) Overview of flood issues in contemporary water management. Marsalek J., Watt W.E., Zeman E. and Sieker, F. (eds.) Flood Issues in Contemporary Flood Management. *NATO Science Series* 71, Kluwer Academic Publishers, Dordrecht/ Boston/London, 1-14.

Marsalek J. (2004) Monitoring of extreme flood events in Romania and Hungary using EO data. Technical Progress Report No. 3, NWRI, Burlington, Ontario, Canada, August.

Muller E., Decamps H. and Dobson M. K. (1993) Contribution of space remote sensing to river studies. *Freshwater Biology* 29, 301-312.

Stancalie G. (2004) Contribution of Earth Observation Data to Flood Risk Analysis. Proc. XXII-nd Conference of the Danubian Countries on the Hydrological Forecasting and Hydrological Bases of Water Management, Brno, Aug. 30- Sep. 3, 2004 (CD-ROM).

Stancalie G., Alecu C., Craciunescu V., Diamandi A., Oancea S. and Brakenridge R. G. (2004) Contribution of Earth Observation data to flood risk mapping in the framework of the NATO SfP "TIGRU" project. Proc. International Conference on Water Observation and Information System for Decision Support, Ohrid, FY Republic of Macedonia, May 25-29, 2004 (CD-ROM).

Stancalie G., Diamandi A., Ungureanu V. and Stanescu V. A. (2003) Sub-system based on remote sensing and GIS technology for flood and related effects management in the framework of the NATO SfP "TIGRU" project. 1st Annual Session of the NIHWM, Bucharest, Sep. 22-25, 2003 (CD-ROM).

Chapter 25

THE APRIL 2000 FLOOD IN THE CRISUL ALB RIVER BASIN – TRANSBORDER COOPERATION ROM-HUN FOR FLOOD MITIGATION

OCTAVIAN STRENG[1], VALENTINA UNGUREANU[2], MITRUT TENTIS[1]
[1]Hydrological Service, Crisuri Rivers Authority, Oradea, Romania,
[2]National Administration "Apele Romane", Bucharest, Romania,

Keywords: Extreme floods, flood defence, transboundary collaboration.

1. INTRODUCTION

Romania is a country that suffered quite often flooding, especially in the last decade, when they were more frequent. Practically, there were years when almost during all months flooding occurred, either in small basins, with torrential character, or at a zonal or basin scale. These floods provoked very important damages that strongly affected the population and various social and economic relevant objectives. There were even losses of human lives registered. The economic consequences of these floods are hard to be compensated in the present context of economic development of Romania.

Figure 1. Crisuri River basins location

J. Schanze et al. (eds.), Flood Risk Management:
Hazards, Vulnerability and Mitigation Measures, 297–304.

298

From the analysis of the rainfall events that led to flooding over the time, it has been found that the most affected areas are those situated in the western part of the country because of their climatic and orographic characteristics. The study zone, Crisul Alb hydrographic basin, is located in this area (Figure 1).

2. DESCRIPTION OF CRISUL ALB RIVER BASIN

The Crisul Alb hydrographic basin (Figure 2) has 4,240 km^2. The largest part of the basin is situated in the contact area of four mountainous massifs from Western Carpathians and, by its main course of 234 km length, flows through some tectonic depressions before it reaches the Tisza Plain. Crisul Alb springs from the western part of Bihor Mountains, at 980 m altitude, under the Certez peak (1,184 m).

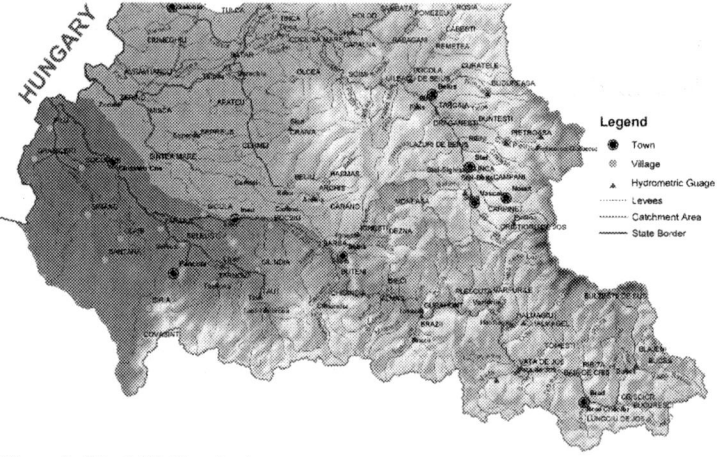

Figure 2. Crisul Alb River basin

The upper part of the Crisul Alb River (till Criscior section), situated in the mountainous area, with only 32 km length and oriented from north to west, is characterized by a cross profile of a tipping sharp V. The hydrographic network is relatively dense (the density coefficient is 0.8 – 1.2 km/km^2) and is characterized by important water discharges due to high precipitation and high flow velocity.

The middle section of the water course (between Criscior and Cociuba), situated in the hill area with a bigger length than the superior course (10 km) and oriented from east to west, is characterized by a more balanced longitudinal profile, with smaller medium slopes (approximately 1.2 ‰). This sector is formed by a succession of tectonic depressions in which the river meanders because of the reduced slopes divided by narrower gorge with higher slopes and flow velocities. These tectonic depressions represent the areas in which the waters (tributaries) are collected and favour flooding.

The inferior course of the river (upstream of Cociuba village), situated in the plain, with 100 km length and oriented from east to north-west, is characterized by smaller medium slopes (0.7 ‰ till Chisinau Cris section and 0.3 ‰ in upstream). In this area, in which the flooding recurrence is high, dams were built in order to protect the agricultural areas. There were also channels built for dewatering and evacuating the excess of water. But some of these constructions are quite old (over 100 years), and are in danger to be damaged, as it was the case of the broken dyke of the Crisul Alb right bank, just in the upstream of the confluence with Cigher river, in April 2000.

From a climatic point of view, the Crisul Alb river basin is characterized by a combination of three influences: Mediterranean, Baltic and continental with important precipitation distributed uniformly during the year, with annual medium temperatures that exceed 10 °C, with frequent warming air during winter, which causes the melting of the snow layer that, combined with liquid precipitation, could generate serious floods.

The coefficient of forestation in Crisul Alb hydrographic basin is of approximately 30 % from the surface of the entire basin. Due to these values that are quite small it could be stated that this coefficient has is important role in floods occurring.

In the depressions situated along the Crisul Alb River there are many inhabitants, very close to the watercourse, in danger to be flooded.

The flood occurrence in Crisul Alb basin is quite frequent: once per 2-3 years the water level could exceed the critical threshold, and, in the last period, this phenomenon occurred even twice a year. As highest floods occurred in this basin, the floods from 1966, 1970, 1974, 1981, 1995, 2000 (in March and April) can be mentioned, when the levels exceeded the danger quote.

The hydrometric network existing in the basin of Crisul Alb consists in approximately 30 hydrometric stations, 6 been situated on the main course. There is a project in progress which stipulates the installation of 5 self-monitoring stations in its framework. Water level and the hourly precipitations will be measured and transmitted.

3. FLOOD IN APRIL 2000

In condition of a moisture soil, resulted from the waterfall as rain in the period between April 1th - 5th, 2000 and a snow pack existed at 1400 m altitude, having about 100 cm thickness, precipitation fell during April 6th - 7th, 2000 with about 30-120 mm allocated over the whole basin. As a result floods occurred in the whole basin with a probability of 2-10 %, which generated a discharge over the assurance of 1 % in the lowest basin at Chisinau Cris.

Flood from April 5th - 7th, 2000 was caused by serial rainfalls which had their maximum in the interval of April 5th - 6th. These rainfalls were close preceded by others fell as far back as 27th February till 5th April that totalized 50 mm and which, on one hand contributed to increase the saturation soil degree and on the other hand produced grows of discharges which kept a high level of the river beds. The biggest amount of rain fell in the north part of the flood totalizing over 100 mm (Figure 3).

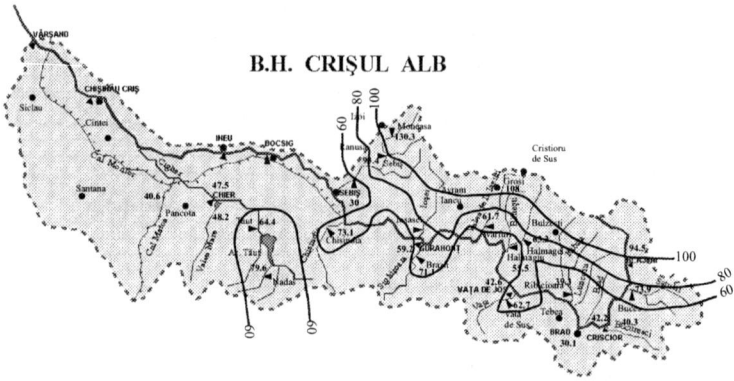

Figure 3. Precipitation in Crisul Alb Basin during 5-6 April 2000

From the analyses of maximum discharges on the main course it can be stated that the rare frequency of these exceptional flood of the Ineu (Bocsig) – Chisinau Cris river reached a probability of once per 35 - 100 years. Rare occurring events are also noted for the Sebis and Crisul Alb rivers at Gurahont. It is remarkable that the flood at the upper stream sector of the Crisul Alb river from March of the same year was higher than at the downstream sector in April 2000 and had frequencies of once per 7-100 years (Vata de Jos and Criscior respectively).

Because of the maximum discharges registered downstream of Ineu (values with exceeding probability of 1 %) the dikes, projected for 2 % assurance, did not resist anymore in Tipari point. In April 7[th,] 07:00 o'clock at Tipari, between 33 - 36 km, in some sectors the right bank was overtopped and from 12.50 o'clock onwards the dike was breaching on about 140 m portion (Figure 4). A volume of about 85 million m^3 flew in the floodplain and covered the area between Crisul Alb and Crisul Negru reaching the left dike of Crisul Negru at Zerind.

Figure 4. Dike brake at Tipari

The severe inundation caused by the dike breach lead to important damages of more than USD 20 millions. These losses include damages of 807 houses, 196 km of roads and railways tracks, 170 bridges, 35 socio-economic objects and 84 hydraulic structures, and loss of 134 domestic animals (Figures 5 and 6).

Figure 5. Flooded houses

302

Figure 6. The Oradea-Arad flooded railway

The reconstructed discharge after the flood at Ineu - Chisineu Cris, was of 826 m³/s – a value superior to the discharge which has been calculated for a 1% assurance (Figure 6).

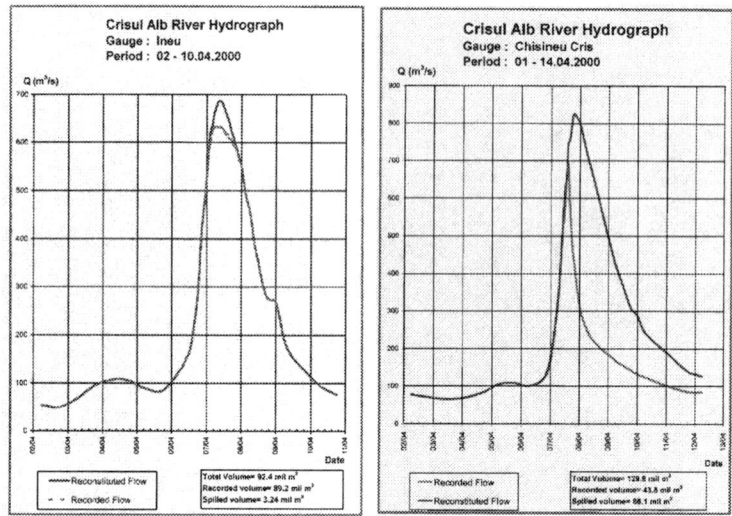

Figure 7. Registered hydrographs in Crisul Alb River basin, at Ineu and Chisineu Cris hydrometric stations

4. ROMANIAN-HUNGARIAN TRANSBORDER COOPERATION

The water was billeted in the frontline area between Crisul Alb and Crisul Negru dikes and the dikes localized at the Hungarian territory.

The water accumulation in the border area had increased the risk of overtopping the dike on the Hungarian side and considerable areas were flooded and important damages recorded.

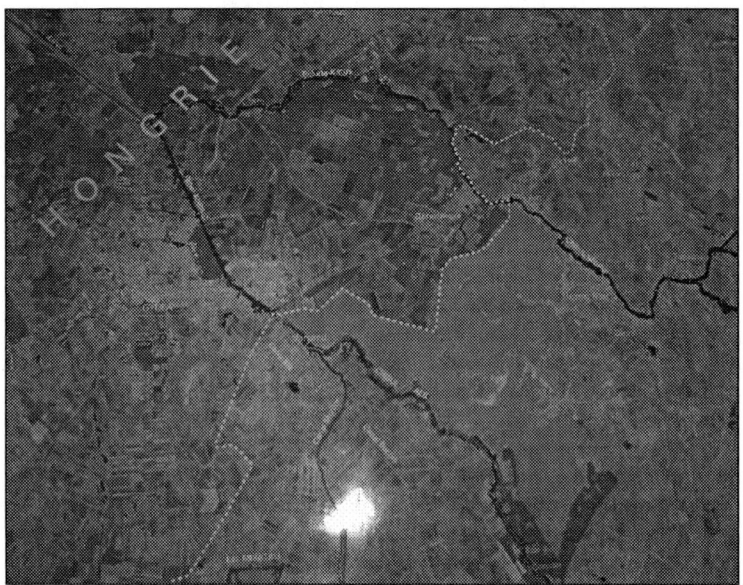

Figure 8. ▨ *The flooded area in April 18th. SPOT Image*

The Romanian-Hungarian collaboration regarding the border waters had operated under the mixed Romanian-Hungarian Hydrotechnical Convention stipulations.

Regarding the situation, the experts from the Hydrotechnical Commission had met immediately in the border area for finding solutions concerning water evacuation and flood mitigation. Because of the pumping stations which did not work due to their inundation, the decision had been taken to create a break in the Crisul Alb right bank dike on the Romanian side close to the border.

The access to the area was easier from the Hungarian side, so it had been decided that the work should be made with the Hungarian technique and specialists. The supervision was done using experts from the two countries. The dike breach was 120 m long (km 0+000 m - km 0+120 m) and the maximum flow capacity was about 40 m^3/sec.

The construction started on the April 11th as soon as the Crisul Alb River level began to lower. The water had returned to the river bad gravitationally. The water evacuation lasted till May 4th. During this period a close connection was kept between the Hydrological Service from Romania in order to prevent any flood.

The dike reconstruction at the initial situation was done on September 13th, 2000. After the work had been finished the repraesentants of the two countries made the project costs evaluation and specified, how the Hungarian side could be compensated. The mutual decision, which had been taken, said that the Romanian side would realise the hydrotechnical structures for flood mitigation, in the same concerning area from the Crisul Alb Basin, which have effects on the Hungarian side, too.

In the Ineu-Chisineu Cris area two non-permanent reservoirs (polders) are being built with a maximum capacity of 12 millions m^3 reducing the flow capacity of the most important floods (reducing pressures on dikes).

5.   CONCLUSIONS

The Crisul Alb basin configuration creates complex problems in flood defence in the two countries. Accordingly supervising and studying thoroughly the floods in the Crisul Alb River is rightful.

Due to the NATO support in the "Science for Peace (SfP)" Program – the Hungarian-Romanian-American project no. 978016, Monitoring of Extreme Flood Events in Romania and Hungarian Using EO Data was realised.

The efficiency in solving the April 2000 issues became possible based on the existence of the legal scheme of transborder Hungarian – Romania cooperation and because of the outstanding collaboration between the specialists from the two countries.

The Romanian-Hungarian-Hydrotechnical Convention started its activity in 1985 and it had created the conditions of an efficient collaboration of the two countries. Inside this convention there were several sub-commissions (for flood defence, qualitative and quantitative water management, hydrometeorology etc. with annual meetings, where the performances and the best ways of collaboration improvement were analysed).

In 2004 the new Romanian-Hungarian-Hydrotechnical Convention was signed. It was actualized and got many improvements. Collaboration inside the Convention is used to provide a small contribution to realise an integrated water management in the Danube River Basin.

PART 7

CONCLUSIONS

Chapter 26

CONCLUSIONS OF THE ADVANCED RESEARCH WORKSHOP ON FLOOD RISK MANAGEMENT

JOCHEN SCHANZE[1], EVZEN ZEMAN[2],
JIRI MARSALEK[3]
[1]*Leibniz Institute of Ecological and Regional Development (IOER),
Member of the Dresden Flood Research Center (D-FRC),
[2]DHI Hydroinform a.s., Prague, Czech Republik
[3]National Water Research Institute, Burlington, Ontario, Canada*

1. FLOOD HAZARD MODELLING (SESSION 1)

Presentations showed various approaches to flood modelling. Flood hazard are assessed from observed data and modelling. Surrogate observations should be approached with caution. The importance of flood hazard determination is increasing, because of changing climate, conflicting demands of water management, and continuing expansion of urban areas (human settlements). Further advances in the modelling of flood protection are recommended and should consider water storage in headwaters. Moreover, the role of groundwater should be further examined.

2. FLOOD FORECASTING (SESSION 2)

The session provided a review of recent developments in flood forecasting. As a result weather radar and earth observations can been seen as successfully incorporated in flood forecasting, but further refinements are needed. In flow routing, 2D models still take too much time. Good progress is being made in the development of flood forecast models. Better sharing of information on models is needed. To avoid duplicity and to assist flood forecasters in selecting appropriate models, an information platform is needed. Further research on weather radar and remote sensing in flood forecasting is recommended.

*J. Schanze et al. (eds.), Flood Risk Management:
Hazards, Vulnerability and Mitigation Measures, 307–309.*
© 2006 *Springer.*

3. MODELLING OF VULNERABILITY (SESSION 3)

The session gave an overview of different approaches to assessing vulnerability. In became evident that the analysis and assessment of vulnerability is an essential part of flood risk analysis. A good understanding of vulnerability with respect to flood characteristics, property characteristics and socio-economic factors is needed. Vulnerability cannot be measured only in monetary terms. Reduction of vulnerability seems to be a crucial approach for flood risk management. In some examples, which have been presented, vulnerability still increases as a result of poor land use planning. Future research should further develop models for the calculation and assessment of vulnerability.

4. FLOOD RISK MITIGATION (SESSION 4)

There are new developments in flood risk mitigation in various countries. The paradigm of flood management is changing from flood protection to flood risk management including risk analysis and risk mitigation. Many countries are in transition between those two approaches. The role of the public has to be recognized for example in terms of negotiating tolerable risks, access to information, uncertainties of modelling results etc. Flood risk maps should form the basis for selection of mitigation measures. While the technical aspects of mitigation measures are important, their acceptance by the public should not be underestimated. Strategic planning approach to risk management is needed. Enforcement of spatial planning policies should prevent certain increase of flood risk by uncontrolled development. Flood risk management should be based on flood risk maps. Moreover, criteria and multi-criteria approaches for allocation of funds to various mitigation projects are an important requirement.

Flood risk management should be dedicated to the whole catchments and explicitly mention the remaining risks. The move from flood defence to the management of flood risk needs to educate and train students and professionals to meet these new challenges.

5. HISTORICAL FLOODS AND TRANSBOUNDARY ISSUES (SESSION 5)

Historical floods represent valuable sources for interpreting recent floods like the 2002 flood in the Vltava/Elbe basin. Based on this, they may contribute to planning future flood risk management.

New international programmes bring attention to transboundary floods and their improved management in the spirit of enhanced co-operation, which was seen as having paramount importance. Further advances have to be made in the field of data harmonization and transboundary access.

6.   FINAL OBSERVATIONS AND OUTLOOK

Hydrometeorological extremes of recent years brought additional attention to floods. A shift in approaches to coping with floods from protection to flood risk management is obvious. There is good progress in integrated measurements and observations referring to data sensing, flood forecast, etc. Successful risk reduction requires more information about socio-economic factors, if for no other reasons, then at least for raising the public awareness of risks. Mitigation measures encompass both structural and non-structural measures, and require public financial support based on priority ranking. Therein, regional and international co-operation is of paramount importance. To facilitate the communication between scientific fields and between various countries a common language in flood risk management is required, as it will be provided from the European FLOODsite research project.

Overall, the format of the ARW turned out to be a very fruitful one. Especially the meeting of a small group of scientists over a few days was a good prerequisite for a more detailed exchange. All participants agreed on the importance of continuation of this discussion in the near future.

LIST OF PARTICIPANTS

DIRECTORS

Schanze, Jochen

Leibniz Institute of Ecological and
Regional Development (IOER),
Weberplatz 1, 01217 Dresden
j.schanze@ioer.de
GERMANY

Zeman, Evzen

DHI Hydroinform a.s.,
Na Vrsich 5, 100 00 Praha 10
e.zeman@dhi.cz
CZECH REPUBLIC

SPEAKERS

Arman, Hasan

Sakarya University, Engineering
Faculty, Dpt of Civil Engineering,
Esentepe Campus, 54187 Sakarya
harman@sakarya.edu.tr
TURKEY

Babiakova, Gabriela

Slovak Hydrometeorological Institute,
Hydrological Forecasts and Warnings
Dpt, Jeséniova 17, 83315 Bratislava 37
gabriela.babiakova@shmu.sk
SLOVAKIA

Balint, Zoltan

Upper Tisza Water Authority, Water
Management Department, Szechenyi
u. 19, H 4400 Nyiregyhaza
balintz@fetikovizig.hu
HUNGARY

Bernhofer, Christian

Technische Universität Dresden,
Institute of Hydrology and
Meteorology
01062 Dresden
bernhofer@frsws10.forst.tu-dresden.de
GERMANY

Brázdil, Rudolf

Masarykova Universita, Geografický
ústav, Kotlářská 2, 611 37 Brno
brazdil@sci.muni.cz
CZECH REPUBLIC

Craciunescu, Vasile

National Institute of Meteorology and Hydrology, Remote Sensing and GIS Lab, 97, Soseaua Bucuresti-Ploiesti, Sector 1, 013686 Bucuresti
vasile.craciunescu@meteo.inmh.ro
ROMANIA

Einfalt, Thomas

einfalt & hydrotec GbR, Wakenitzmauer 33, 23552 Lübeck
thomas@einfalt.de
GERMANY

Hutter, Gérard

Leibniz Institute of Ecological and Regional Development (IOER), Weberplatz 1, 01217 Dresden
g.hutter@ioer.de
GERMANY

Jirásek, Václav

Povodí Labe, s.p., Víta Nejedlého 951, 500 03 Hradec Králové
jirasek@pla.cz
CZECH REPUBLIC

Juza, Bohumil

Baker Engineering NY, Inc., 45-18 Court Square, Suite 503, NY 11101 Long Island City
bjuza@mbakercorp.com
USA

Kohnová, Silvia

Slovak University of Technology, Radlinského 11, 813 68 Bratislava,
kohnova@svf.stuba.sk
SLOVAKIA

Korndörfer, Christian

Environmental Office Dresden 01069 Dresden
umweltamt@dresden.de
GERMANY

Kovacs, Sandor

Middle Tisza Valley District Water Authority (KOTIVIZIG), Sagvari krt 4, H-5000 Szolnok
drkovacz@kotikovizig.hu
HUNGARY

Kubát, Jan

CHMU Czech Hydrometeorological
Institute, Na Šabatce 17,
143 06 Praha 4 – Komořany
kubat@chmi.cz
CZECH REPUBLIC

Kukharchyk, Tamara

National Academy of Belarus, Institute
for Problems of Natural Resources Use
& Ecology, 10 Staroborisovsky tract,
220114 Minsk
kukharchyk@mail.ru
BELARUS

LeMarquand, David

Public Security and Emergency
Preparedness Canada, Research &
Development, 340 Laurier, 12th floor
K1A 0P8 Ottawa, Ontario
david.lemarquand@psepc-sppcc.gc.ca
CANADA

Loza, Irina

State Agrarian University, Plant
Psychology & Botany Department,
Voroshilov str. 25,
49027 Dniepropetrovsk
irinaloza@hotmail.com
UKRAINE

Lukáč, Iroslav

Water Research Institute, Department
of Hydrology and Hydraulics, nabr.
arm. gen. L. Svobodu 5,
812 49 Bratislava
lukac@vuvh.sk
SLOVAKIA

Marsalek, Jiri

UWMP, National Water Research
Institute, 867 Lakeshore Rd.,
Burlington, Ontario L7R 4A6
jiri.marsalek@ec.gc.ca
CANADA

Messner, Frank

UFZ Centre for Environmental
Research Leipzig-Halle,
Permoserstr. 15, 04318 Leipzig
messner@alok.ufz.de
GERMANY

Mic, Rodica Paula

National Institute of Hydrology and
Water Management, Sos. Bucuresti –
Ploiesti 97, Sector 1, 013686 Bucuresti
rodica.mic@hidro.ro
ROMANIA

Pagliara, Stefano

Universita di Pisa, via Gabba 22,
56100 Pisa
s.pagliara@ing.unipi.it
ITALY

Precht, Elimar

DHI, Water & Environment,
DHI Büro Deutschland,
Krusenberg 31, 28857 Syke
e.precht@dhi-umwelt.de
GERMANY

Punčochář, Pavel

Ministerstvo zemědělství, Water
Management Sector,
Těšnov 17, 117 05 Praha 1
puncochar@mze.cz
CZECH REPUBLIC

Roos, Alex

Ministerie van Verkeer en Waterstaat,
Rijkswaterstaat, Van der Burghweg 1,
2600 GA Delft
a.roos@dww.rws.minvenw.nl
THE NETHERLANDS

Samuels, Paul

HR Wallingford, Water Management
Dpt, Howbery Park,
Wallingford, OX 8BA Oxon
p.samuels@hrwallingford.co.uk
UNITED KINGDOM

Spatka, Jan

DHI Hydroinform a.s.,
Na Vršich 5, 100 00 Prague 10
j.spatka@dhi.cz
CZECH REPUBLIC

Stancalie, Gheorghe

National Institute of Meteorology and
Hydrology, Remote Sensing and GIS
Lab, 97, Soseaua Bucuresti-Ploiesti,
Sector 1, 71552 Bucharest
Gheorghe.Stancalie@meteo.inmh.ro
ROMANIA

Streng, Octavian	Crisuri Rivers Authority Oradea, 35 Ion Bogdan Street, 410125 Oradea octavian.streng@dac.rowater.ro ROMANIA
Toth, Sandor	National General Directorate for Environment, Flood Control Department, Márvány ut. 1/c H 1013 Budapest toth.sandor@ovf.hu HUNGARY
Trak, Baris	Council of Development Bank, 55 Avenue Kléber, 75116 Paris baris.trak@coebank.org FRANCE
Valdhans, Jiři	VRV, a.s., Nábřežní 4, 150 56 Praha 5 – Smíchov valdhans@vrv.cz CZECH REPUBLIC

OTHER PARTICIPANTS

Báča, Václav	Povodí Vltavy, s.p., Holečkova 8, 154 21 Praha 5 baca@pvl.cz CZECH REPUBLIC
Cabmoch, Jiří	VRV, a.s., Nábřežní 4, 150 56 Praha 5 – Smíchov CZECH REPUBLIC
Hajtášová, Katarina	Slovak Hydrometeorological Institute, Jeséniova 17, 83315 Bratislava 37 Katarina.hajtasova@shmu.sk SLOVAKIA
Jiřinec, Petr	DHI Hydroinform a.s., Na Vrších 5, 100 00 Prague 10 p.jirinec@dhi.cz CZECH REPUBLIC
Kiss, Attila	Koros-valley District Water Authority (KOVIZIG), Varoshaz u.26, H-57 Guyla kiss.attila@kovizig.hu HUNGARY

316

Mahriková, Ivana

Slovak University of Technology,
Radlinského 11, 813 68 Bratislava
mahrikov@svf.stuba.sk
SLOVAKIA

Seegert, Jörg

Technische Universität Dresden,
Institute of Hydrology and
Meteorology,
01062 Dresden
joerg.seegert@mailbox.tu-dresden.de
GERMANY

Socher, Martin

Sächsisches Staatsministerium für
Umwelt und Landwirtschaft,
Archivstr. 1, 01097 Dresden
martin.socher@smul.sachsen.de
GERMANY

Szekeres, Janos

Water Resources Research Centre plc.
(VITUKI),
Kvassay J.u. 1, H-1095 Budapest
szekeres@vituki.hu
HUNGARY

Tentis, Mitrut

Romanian Waters Authority, Crisuri
Rivers Authrity Oradea, Str. lullu
Manlu nr2 ap.1, 410104 Orades
director@oradea.rowater,ro
ROMANIA

INDEX

Printed in the United Kingdom
by Lightning Source UK Ltd.
128789UK00001B/75/A